Ecological Studies

Analysis and Synthesis

Volume 224

Ecological Studies is Springer's premier book series treating all aspects of ecology. These volumes, either authored or edited collections, appear several times each year. They are intended to analyze and synthesize our understanding of natural and managed ecosystems and their constituent organisms and resources at different scales from the biosphere to communities, populations, individual organisms and molecular interactions. Many volumes constitute case studies illustrating and synthesizing ecological principles for an intended audience of scientists, students, environmental managers and policy experts. Recent volumes address biodiversity, global change, landscape ecology, air pollution, ecosystem analysis, microbial ecology, ecophysiology and molecular ecology.

More information about this series at http://www.springer.com/series/86

Thomas R. Horton
Editor

Mycorrhizal Networks

Springer

Editor
Thomas R. Horton
Department of Environmental and Forest Biology
State University of New York – College of Environmental Science and Forestry
Syracuse, NY
USA

ISSN 0070-8356 ISSN 2196-971X (electronic)
Ecological Studies
ISBN 978-94-017-7394-2 ISBN 978-94-017-7395-9 (eBook)
DOI 10.1007/978-94-017-7395-9

Library of Congress Control Number: 2015951398

Springer Dordrecht Heidelberg New York London

Printed on acid-free paper

Springer Science+Business Media B.V. Dordrecht is part of Springer Science+Business Media (www.springer.com)

I dedicate this book to Michael Booth. Michael was an incredible scientist whose Ph.D. on mycorrhizal networks took us to a new level with respect to field experiments on mycorrhizal networks. I was fortunate enough to be on his committee and see that his energy and intellect combined to give him an edge that few have at that stage. He agreed to contribute to this book and was working on the chapter with Jason Hoeksema when we lost him. His influence on the field was large and he was just getting started.

Foreword

In 1842, Carlo Vittadini, Professor of Medicine at the University of Milan and eminent mycological taxonomist, described the husk of tree feeder rootlets enclosing mature sporocarps of the mycorrhizal hypogeous genus *Elaphomyces.* He pointed out that the rootlets were mantled by hyphae growing out of such sporocarps, yet showed no sign of disease. He bravely stated "extra omne dubium," beyond all doubt, that the rootlets were nourished by the *Elaphomyces* hyphae. Bravely, because at that time fungi were universally regarded either as causing disease or rotting organic matter. Buried in his monograph of *Elaphomyces,* that startling assertion was overlooked by mycorrhiza researchers for more than a century, and Vittadini did not pursue the topic again. Nonetheless, it was the first hypothesis later proven true that led to the real meaning of mycorrhizae and the translocating function of the mycorrhizal fungi.

The ensuing 45 years witnessed considerable discussion about the fungal colonization of rootlets of achlorophyllous plants such as *Monotropa* spp. and its connection with tree roots. The nature of that connection was debated, but no firm conclusions reached: Is the fungus a parasite on the achlorophyllous plant, or is the plant "humicolous," obtaining its nutrients directly from soil organic matter via the fungi? Feeder rootlets of ectomycorrhizal trees were noted to be among those of the achlorophyllous plants, but the sharing of those fungi was mentioned in passing or not at all. The Russian botanist F. Kamienski speculated in 1882 that *Monotropa* might be nourished by overstory trees through shared hyphae, or perhaps the fungus was humicolous. His rather convoluted discussion leaves the reader in doubt exactly what he hypothesized.

Then, the German plant pathologist A.B. Frank publishes his epic paper in 1885, in which he accurately describes with clarity the morphology of ectomycorrhizae, coins the term "mycorrhiza" for them, and correctly interprets their mutualistic symbiotic nature. Given the confusion about the phenomenon at the time, Frank's paper was an amazing *tour de force.* But, the puzzle of achlorophyllous plants remained.

In the early half of the twentieth century, a rather desultory interest in mycorrhizae continued, albeit a few important contributions appeared, such as that on the role of mycorrhizae in mineral nutrition of plants by the American A.B. Hatch in 1937. After World War II, Prof. Elias Melin and his students at the University of Uppsala, Sweden, became the first mycorrhiza researchers to enlist isotopes in study of translocation via mycorrhizal hyphae of elements and compounds from external sources to mycorrhizal seedlings and vice versa. Melin's group performed the epochal experiments that confirmed Frank's hypotheses of some 65 years earlier, thus laying the methodological foundation for study of common mycorrhizal networks (CMN). Eric Björkman, also Swedish, reported his results from experiments with ^{14}C-labelled glucose and ^{32}P-labelled phosphate injected into pine and spruce trees under which grew achlorophyllous *Monotropa* plants 1–2 m from the tree trunks. In 5 days, actively growing *Monotropae* had concentrated the isotopes, which did not appear in other nearby Ericaceae. This would seem to be the 1st experimental demonstration of CMN in Nature.

Little attention was then paid to CMN until the 1980s, when Roger Finlay and David Read of the University of Sheffield experimentally demonstrated movement of ^{14}C and ^{32}P to tree seedlings linked through CMN. From then to the present, research on the role of CMN in plant nutrition in laboratory and field has blossomed. As shown in the chapters of this book, CMN can have immense nutritional, ecological, agricultural, and forestry consequences undreamed of 40 or even 20 years ago. New terms have entered the lexicon of mycorrhiza research, e.g., CMN, infochemicals, and network theory. So, readers, feast your eyes and minds on the pages that follow.

Jim Trappe
Department of Forest Ecosystems and Society
Oregon State University U.S. Forest Service
Pacific Northwest Research Station
Forestry Sciences Laboratory

Preface

The overwhelming majority of the world's plant species are associated with mycorrhizal fungi in nature. As the term mycorrhiza implies, the association involves fungal *hyphae* interacting in the *roots* of a plant. Importantly, the hyphae do not penetrate the cell membranes of the root cells, although they may penetrate the cell wall. Further, the plant does not reject the fungus as a parasite or pathogen. The hyphae extend away from the roots into the soil where they take up nutrients and transport them through the mycelium and to colonized roots. Multiple hyphae connect plant hosts into what has become known as a mycorrhizal network. Mycorrhizal networks are below ground and cryptic. As such, plant and ecosystem ecologists in the past had to largely "black-box" the role of mycorrhizal networks in plant community and ecosystem dynamics. Björkman was the first to report field evidence of a mycorrhizal network in his work on the nutritional mode of the mycoheterotrophic plant *Monotropa hypopitys* (Björkman 1960). Newman (1988) provided a thorough review of the structure and function of mycorrhizal networks. Since Newman's initial review there has been an impressive amount of work on the topic using advanced methods such as isotopic labeling and PCR-based identification methodologies and additional reviews have followed (Simard and Durall 2004; Selosse et al. 2006; Horton and Van der Heijden 2008; Peay et al. 2008; Van Der Heijden and Horton 2009; Bahram et al. 2014).

The textbook explanation for the benefit to plants of associating with mycorrhizal fungi is that fungi provide plants with a physical extension of the root system. Importantly, the fungi also produce unique enzymes that give plants access to pools of nutrients with limited availability to the plants alone (e.g., organic nitrogen, phosphorus). However, an individual mycorrhizal plant colonized by a single fungus individual is too simple. The mycorrhizal condition is much more complex and interesting. For example, an ectomycorrhizal tree may support tens of fungus species on its roots and multiple genotypes of each (Bahram et al. 2011). And each genetic entity (genet) may be isolated into multiple independent ramets. Further, this does not address the increasingly recognized role of mycorrhizal fungi in forming intra- and inter-specific plant connections, the mycorrhizal networks.

It has become clear that most mycorrhizal fungi colonize and provide nutrients to multiple plant species. This has important implications for plant competition for soil nutrients, seedling establishment, plant succession, and plant community and ecosystem dynamics. Schimel and Bennett (2004) suggested a paradigm shift was occurring in ecosystem science based on the realization that microbes, including mycorrhizal fungi, acquire organic N through depolymerization of N containing polymers. Plant ecologists have accumulated a rich body of knowledge regarding nutrient acquisition by plants. Much of the work is based on hydroponic systems replete with nutrients and, importantly, without mycorrhizal fungi. However, plants do not typically grow in nutrient-rich soils, or without mycorrhizal fungi. Plant competition for nutrients is not strictly a function of interactions between plants. Rather, mycorrhizal fungi compete for soil nutrients that then become available to multiple plant hosts through mycelial networks. Competition for soil nutrients may therefore involve compatibility interactions between fungi and hosts as much as plant–plant competition. This is a new paradigm for plant ecologists that may be as important as the one highlighted for ecosystem ecologists by Schimel and Bennett (2004).

A mycorrhizal fungus network reduced to its simplest form is a single fungal individual (a thallus) that has colonized multiple root tips of a plant individual (Fig. 1a). The connected mycelium of an individual fungus colonizing roots of multiple plants of the same or different species is a *common mycelial network* (Fig. 1b). Multiple individuals of multiple fungus species colonizing multiple plant species make up a *common mycorrhizal network* (Fig. 1c). The reader should be careful to understand when an author uses CMN to denote a common mycelial network usually with a limited spatial extent versus a common mycorrhizal network that may be much more extensive and involve multiple trees and their fungi.

It is reasonable to assume that an individual plant interacts with a rich assemblage of mycorrhizal fungi as indicated in Bahram et al. (2011) on a single poplar tree, and this level of complexity is ramified by the addition of other local plant hosts. Individual fungi in such a mycorrhizal network have direct and indirect impacts on each other and their hosts through competition for soil resources and compatibility interactions with common hosts. Therefore, a community of mycorrhizal plants and fungi interact not as a superorganism *sensu* Clements (1936) and Phillips (1935) but individualistically as independent organisms *sensu* Gleason (1926).

Following nutrient dynamics in a mycelial network of a single fungus is daunting, especially in a field setting. Following nutrient dynamics in a complex mycorrhizal network, with many interacting fungal and plant individuals, is all the more difficult. Modeling these interactions is helping to overcome the limitations of an in vivo system and is ripe for new research efforts (see Bahram et al. 2014). However, these approaches will only yield accurate results with more and better data on the role each species plays in network dynamics to feed into the models (e.g., the kind and amount of nutrients acquired and transported, compatibility interactions with various hosts, competitive interactions between the fungi, and how plants and fungi recognize and reward good symbionts). It is my hope that the

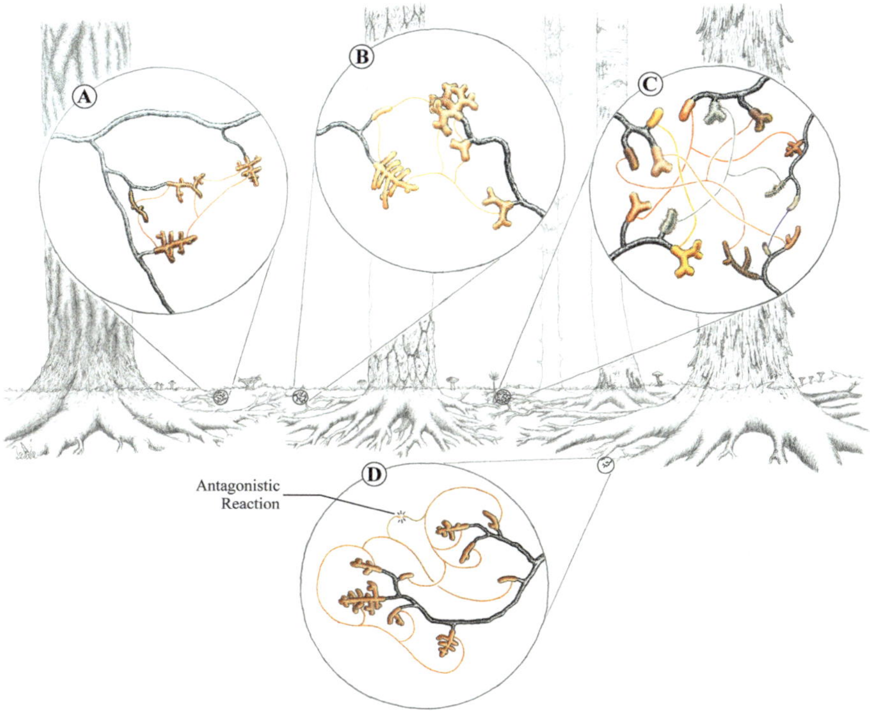

Fig. 1 Schematics of mycorrhizal networks. Mycelial connections can vary from one fungus individual connecting root tips of one plant individual (*A*), to one fungus individual connecting root tips of two plant individuals (*B*), to multiple fungi interacting on multiple plant species (*C*). We also highlight an antagonistic interaction between different thalli of the same species (*D*); the hyphae recognize nonself tissue and reject the attempt to anastomose. Figure drawn by Sam Tourtellot

chapters in this book lay the next foundation for research on mycorrhizal networks and point the way to areas for research needs and opportunities.

The book is organized into three sections: network structure, nutrient dynamics, and the mutualism–parasitism continuum. A necessary requirement for the development of a mycelial network is compatibility between a fungus and a plant. Molina and Horton review specificity of ectomycorrhizal symbionts and its role in plant communities in Chap. 1. A lot of work has been conducted since Molina et al. (1992) provided a comprehensive review of specificity. Chapter 1 includes an updated list of terms and their definitions that should prove useful in communicating about specificity phenomena and mycorrhizal networks. Predictions in the earlier review about the role of specificity phenomena in plant community dynamics have been supported in numerous field studies using molecular techniques. However, difficulties with sampling ectomycorrhizal fungi that are infrequently encountered and belowground continue to be a problem when investigating large-scale patterns of host preference and specificity.

Networks of mycorrhizal fungi involve interactions between symbionts but also interactions between fungal individuals, a topic explored by Giovannetti et al. in Chap. 2. When hyphae from the same genotype come into contact, they can anastomose, or fuse, into a continuous thallus even if the two hyphae were from different ramets of the same genet. Ectomycorrhizal fungi in the Basidiomycota and the Ascomycota have genetic systems for recognizing and rejecting nonself tissue, preventing anastomosis between different genotypes in a mycorrhizal network (Fig. 1d). This so-called vegetative incompatibility system is analogous to our own immune system. The vegetative incompatibility system in ectomycorrhizal fungi is why I suggested above that a community of mycorrhizal fungi functions individualistically rather than as a superorganism. However, as Giovannetti et al. review in Chap. 2, interactions between arbuscular mycorrhizal fungi may be different. These fungi are thought to be strictly clonal over the course of their greater-than-400-million-year history, yet there is evidence for recombination in the group. Thalli of Glomeromycota have few septae, making a mycelial network essentially a single cell with hundreds and even thousands of nuclei. Germinants of single spores contain multiple genotypes but whether the genetic diversity occurs within the nucleus that is mitotically propagated and packaged into new spores (homokaryosis model; Pawlowska and Taylor 2004) or across multiple genetically distinct nuclei, each with their own mitotic fate and possibility for packaging into new spores (heterokaryosis model; Sanders et al. 1995) is hotly debated. The vegetative incompatibility system may not be as active in the arbuscular mycorrhizal fungi as it is in the ectomycorrhizal fungi and anastomosis between thalli derived from different genets may be a way of generating and maintaining a mosaic of genetic types in a thallus. Interestingly, Tisserant et al. (2012) report finding meiosis-specific genes in the transcriptome of *Glomus interadices* (now *Rhizophagus irregularis*) raising the possibility for some mechanism supporting recombination in the group other than heterothallic anastomosis. This is an exciting area with great potential for additional lessons about the unique genetic system in Glomeromycota and interactions between symbionts in mycorrhizal networks.

The second section of the book focuses on nutrients and their movement through networks. Wallander and Ekblad begin the section with Chap. 3 and their coverage of extramatrical mycelium in ectomycorrhizal fungi (extramatrical being mycelia beyond the root tips). They focus on carbon and nitrogen. Plants allocate an estimated 15–20 % of the carbon they fix to their mycorrhizal fungi (Hobbie and Hobbie 2006). The fungi use this carbon in part to produce mycelial networks in soils to gain access to limiting resources. It is well known that fertilized plants allocate fewer resources belowground and as a result, fewer resources to mycorrhizal fungi. Conversely, if a plant is growing under nutrient limitation, more carbon is allocated belowground, supporting mycelia of fungi that provide access to the limiting nutrient (Werner and Kiers 2015). Network production is then a function of both carbon availability and nutrient availability. It is still hard to quantify the rate of extramatrical mycelium turnover, but Clemmensen et al. (2015) have shown that dark septate root endophytes may support long-term sequestration of host carbon in boreal forests. As Wallander and Ekblad suggest, there is increasing evidence that ectomycorrhizal fungus networks should be included in soil carbon models.

In Chap. 4, Jakobsen and Hammer review the influence of mycorrhizal networks on outcomes of plant competition in arbuscular mycorrhizal plant communities. Observations that plants do not always benefit from associating with mycorrhizal fungi have led to the idea that the symbiosis exists on a mutualism–parasitism continuum (Johnson et al. 1997), a topic that will be explored further in Section 3 of this book. Clearly, plant hosts allocate carbon to their symbionts, and this allocation reduces the carbon available for their own growth. This reduced growth may, in situations of high nutrient availability or intense competition, reduce the competitive outcome and fitness of the host. Jakobsen and Hammer suggest that nutrient movement in networks moves toward carbon sources and larger plants. As a result, these authors suggest mycorrhizal networks accentuate competitive outcomes rather than relaxing them. While it is known that seedlings can experience reduced growth when connected to mycorrhizal networks, the negative effect of mycorrhizal networks on seedling establishment may be temporary, especially when larger hosts become less significant carbon sources through senescence. Still, Jakobsen and Hammer predict that mycorrhizal networks help plants that are already larger than others in the network.

Simard and colleagues have used labeled isotopes to follow transfer of resources between plants through ectomycorrhizal fungus networks. In Chap. 5, Simard et al. review the literature on nutrient movement between plants through networks with a focus on the magnitude, fate, and importance of mycorrhiza-derived nutrients in ectomycorrhizal plants. It is clear that many mycoheterotrophic hosts are dependent on network fungi to supply carbon, and that carbon comes from autotrophic hosts in the network. While there are data showing a low level of carbon can be transported to autotrophic hosts, it remains controversial whether the amounts are ecologically significant. Other nutrients such as nitrogen and water are transported through networks and contribute to plant survival. Phosphorus transport remains difficult to trace. In summary, resource fluxes through ectomycorrhizal networks can contribute to plant establishment and survival, but the level of the effect is context dependent.

Section three focuses on studies that investigate mycorrhizal fungi as mutualists or parasites and the implications of those two symbiotic types for plant community dynamics. Nara opens this section with Chap. 6 and a review of his work on the role of ectomycorrhizal networks on seedling establishment in a primary successional habitat. Networks of ectomycorrhizal fungi associated with pioneer *Salix* support establishment of conspecific *Salix* seedlings, but also seedlings of successional hosts such as *Picea* and *Betula*. Plant hosts planted away from *Salix* patches are not colonized by ectomycorrhizal fungi, suggesting spores are not functioning as inocula as much as *Salix*-associated mycelial networks. Using microsatellite markers, Nara shows that some species such as *Laccaria* produce small thalli generally less than a meter in extent and short lived. However, other species produce longer-lived thalli that are up to 10 m in extent. These data are consistent with other studies from a variety of successional settings suggesting most individuals of ectomycorrhizal species are typically less than 3 m in extent, but some individuals of some species can be much larger (Douhan et al. 2011). Although the degree of benefit to the plant host varied by fungus species, ectomycorrhizal fungus networks are important for facilitating seedling establishment in this system.

Wagg and colleagues focus Chap. 7 on facilitation and antagonism in arbuscular mycorrhizal networks. They suggest arbuscular fungal communities might contribute to greater plant performance through functional complementarity or niche specialization. They argue that allocation of resources through mycorrhizal networks alters competitive outcomes among the plant species, an idea that follows a model proposed by Bever (2003). By considering antagonistic as well as facilitation in the mycorrhizal mutualism, the authors give a more complete framework for understanding how networks function in plant community dynamics.

Kennedy et al. explore the unique networking dynamic of *Alnus* in Chap. 8. Most ectomycorrhizal plant hosts are known to associate with multiple fungal species and vice versa, resulting in a high potential for complex ectomycorrhizal networks in a forest setting. However, exceptions to this pattern are known, with *Alnus* being the most commonly cited exception. *Alnus* forms isolated networks with little direct connections to networks of other host species in a forest. This seems to put both the fungus and plant at a disadvantage given the fungus has fewer sources of carbon and the plant has less access to nutrients than if they were generalist symbionts. In addition to reviewing the literature on the specificity observed in *Alnus*, the authors discuss why having isolated networks is an advantage, and how the high specificity in the genus is maintained.

It should be clear from the coverage of these chapters that nutrient transfer between plants via mycorrhizal networks and its effects on plant community dynamics remains controversial or at least context dependent. Hoeksema argues in Chap. 9 that most experiments have not adequately tested the role of mycorrhizal networks on plant community dynamics. He suggests more tests should be conducted to rule out alternative hypotheses to carbon movement between plants, especially those that include experimental manipulations of the mycorrhizal networks. This is obviously not a trivial request considering the fact that mycorrhizal networks function largely belowground and are difficult to observe. However, Hoeksema's recommendations for future studies (and those of the authors of all the chapters) point to the exciting possibilities for additional research on mycorrhizal networks.

I close with my acknowledgements to all the people that have helped make this book possible. First and foremost are my coauthors who accepted the invitation to write a review chapter focused on their research. A book on a topic such as mycorrhizal networks will already seem rather narrowly focused to some, yet as each chapter shows, the topic can be broken down even further. I have learned a lot from these authors in the past and even more from this project. I thank all the authors for their patience with the delays in my completing the book. I also thank the anonymous reviewers who helped make the book stronger with their constructive comments and edits. I thank Valeria Rinaudo for her interest and enthusiasm for a book on mycorrhizal networks and shepherding the proposal through

Springer. Thank you also to Takeesha Moerland-Torpey who helped move the book forward. And I thank the subject editor Dr. Hal Mooney for his help and encouragement as I wrapped up the project.

References

Bahram M, Harend H, Tedersoo L (2014) Network perspectives of ectomycorrhizal associations. Fungal Ecol 7:70–77

Bahram M, Põlme S, Kõljalg U, Tedersoo L (2011) A single European aspen (*Populus tremula*) tree individual may potentially harbour dozens of *Cenococcum geophilum* ITS genotypes and hundreds of species of ectomycorrhizal fungi. FEMS Microbiol Ecol 75:313–320

Bever JD (2003) Soil community feedback and the coesxitence of competitors: conceptual frameworks and emprical tests. New Phytol 157:465–473

Björkman E (1960) *Monotropa hypopitys* L.—an epiparasite on tree roots. Physiol Plant 13:308–327

Clements FE (1936) Nature and structure of the climax. J Ecol 24:252–284

Clemmensen KE, Finlay RD, Dahlberg A, Stenlid J, Wardle DA, Lindahl BD (2015) Carbon sequestration is related to mycorrhizal fungal community shifts during long-term succession in boreal forests. New Phytol 205:1525–1536

Douhan GW, Vincenot L, Gryta H, Selosse M-A (2011) Population genetics of ectomycorrhizal fungi: from current knowledge to emerging directions. Fungal Biol 115:569–597

Gleason HA (1926) The individualistic concept of the plant association. Bull Torrey Bot Club 53:7–26

Hobbie JE, Hobbie EA (2006) ^{15}N in symbiotic fungi and plants estimates nitrogen and carbon flux rates in arctic tundra. Ecology 87:816–822

Horton TR, Van der Heijden MGA (2008) The role of symbiosis in seedling establishment and survival. In: Leck M, Parker VT, Simpson B (eds) Seedling ecology and evolution. pp 150–171. Cambridge University Press. Cambridge

Johnson NC, Graham JH, Smith FA (1997) Functioning of mycorrhizal associations along the mutualism-parasitism continuum. New Phytol 135:575–585

Molina R, Massicotte H, Trappe JM (1992) Specificity phenomena in mycorrhizal symbioses: community-ecological consequences and practical implications. In: Allen MF (ed) Mycorrhizal functioning an integrative plant-fungal process, pp 357–423. Chapman and Hall, New York

Newman EI (1988) Mycorrhizal links between plants: their functioning and ecological significance. Adv Ecol Res 18:243–270

Pawlowska TE, Taylor JW (2004) Organization of genetic variation in individuals of arbuscular mycorrhizal fungi. Nature 427:733–737

Peay KG, Kennedy PG, Bruns TD (2008) Fungal community ecology: a hybrid beast with a molecular master. Bioscience 58:799–810

Phillips JP (1935) Development, the climax, and the complex organism: an analysis of concepts: Part III. The complex organism: conclusions. J Ecol 23:488–508

Sanders IR, Alt M, Groppe K, Boller T, Wiemken A (1995) Identification of ribosomal DNA polymorphisms among and within spores of the Glomales: application to studies on the genetic diversity of arbuscular mycorrhizal fungal communities. New Phytol 130:419–427

Schimel JP, Bennett J (2004) Nitrogen mineralization: challenges of a changing paradigm. Ecology 85:591–602

Selosse M-A, Richard R, He X, Simard SW (2006) Mycorrhizal networks: des liaisons dangereuses. Trends Ecol Evol 21:621–628

Simard SW, Durall DM (2004) Mycorrhizal networks: a review of their extent, function, and importance. Can J Bot 82:1140–1165

Tisserant E, Kohler A, Dozolme-Seddas P, Balestrini R, Benabdellah K, Colard A, Croll D, Da Silva C, Gomez SK, Koul R, Ferrol N, Fiorilli V, Formey D, Franken P, Helber N, Hijri M, Lanfranco L, Lindquist E, Liu Y, Malbreil M, Morin E, Poulain J, Shapiro H, van Tuinen D, Waschke A, Azcón-Aguilar C, Bécard G, Bonfante P, Harrison MJ, Küster H, Lammers P, Paszkowski U, Requena N, Rensing SA, Roux C, Sanders IR, Shachar-Hill Y, Tuskan G, Young JPW, Gianinazzi-Pearson V, Martin F (2012) The transcriptome of the arbuscular mycorrhizal fungus Glomus intraradices (DAOM 197198) reveals functional tradeoffs in an obligate symbiont. New Phytol 193:755–769

Van Der Heijden MGA, Horton TR (2009) Socialism in soil? The importance of mycorrhizal fungal networks for facilitation in natural ecosystems. J Ecol 97:1139–1150

Werner GDA, Kiers ET (2015) Partner selection in the mycorrhizal mutualism. New Phytol 205:1437–1442

Contents

Chapter 1
Mycorrhiza Specificity: Its Role in the Development and Function of Common Mycelial Networks

Randy Molina and Thomas R. Horton

Abstract The establishment of common mycelial networks by mycorrhizal fungi shared between host plants depends on the ability of neighboring plants to enter into mycorrhizal associations with compatible fungal species. Such compatibility is governed by the potential mycorrhiza specificities of the symbionts. Mycorrhiza specificities exist along a continuum from low specificity (association with multiple partners) to high specificity (association with one or few partners). Although the ability of symbionts to form mycorrhizas may be largely governed by host-fungus gene interactions as influenced by co-evolutionary events, mycorrhizal associations in natural ecosystems can also be influenced by environmental factors (e.g. soil) and biological factors (e.g. different neighboring host species), phenomena referred to as "ecological specificity." For example, in natural settings, mycorrhizal fungi often express "host preference" wherein fungi may be more common on a particular host in mixed-host settings than would be expected by random species assemblage within the fungal and plant communities. Mycorrhiza specificity phenomena significantly influence plant community dynamics, particularly plant succession. Early seral plants can positively affect the establishment of later-seral plants by maintaining commonly shared mycorrhizal fungi, and thus affecting the function of common mycelial networks over time. Such knowledge provides guidance for ecosystem managers to maintain "legacy" early -seral plants that benefit later-seral plants via shared mycorrhizal fungus species. Understanding specificity phenomena is also crucial for predicting the successful migration of plants and compatible mycorrhizal fungi during climate change. We review mycorrhiza specificity terminology and types of specificity phenomena, and suggest use of common terms to

R. Molina
Mycorrhiza, 620 SE 14th Court, Gresham, OR 97080, USA
e-mail: r.molina@comcast.net

T.R. Horton (✉)
Department of Environmental and Forest Biology, State University of New York – College of Environmental Science and Forestry, 246 Illick Hall, 1 Forestry Drive, Syracuse, NY 13210, USA
e-mail: trhorton@esf.edu

T.R. Horton (ed.), *Mycorrhizal Networks*,
Ecological Studies 224, DOI 10.1007/978-94-017-7395-9_1

provide consistency in addressing this research topic. We also provide extensive examples from diverse ecosystems on the ability (or inability) of neighboring plants to develop common mycelial networks.

Keywords Host specificity • Mycorrhiza specificity • Mycorrhiza compatibility • Host preference • Ectomycorrhiza

1.1 Introduction

Common mycelial networks (CMNs) of mycorrhizal fungi connecting neighboring host plants affect ecosystem processes and community dynamics including seedling establishment, plant succession, and ecosystem resiliency (Simard et al. 2002, 2012; Simard and Durall 2004; Simard and Austin 2010; Selosse et al. 2006; Horton and van der Heijden 2008; van der Heijden and Horton 2009). Requisite to the establishment and function of CMNs is the ability of neighboring plants to be colonized by shared mycorrhizal fungi, or more specifically, individuals with continuous mycelial systems. Formation of linkages via compatible mycorrhizal fungi is governed in large part by the potential mycorrhiza specificities of the symbionts (i.e., host range of fungus, fungus range of host).

Molina et al. (1992) comprehensively synthesized concepts, phenomena, and ecological implications of mycorrhiza specificity. In the ensuing 20 years, many researchers have expanded upon those ideas to support an overarching concept: the degree of specificity displayed by both plant and fungal symbionts varies along a continuum from low specificity (associate with many symbiotic species) to high specificity (associate with one or a few species) (Fig. 1.1).

Several general terms are used to express where the symbionts lie along this continuum. For example, fungi only known to associate with a particular host species, or, more commonly, a host genus or family, are called "host specialists" for that host taxon. Those fungi that show less or no restriction to a taxonomic group of hosts are commonly called "generalists". Taylor et al. (2002) note that the "degree of specificity is a unique attribute of each partner". Although we can use general terms to describe similarities among fungi and hosts in mycorrhiza specificity

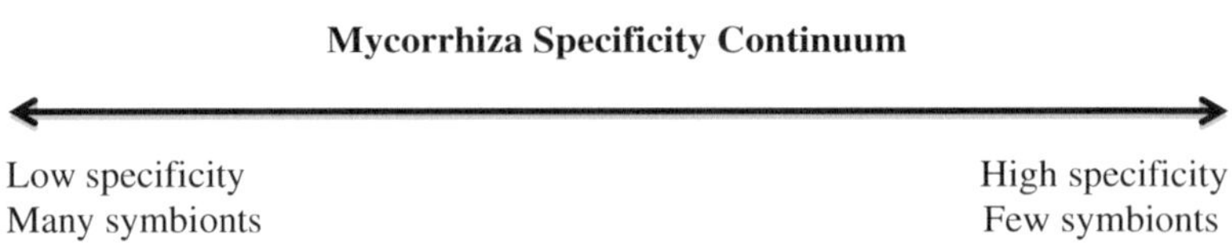

Fig. 1.1 Specificity continuum. Symbiotic plants and fungi fall along a continuum of specificity patterns. These interactions influence the complexity (number and diversity of species) of a mycorrhizal network

attributes, we must recognize that such general terms are relative and require more precise definition within the context of a study or when comparing studies. One of our goals is to bring clarity in the use of terms relevant to describing mycorrhiza specificity phenomena.

Mycorrhiza specificity processes and phenomena are complex due to the high diversity of mycorrhizal symbionts that span a phylogenetically diverse group of plant and fungal taxa, forming a variety of mycorrhizal types (e.g., arbuscular-, ecto-, orchid-, and ericoid mycorrhiza) with varied evolutionary histories (Hibbett and Matheny 2009; Tedersoo et al. 2010a; van der Heijden et al. 2015). Such complexity challenges our ability to generalize about global patterns of mycorrhiza specificity because we continually discover exceptions as we explore new ecosystems. A commonly cited generalization from the Molina et al. (1992) review, for example, is that most mycorrhizal fungi display a broad host range (or that little specificity is expressed in mycorrhizal associations). While this generalization may be true in a numerical sense, those same authors also provided numerous examples of fungi that showed variously restricted levels of host specificity and cautioned about over generalizing about these patterns. Since that 1992 review several papers support both the widespread nature of generalists and specialists among ectomycorrhizal (EM) fungi, and shed light on how host community composition and ecological conditions at a site impacts the relative abundance and function of EM fungi along the specificity continuum.

Concepts of mycorrhiza specificity go beyond a simple understanding of whether particular plants and fungi can enter into a mycorrhizal symbiosis. Harley and Smith (1983) coined the term "ecological specificity" to emphasize that the ability of plants and fungi to form mycorrhizae in the field may differ from that demonstrated in experimental syntheses wherein the symbionts are brought into contact under controlled conditions. Ecological specificity may be thought of as a variation on the ecological concepts of fundamental and realized niches. Molina et al. (1992) expanded this concept to include how a diversity of abiotic and biotic factors may affect the ability of plants and fungi to develop mycorrhizae in nature. This is analogous to applying the disease triangle in plant pathology (Fig. 1.2). The presence and composition of neighboring plants can affect the ability of some mycorrhizal fungi to develop mycorrhizae with particular hosts (Massicotte et al. 1994; Molina et al. 1997; Simard et al. 1997a; Kohout et al. 2011). As will be discussed in more detail, several studies have shown that in spite of the presence of a selection of potential host plants, certain fungi only form mycorrhiza with particular hosts. Such "host preference" appears widespread in both EM (Kranabetter et al. 1999; Cullings et al. 2000; Kernaghan et al. 2003; Ishida et al. 2007; Tedersoo et al. 2007a, 2008a, 2010b, 2011; Morris et al. 2008, 2009; Cavender-Bares et al. 2009; Smith et al. 2009, 2011; Diédhiou et al. 2010; Wolfe and Pringle 2011) and arbuscular mycorrhizal (AM) systems (Helgason et al. 2002, 2007; Hart et al. 2003; Bever et al. 2009; Hausman and Hawkes 2009, 2010; Kiers et al. 2011; Davidson et al. 2011). Hart and Klironomos (2002) describe such preferences in AM systems as a type of "functional specificity" wherein fungi display differential benefit to neighboring plants. Recent research demonstrates that some hosts can selectively

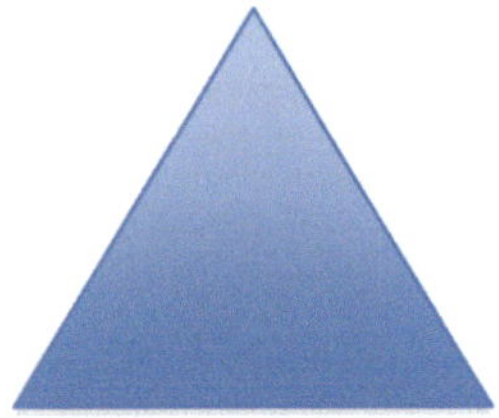

Fig. 1.2 The disease triangle. Compatible interactions between EM plants and fungi are influenced by the two symbionts (e.g., age of the seedling, carbon cost of the fungus, etc.) and by the environment. The environment includes the light environment, soil and nutrient type or availability (e.g., organic or inorganic N), as well as neighboring plant and fungal species in the networks

allocate photosynthate to a beneficial versus a non-beneficial fungus, and some fungi can discriminate among roots for carbon supply (Bever et al. 2009; Kiers et al. 2011). As we will discuss in more detail, these types of interactions and responses by fungi and plants associated with a mycelial network can significantly affect mycorrhizal community dynamics.

Early research on patterns of mycorrhiza specificity relied on associations of sporocarps (or spores for AM fungi) with particular plant hosts in nature or inoculation experiments such as the elegant pure culture syntheses of ectomycorrhiza by Melin (1922, 1923) or early pot studies by many researchers on arbuscular mycorrhiza. For mycorrhizal fungi that develop macroscopic reproductive structures (e.g. mushrooms and truffles) or easily retrievable spores (certain AM fungi), the resulting patterns of host-fungus associations provided valuable clues on mycorrhiza specificity, evolution, and host-fungus migration (Trappe 1962; Newton and Haigh 1998; Halling 2001; den Bakker et al. 2004; Vellinga et al. 2009; Wilson et al. 2012). Unfortunately not all mycorrhizal fungi produce showy and easily identified reproductive bodies, especially not in controlled settings, and field associations are not absolute proof of mycorrhizal relationships, particularly in stands composed of multiple host species.

The advent of PCR-based molecular techniques revolutionized our ability to identify plant and fungal symbionts on colonized roots, to derive phylogenetic relationships among mycorrhizal fungi, and thus to address concepts of mycorrhiza specificity with enhanced precision (reviewed in Horton and Bruns 2001; Peay et al. 2008). The patterns of presence and prevalence of multi-host fungi versus host specific fungi, or those showing host preference, have been fine-tuned with the application of molecular techniques in the field (Kennedy et al. 2003; Ishida et al. 2007; Twieg et al. 2007; Diédhiou et al. 2010; Smith et al. 2011; Polme et al. 2013; Roy et al. 2013). Molecular studies have also shed new light on the abilities of some fungi to form multiple types of mycorrhizae. For example, many EM fungi also form orchid, arbutoid, or monotropoid mycorrhizae (Taylor et al. 2002; Bidartondo et al. 2004), and some may even form ericoid mycorrhizae (Bergero et al. 2000; Allen et al. 2003; Villarreal-Ruiz et al. 2004; Grelet et al. 2009, 2010). Molecular studies of mycoheterotrophic plant roots have also revealed extreme fungal specificity and exploitive (parasitic) associations within mycorrhizal symbioses (Taylor et al. 2002; Bidartondo 2005). Molecular studies continue to provide a wealth of new information and approaches to determine the potential for linkage between neighboring plants, resulting in the formation of CMNs.

Many ecological and management implications flow from our understanding of CMNs as structured by mycorrhiza specificity processes and phenomena. Molina and Trappe (1982) first used "common mycelial networks" in a discussion of mycorrhiza specificity expressed by the arbutoid mycorrhizal plants *Arbutus menziesii* and *Arctostaphylos uva-ursi*; the authors hypothesized that these plants maintain EM fungus diversity in forest ecosystems following disturbance and benefit seedling establishment of later-seral Pinaceae because seedlings can exploit the mycorrhizal networks supported by these arbutoid plants. Similar examples of mycorrhiza specificity affecting formation and function of CMNs in diverse ecosystems abound and will be discussed later. The introduction of exotic, and potentially invasive, mycorrhizal fungi and plants worldwide is also influenced by compatible symbionts and mycorrhiza specificity processes (Vellinga et al. 2009; Nuñez et al. 2009; Dickie et al. 2010; Hynson et al. 2013; Karst et al. 2014; Hayward et al. 2015a, b). Similarly, as climate changes, the successful migration of plants and fungi into new habitats shaped by changing environmental conditions will be affected by compatibility of existing mycelial networks and the co-migration of compatible symbionts during migration.

We do not comprehensively review mycorrhiza specificity in this chapter; instead, we focus on what has been learned since the review by Molina et al. (1992). Our main goal is to provide a clear understanding of how mycorrhiza specificity influences the development and function of CMNs by addressing the following objectives: (1) refine a working lexicon of terms for mycorrhiza specificity research, (2) provide an updated overview of patterns of specificity seen in EM associations, and (3) exemplify the ecological consequences of how these specificity phenomena influence community dynamics, ecosystem resiliency, and the functions of CMNs. Given our expertise, we focus on EM symbioses but draw upon examples of other mycorrhiza types as appropriate.

1.2 The Lexicon

As ecologists we are acutely aware of the use and abuse of ecological terms (Tansley 1935), and those applicable to mycorrhiza specificity are no exception. In this section our objective is to clearly define our use of terms in this chapter. Some of the terms are drawn from the ecological literature, some from plant pathology, and some are unique to mycorrhizal symbioses. We provide definitions in Table 1.1, and elaborate on these in the following section.

Table 1.1 Definitions of terms used when discussing mycorrhiza specificity

Word or phrase	Definition
Symbiosis	Literally, living together. Can be a pathogenic, parasitic or mutualistic interaction
Mutualism	A symbiosis in which both organisms increase their fitness through the interaction. Increased uptake or allocation of resources, and improved growth are typically surrogates for fitness in mycorrhizal mutualisms
Fitness	The genetic contribution by an individual's descendants to future generations of a population. While biomass is often used as a proxy for fitness, it does not directly account for reproductive success
Mycorrhiza, mycorrhizas, mycorrhizae, mycorrhizal When to use these?	It is believed that the term mycorrhizae was first coined by A.B. Frank. He combined two Greek roots, mycor- for fungus and -rhiza for root (plant root). He used mycorrhiza for singular, mycorrhizae for plural, and most researchers followed that approach. But actually there was a problem. The—ae ending is Latin, and so Frank combined two Greek roots with a Latin ending. Many today argue for an English plural ending, so we get mycorrhizas. Neither uses a Greek ending, so neither is actually 'correct'.
	The use of the adjective mycorrhizal is perhaps more interesting (for those interested in this kind of thing!). What follows is a personal communication from Dr. Jim Trappe on whether to use mycorrhiza or mycorrhizal with specificity: 'Mycorrhiza' is a noun, but unlike most languages, English lets us use nouns as adjectives. When we do, always or maybe nearly always it's a short-handed way of saying the possessive case, i.e. 'mycorrhiza specificity' = 'of mycorrhiza' or 'of mycorrhizae', depending on the context. 'Mycorrhizal' is strictly an adjective, meaning the property of forming or being part of a mycorrhiza, e.g. 'a mycorrhizal host' or a 'mycorrhizal fungus'. Having counted how many angels can dance on the head of a pin, I pound my gavel on the judge's bench and pronounce: "specificity" cannot be mycorrhizal, that's like saying 'specificity forms a mycorrhiza' or 'specificity is a participant in forming a mycorrhiza'. 'Mycorrhiza specificity' means "specificity of mycorrhizae". This may not be strictly accurate, because we really mean host or fungus specificity, but in my opinion (that's what judges do, give opinions), the term implies specificity of either or both components of a 'mycorrhiza'
	Note from TRH: It seems to me that the issue is not whether to use mycorrhizal or mycorrhiza with specificity if both can connote an adjective and both suffer from the suggestion that specificity can form a mycorrhizal root which it cannot. We probably will continue to use mycorrhiza specificity (or mycorrhizal specificity) even though, as Judge Trappe suggests, we mean the specificity of the fungus and/or plant forming the mycorrhizal root tip. Note that the titles of multiple publications by Molina, Horton and Trappe in the literature citation list for this chapter used the –al form when referring to specificity, symbioses, inoculation, ecology and networks, none of which can be mycorrhizal according to judge Trappe! So much for clarity

(continued)

Table 1.1 (continued)

Word or phrase	Definition
Mycorrhiza specificity	An umbrella term that refers to the range of symbionts with which a fungus or plant develops a mycorrhizal symbiosis and the influences that contribute to the compatibility of the symbionts. Narrow range species necessarily associate with fewer symbiotic partners compared to broad range species, but we emphasize the phylogenetic breadth over richness
Mycorrhiza compatibility	The ability of a fungus and plant to form an anatomically defined mycorrhiza (Refer to Peterson et al. (2004) for anatomical definitions and images of the various mycorrhizal types)
Degree of mycorrhiza specificity	The breadth of taxonomic diversity with which a mycorrhizal species associates (synonymous with degree of host specificity)
Host range of the fungus:	Host range displays a continuum from narrow (associated with closely related hosts such as members of a single genus) to broad (associated with unrelated hosts such as both angiosperm and conifer species). Mycorrhizal fungi with narrow host ranges are often called "specialists" while those with diverse hosts are called "generalists." Specialist and generalist are relative terms and should be carefully defined within the context of the study
Fungus range of the host:	The breadth of fungus taxonomic diversity with which a plant species associates. Analogous to host range, fungus range displays a continuum from narrow (fungal associates are closely related) to broad (fungal associates are a phylogenetically diverse group)
Fidelity to a mycorrhizal type	The ability of a plant or fungus to form one or more mycorrhizal types (type in this case refers to the anatomically defined categories of mycorrhiza)
Ecological specificity	The influence of biological or environmental factors on the ability of a plant and fungus to form a compatible mycorrhiza in soil. Based on the disease triangle in plant pathology
Host preference/selectivity	Consistent patterns of nonrandom assemblages between plant and fungal species are observed more or less frequently than expected by chance, despite an absence of compatibility limitations between the symbionts. The mechanisms behind these patterns are not well understood, including whether the plant or fungus or both control the association frequency
Host shift	An evolutionary process wherein a fungus colonizes a new plant species when its primary host species is going locally extinct or no longer available

1.2.1 Symbiosis

Frank originally coined the term "symbiotism" in 1885 (Trappe 2005) to encompass the full range of interspecific interactions from parasitism to mutualism. More recently symbiosis, a variation of the term coined by de Bary in 1879, is widely used almost synonymously with mutualism. We find Frank's original meaning better, in part because it allows for multiple outcomes along the parasitism—mutualism continuum.

1.2.2 Mutualism

Mutualisms are symbioses in which both organisms benefit from the interaction. Widespread examples include mycorrhizal plants and fungi, lichens and plant pollinator systems. An organism that is generally labeled as a mutualist in relation to a second organism may still reduce the growth or even fitness of the second organism under certain conditions, a feature of the mutualism-parasitism continuum noted in mycorrhizal symbioses (Johnson et al. 1997).

1.2.3 Fitness

The contribution of an individual's genotype to succeeding generations. While biomass and relative growth rate (RGR) may provide good proxies for fitness, they are not direct measures. As a result, a plant may sustain a negative RGR at the seedling stage as it allocates carbon to its mycorrhizal fungi while actually increasing its fitness through increased survival and reproductive output in the future (Stanley et al. 1993).

1.2.4 Mycorrhiza Specificity

Phylogenetic range of the symbionts known to form mycorrhizal associations with a particular plant or fungal species. This general term takes neither a fungus- or plant-centric view. Mycorrhiza specificity may be narrow, as in the restriction of *Suillus* species to members of Pinaceae (Kretzer et al. 1996), or broad as in the 2000 species of EM fungi known to associate with Douglas-fir (*Pseudotsuga menziesii*, Trappe 1977). This umbrella term serves as a construct to discuss various specificity phenomena expressed in mycorrhizal symbioses.

1.2.5 Mycorrhiza Compatibility

The ability of a fungus and plant to form an anatomically defined mycorrhiza. This definition does not necessitate having data on the physiological nature of the symbiosis or "functional compatibility" as defined by Gianinazzi-Pearson and Gianinazzi (1983), i.e. physiological exchange of materials that point to a mutualistic symbiosis.

Given that the functional nature of mycorrhizal interactions varies tremendously between different plant and fungus associations (from one-sided parasitism to obligate mutualism), and that it is difficult to measure functional interactions, we use an anatomical definition of compatibility. Anatomically, the arbuscule (AM), Hartig net (EM), coil (ericoid, arbutoid, AM), hyphal peg (monotropoid), and peloton (orchidoid) are sites of nutrient transfer that we consider hallmarks of compatible associations. There may also be indications of incompatible associations, such as cortical cell disruption and phenolic compounds in colonized plant roots. The fact that hyphal cell disruption occurs in monotropoid hyphal pegs and orchidoid pelotons is an indication that these associations may be closer to, if not at, the parasitic end of the mutualism-parasitism continuum. We recognize that physiological interactions may be integral to the expression of mycorrhiza specificity phenomena, but leave functional aspects to a more thorough taxonomic survey based on experimental testing and genome surveys of functional genes.

1.2.6 Degree of Mycorrhiza Specificity

Prior to the use of extensive morphotyping in concert with molecular tools, the degree of mycorrhiza specificity in EM associations was based either on sporocarp-host observations, visually linking mycelium from sporocarps to EM roots of hosts in the field, or on pure culture synthesis experiments in the laboratory. The application of molecular tools on field-collected ectomycorrhizal root tips has provided greater data with more precision regarding fungal identity and the ability to test many of the assumptions and hypotheses put forth by earlier methods. Knowledge on specificity patterns represents hypotheses that are necessarily upheld or modified as new data are generated.

Host range of the fungus: Molina et al. (1992) described how EM fungi associate with host species along a spectrum from narrow to broad, and, for simplicity, divided the spectrum into three categories: narrow host range (restricted to a single host species or genus), intermediate host range (restricted to a host family, or a single taxonomic grouping, such as conifers or angiosperms), and broad host range (mostly unrestricted, associated with many host families, including both conifer and angiosperm). Species in the hypogeous EM genus *Rhizopogon* are classic examples that express primarily narrow specificity, associating with single genera within Pinaceae (e.g., *R. vinicolor* and *Pseudotsuga*); some *Rhizopogon* species, however, are intermediate and associate with several host genera within Pinaceae (e.g., *R. salebrosus*) (Molina et al. 1999). *Cenococcum geophilum*, with its cosmopolitan range and association with most EM hosts, exemplifies a broad host range EM fungus. Although these terms have been used in several publications, the terms "specialists" and "generalists" have proven more common as they are used in other ecological contexts.

These are all relative terms, however, and require definition within the context of the study, e.g., *Quercus* specialist, Fagaceae specialist, or angiosperm specialist.

Alternatively, one may directly state the degree of host restriction such as being *Quercus* specific, Fagaceae specific, or angiosperm specific.

"Multi-host fungus" is another commonly used phrase in EM fungal community studies in stands of mixed hosts and reflects the observed host range in that particular location. "Multi-host" is also a relative term that requires careful definition within the context of the study or when comparing results from different studies. For example, if one examines a stand with three genera of Pinaceae (e.g., *Pinus*, *Picea*, *Tsuga*) and three genera of angiosperms from two families (e.g., *Quercus*, *Betula*, *Fagus*), fungi that form ectomycorrhiza with all three Pinaceae hosts, all three angiosperm hosts, or a combination of Pinaceae and angiosperm hosts would all be considered "multi-host" fungi, yet could differ significantly in their host ranges and thus mycorrhiza specificity. Similarly, all "generalists" are not the same. For instance, although Wolfe and Pringle (2011) found that the EM fungus *Amanita phalloides* expresses broad compatibility with new hosts since it was introduced into North America, they did not consider it a true "generalist", because it did not associate with all EM hosts within its new range and showed strong host preferences in different locations. It is also important to keep in mind that genetic data has shown that some generalists are actually species complexes with individual species showing a higher degree of specialization (ecological or plant host) than expected based on a morphological species concept (Martin et al. 2002, 2008; Douhan et al. 2007; Geml et al. 2008).

Fungus range of the host: Ectomycorrizal hosts differ in the phylogenetic breadth of associated fungi. Molina et al. (1992) termed this fungus specificity phenomenon "host receptivity," defined simply as the number of fungal species associated with a host. Note that this is without reference to the phylogenetic breadth of the associates. Host receptivity has not been widely used since the 1992 paper. To make the fungus range of the host analogous to the host range of the fungus, we suggest using the degree of phylogenetic breadth of the fungi associated with hosts ranging from narrow (as seen in *Pterospora andromedea*, which only associates with several closely related *Rhizopogon* spp.) to broad (as seen in *Pseudotsuga menziesii*, known to associate with thousands of fungal species from numerous and distantly related phylogenetic groups). As noted for fungal host range, it is best to define the context when referring to the fungus range of the host.

1.2.7 Symbiont Fidelity to a Mycorrhiza Type

The term "fidelity" is used in early plant community ecology literature in reference to the constancy a plant species exhibits in a particular community association. We use it similarly but in reference to whether plant or fungal species are constant in the type of mycorrhiza they form. It is a valuable concept in describing mycorrhiza specificity phenomena, because it implies wide ranges of compatibility among diverse plants and fungi and indicates the potential for CMNs among plants. Most mycorrhizal plants and fungi express fidelity to one mycorrhiza type. There are many exceptions, however, and molecular tools have raised several questions in this regard. Well known exceptions include a large group of hosts that form both ectomycorrhizae and arbuscular mycorrhizae (e.g., several species in Fagaceae, *Eucalyptus*, *Populus*, *Salix*); these hosts often form arbuscular mycorrhiza early as seedlings but become predominately EM as mature plants (see Molina et al. 1992; Brundrett 2004; Smith and Read 2008 for lists of these plants and more through discussion). AM colonization has also been noted in typically non-AM hosts, such as the *Pseudotsuga* (Cázares and Smith 1995), and *Pinus* (Horton et al. 1999). Koske et al. (1990) reported AM colonization along with ericoid mycorrhiza in several Hawaiian Ericaceae. We will probably continue to discover more incidences of AM colonization of typically EM or ericoid hosts as we explore this phenomenon further, and it remains to be seen whether these result from opportunistic colonization in roots lacking a fungal mantle with no obvious fitness enhancement to the plants (perhaps following disturbance as seen in Horton et al. 1999) or functional mutualisms as evidenced by increased P uptake with AM colonization in *Pseudotsuga* (Cázares and Smith 1995).

Molecular tools have shed considerable new light on the interactions of EM fungi with arbutoid hosts, such as *Arctostaphylos* spp. (covered below), and mycoheterotrophic plants in Orchidaceae and Ericaceae. EM fungi of forest trees are the main mycobionts of mycoheterotrophic plants, and mycotrophic plants typically have a very narrow fungus range such as a single species or a group of closely related species (Taylor et al. 2002; Bidartondo 2005). Although debate continues on whether these are mutualistic or parasitic symbioses, they are anatomically referred to as mycorrhiza (Peterson et al. 2004), and certainly lie within the parasitism-mutualism continuum recognized for all mycorrhizal symbioses (Johnson et al. 1997). Several recent molecular studies have also shown that some EM fungi can form both ectomycorrhiza and ericoid mycorrhiza (Bergero et al. 2000; Allen et al. 2003; Villarreal-Ruiz et al. 2004; Grelet et al. 2009, 2010). Such findings led Vrålstad (2004) to entertain the possibility of EM and ericoid fungi operating within a "common guild" and potentially developing CMNs that may yield ecologically significant interactions between overstory EM trees and understory Ericaceae.

Molecular studies and further root sampling worldwide will continue to clarify the lines of mycorrhiza fidelity. For example, the ericoid mycorrhizal fungus *Rhizocyphus ericae* is widespread in the leafy liverwort *Cephaloziella various* in Antarctica, and an isolate from the liverwort formed typical ericoid mycorrhizae with *Vaccinium macrocarpon* seedlings upon inoculation (Upson et al. 2007). Several fungi that form ericoid mycorrhizae with *Woolsia pungens* were also isolated from 17 plants in a southeastern Australian forest (Chambers et al. 2008). Members of the Sebeniales also blur the fidelity line, as several species are involved in ecto-, orchid, and ericoid mycorrhiza (Selosse et al. 2002a, b; Allen et al. 2003; Urban et al. 2003; Setaro et al. 2006). These patterns reveal an interesting line of research: who is in control of a compatible interaction and characteristic anatomical features of each mycorrhizal type, the plant, fungus, or both?

1.2.8 Ecological Specificity

Harley and Smith (1983) used this concept in reference to the ability of plants and fungi to express different mycorrhiza compatibility under natural conditions compared to laboratory conditions, such as pure culture syntheses, and it has its roots in the disease triangle (Fig. 1.2).

Molina et al. (1992) expanded the definition to include the influence of biological or environmental factors on the ability of a plant and fungus to form a compatible interaction in soil. This concept emphasizes the point that factors beyond potential genetic compatibility of host and fungus can influence whether mycorrhizas develop under natural conditions, as indicated in the disease triangle. For mycorrhizal networks, the environment includes neighboring plants that can influence whether a particular fungus forms mycorrhiza with other co-occurring plant species.

1.2.9 Host Preference and Selectivity

In field studies with experimental designs that allow researchers to rule out random affects, consistent patterns of associations between plant and fungal species are observed more or less frequently than expected by chance, despite an absence of compatibility limitations between the symbionts. Further, in field studies of EM fungal communities involving multiple neighboring host species, some fungi occur more frequently on one host compared to a different neighboring host species (Kranabetter et al. 1999; Cullings et al. 2000; Kernaghan et al. 2003; Ishida et al. 2007; Tedersoo et al. 2007a, 2008a, 2010b, 2011; Morris et al. 2008, 2009; Cavender-Bares et al. 2009; Smith et al. 2009, 2011; Diédhiou et al. 2010; Wolfe and Pringle 2011). This phenomenon is called "host preference" (or selectivity) and is widespread in EM and AM systems (several detailed examples are provided in a later section). The mechanisms and ecological processes that yield patterns of host preference are largely unknown but likely include a complex of factors, such as competitive interactions among fungi, phylogenetic and physiological differences among hosts, and preferential allocation of resources between symbionts (Dickie 2007; Bever et al. 2009; Tedersoo et al. 2010a; Kiers et al. 2011). Host preference can influence fungal and plant community dynamics, as well as the structure and function of CMNs.

1.2.10 Host Shift

Host shift (or switch) is viewed in the pathology literature as an evolutionary process wherein a fungus becomes relatively more abundant on (shifts to) a new host when the original primary host is declining or no longer available. This allows the fungus to persist in its current range or even expand its range. The concept has also been applied in the evolution of EM fungi. As examples, Wilson et al. (2012) provided evidence for host shifts of *Scleroderma* species from ancestral Pinaceae to various angiosperms, as well as between angiosperms (Myrtaceae and Fagaceae). Within EM *Leccinum*, specific to *Betula* hosts, den Bakker et al. (2004) describe likely host shifts from *Betula* to *Populus* to Arbutoidae (that is, associations with Betulaceae to Salicaceae to subfamily Arbutoideae in Ericaceae). Given the expression of strong host specificity by several linages of EM fungi associated with *Alnus*, Tedersoo et al. (2009) suggest "multiple, independent host shifts". An exact event of host shift is difficult to distinguish, because the process is likely to unfold over a long period of time wherein the fungus may associate with both hosts until the original host disappears. An exception may be when fungi are introduced into new locations far distant from their original hosts. For example, Wolfe and Pringle (2011) note the shift by *Amanita phalloides* to several EM hosts when introduced into North America, an ability likely made possible by ancestral compatibilities with a diverse range of EM hosts in Europe.

Host shift has also been used to describe the movement of native EM fungi onto introduced hosts (Tedersoo et al. 2007b). However, we do not support this use of the term because this does not describe an evolutionary process, and the original local hosts are present and not declining. Cases of introduced hosts forming ectomycorrhiza with native EM simply represent an expansion of the host range for those fungi, again likely to be brought about by broad ancestral host compatibility. For example, Bahram et al. (2013) found that several EM fungi associated with native Fagaceae and Betulaceae in Iran formed ectomycorrhiza on introduced *Pinus sylvestris*. They hypothesized that, although *P. sylvestris* is not native to Iran, it occurs sympatrically with Fagaceae and Betulaceae in Europe, and compatibility of fungi associated with native angiosperm hosts with introduced pines in Iran may reflect ancestral EM fungus compatibility with Pinaceae.

1.3 Ecological Specificity and Host Preference

How do biotic and abiotic factors influence fungal and plant community dynamics and the role of mycorrhiza specificity phenomena? Among biotic factors, neighboring plants can exert significant effects on how fungi develop mycorrhizae within a mixture of potential hosts. For example, when various fungus and host species are grown in plant mixture and monoculture experiments, a fungus may only develop mycorrhizae on a particular host in specific treatment combinations, or differ in the degree of colonization depending on the presence of different hosts. Massicotte et al. (1994) and Molina et al. (1997) found that following spore inoculation, several *Rhizopogon* species (Pinaceae specialists) formed arbutoid mycorrhizae with arbutoid hosts (*Arbutus* or *Arctostaphylos* spp.) when grown in bioassays with their typical Pinaceae host species, but not when grown in an arbutoid host monoculture. Similarly, Douglas-fir specialist *Rhizopogon* species can also colonize western hemlock (*Tsuga heterophylla*) seedlings when grown in dual host bioassays with Douglas-fir seedlings (Smith et al. 1995). Massicotte et al. (1999) grew several mixtures and monocultures of EM hosts in forest soil and found that some EM fungi only formed ectomycorrhizae with particular hosts in mixed cultures. Again, this included the colonization of *Arbutus menziesii* by Pinaceae specialists when grown in host mixtures. In a greenhouse forest soil bioassay that examined shared compatibility between Douglas-fir and *Betula papyrifera*, Simard et al. (1997a) found that Douglas-fir only formed ectomycorrhizae with *Tuber* spp. when grown in mixture with *Betula*, and the mixture treatment also affected frequency and abundance of retrieved morphotypes.

Host neighbor effects also occur under natural field conditions. Jones et al. (1997) found that the evenness of the EM fungal community on Douglas-fir was greatest when seedlings were planted in mixture with *Betula papyrifera* seedlings in clearcuts, although overall richness of EM types was not affected. Nara (2006a, b) detected the *Larix* specialist *Suillus larcinus* on a *Betula* seedling when growing next to a *Larix* sapling. Similarly, Horton (unpublished data) found the *Arctostaphylos* specialist *Leccinum manzanitae* on neighboring *Pinus contorta* roots in a sand dune habitat, corroborating earlier pure culture synthesis results of Molina and Trappe (1982). Presence of ericaceous plants can influence the EM fungal community of forest trees. Kohout et al. (2011) report that neighboring *Vaccinium* significantly promoted abundance of *Rhizopogon salebrosus* and inhibited *Thelephora terrestris* on pine; *Wilcoxina* only occurred on pine when *Vaccinium* was present.

Abiotic factors such as soil composition, chemistry, and soil moisture can influence the growth and establishment of different EM fungal species (Baar and de Vries 1995; Koide et al. 1998; Conn and Dighton 2000; Dighton et al. 2000; Cullings et al. 2003; Cavender-Bares et al. 2009). For example, the litter of some plants and resulting decomposition products may influence the EM fungal community found on adjacent hosts. Aponte et al. (2010) examined the EM fungal community of two co-occurring Mediterranean oaks, and of 69 OTUs (operational taxonomic units) recovered, only 13 were found on both oak species; 29 were exclusive to *Quercus canariensis* and 27 only on *Q. suber*. They found that Ca content was highest under the winter deciduous *Q. canariensis* and that differences in EM fungal communities were correlated with Ca content in the soil. Morris et al. (2009) also suggest that differences in litter quality affected host preferences between two *Quercus* species in a tropical cloud forest in southern Mexico.

Most field reports of ecological specificity phenomena are based on observational data of host-fungus associations and frequency of occurrence. Tests of hypotheses regarding expression of ecological specificity are rare and needed to improve our understanding of the factors influencing mycorrhiza specificity. Hayward and Horton (2012), for example, tested whether the host specific nature of EM fungi associated with *Pisonia grandis* on the Pacific island of Rota was due to soil or host factors. *Pisonia grandis* associates with a restrictive set of EM fungi and throughout much of its range occurs in habitats rich in guano. Cairney et al. (1994),

Chambers et al. (2005), and Suvi et al. (2010) suggest that this unique habitat has shaped the EM fungal associates of *P. grandis*, i.e., an expression of ecological specificity. Hayward and Horton (2012) found that *P. grandis* formed ectomycorrhizae with the same set of EM fungi on guano rich and guano poor habitats, and that several EM fungi on neighboring EM hosts (*Instia bijuga* and *Casaurina equisitifolia*) were not observed with *P. grandis*. They concluded that edaphic factors (i.e., ecological specificity phenomena) did not explain the host specialist fungal associations of *P. grandis*, and that specificity may be due to derived or ancestral characters within *Pisonia*.

1.3.1 Host Preference

When Newton (1991) grew *Quercus robur* and *Betula pendula* seedlings in a variety of soils in England, he found dissimilar EM fungal communities; although the two most common fungi were found on each host, they differed in abundance. He stated that this type of "ecological specificity" accounted for the distinct EM fungal communities of oak and birch, and should more accurately be termed "host preference". With the advent of molecular tools to identify host-fungus associations of field collected roots and statistical testing for host association, several field studies have subsequently demonstrated the widespread prevalence of host preference in EM systems (Kranabetter et al. 1999; Cullings et al. 2000; Kernaghan et al. 2003; Ishida et al. 2007; Tedersoo et al. 2007a, 2008a, 2010b, 2011; Morris et al. 2008, 2009; Cavender-Bares et al. 2009; Smith et al. 2009, 2011; Diédhiou et al. 2010; Wolfe and Pringle 2011). Given the difficulty of demonstrating absolute host-fungus specificity in the field (Taylor 2002; Dickie 2007; Tedersoo et al. 2010b) and the widespread nature of host preference, Dickie and Moyerson (2008) state that host preference (rather than host specificity) may be considered "more the rule rather than the exception" in diverse EM fungal communities. Although host preference is often displayed among taxonomically distant hosts (e.g., angiosperms versus conifers) or at the family level, it also occurs between closely related taxa such as co-occurring *Quercus* species (Morris et al. 2008, 2009; Cavender-Bares et al. 2009; Aponte et al. 2010).

Although widespread, the degree of host preference exhibited in different EM fungal communities can vary from high to low. For example, in a neotropical forest of the western Amazonia, Tedersoo et al. (2010b) found that two thirds of the EM fungi preferred one of three hosts examined, and four of the six most frequent EM fungi showed statistically significant host preference at the host genus level but not at the species level. Tedersoo et al. (2008a) similarly found strong host preference of EM fungi in a Tasmanian sclerophyll forest. In a mixed boreal forest in Canada, Kernaghan et al. (2003) found dissimilar EM fungal communities between angiosperm and conifer hosts, with some fungal species showing a preference for *Abies/Picea* and others for *Populus/Betula*; overall, 30 % of the most abundant EM fungi expressed host specificity and 25 % expressed various levels of host preference. Similarly, Ishida et al. (2007) showed a high degree of host specificity and preference among eight EM hosts in a mixed conifer-broadleaf forest in Japan; host preference was most common at the host family level. In contrast, Tedersoo et al. (2011) found low levels of host preference in wooded savannahs and rain forests of Africa, while Smith et al. (2013) report low levels of host preference between distantly related ectomycorrhizal hosts in neotropical highlands of the Guiana shield in Guyana. Smith et al. (2011) detected no host preference for EM fungi associated with three co-occurring leguminous host trees in a neotropical rainforest (a sharp difference to results of Tedersoo et al. (2010b) in neotropical western Amazonia). EM fungi that display host preference are not necessarily restricted to those found in low abundance or with a restricted host range. Several studies show that many of the most common fungi in EM fungal communities, including many multi-host fungi, can display host preference (Kranabetter et al. 1999; Kernaghan et al. 2003; Tedersoo et al. 2008a, 2010b). As with general specificity phenomena, host preference is likely to be influenced by environmental conditions. Also, as nicely argued in Taylor (2002), the typically high species richness in EM fungal communities on root tips and their cryptic nature makes it very difficult to sample the number of root tips needed to adequately sample the mycorrhizal root types of all the plant species in a plot. This issue has implications for our ability to fully document host preference, especially for species that are observed on a limited number of samples. Further, environmental factors that favors both host and fungal species may give a false impression of host preference at the root tip level. It is critically important that future field studies use robust sampling methods that provide strong statistical inferences regarding the interpretation of host preference patterns.

Many factors determine how an EM fungus responds to new hosts when introduced into a novel geographic range, including the degree of host specificity displayed in the native range (i.e. generalist to specialist tendencies), compatibility with newly encountered hosts, niche availability, interactions with the native fungal flora, soil conditions and other biotic factors (Molina et al. 1992; Vellinga et al. 2009; Wolfe and Pringle 2011). The introduction and spread of *Amanita phalloides* into North America (Wolfe and Pringle 2011) provides a robust example of how these factors interact with relevance to ecological specificity and host preference. Wolfe and Pringle (2011) conducted an extensive survey of the geographic distribution of *A. phalloides* across its native range in Europe and expanded range in N. America, and tested for host selectivity and niche shifts. In Europe *A. phalloides* primarily associates with *Quercus* and other Fagaceae, but rarely with Pinaceae. In N. America, *A. phalloides* associates primarily with Pinaceae in the East Coast, rarely spreading into natural forests. On the West Coast it is more widespread, occurs in native forests, and as it is in Europe, is most commonly associated with *Quercus*. Although the 11 documented novel host associations (host shifts) in N. America is indicative of broad host compatibility, the authors did not consider *A. phalloides* a true "generalist." Instead, they state that *A. phalloides* exhibits "geographically structured host specificity." In California, for example, *A. phalloides* "selectively" associates with

Quercus agrifolia, an evergreen oak, and the distribution of the fungus strongly correlates with the distribution of this oak species. Wolfe and Pringle (2011) suggest that association with *Q. agrifolia* may provide a competitive advantage of *A. phalloides* over the local EM fungal flora and allow it to persist and spread, as an outcome of ecological specificity and host preference. They conclude by stating "specificity in local habitats can influence the success of introduced mutualist species even when the species otherwise appears a generalist."

A complex of factors contribute to the expression of host preferences and differences among hosts and fungi. Tedersoo et al. (2010b) list historical factors, specialized habitat, partial autotrophy, as well as phylogenetic and physiological differences among hosts as important contributors to host preference expression. In the absence of absolute host specificity as determined by genetic factors, Dickie (2007) hypothesizes that host preference is in essence an expression of "realized niche" that may be driven by competitive interactions among the EM fungi in the community, or alternatively, by direct host selection of a particular fungus, i.e., that a host selectively provides resources (e.g. photosynthates) to a preferred fungal species that is highly beneficial to the host. Discussion of "partner choice" has received substantial attention in the recent AM literature, including how such host-fungus interactions may yield significant host preference and stability to mycorrhizal symbioses (Kiers and van der Heiden 2006; Kiers et al. 2011; Bever et al. 2009). Bever et al. (2009) demonstrated preferential allocation of photosynthate by *Allium* to a mutualistic *Glomus* species rather than non-beneficial *Gigaspora margarita* when the plant was mycorrhizal with each fungus growing in separate split root compartments; the preferential C allocation also increased fitness (spore number) of *Glomus* under these growth conditions. Kiers et al. (2011) also report how significant host preference between *Medicago truncatula* and three *Glomus* species resulted in both preferential carbon allocation to the most beneficial fungus and the ability of the cooperative fungi to transfer more P to those roots providing greatest access to photosynthate (i.e. that fungi can discriminate among carbon supply by different hosts). Fungi were not separated into compartments in these experiments. Kiers et al. (2011) suggest that such reciprocal rewards in mycorrhizal host-fungus interactions contribute to stability of the mycorrhizal mutualism.

Selective allocation of resources to differentially beneficial fungi in EM systems has yet to be demonstrated. Some host specific fungi, however, can provide greater benefit to their specific host than generalist fungi. For example, Gorissen and Kuyper (2000) found that pine seedlings inoculated with the host-specialist fungus *Suillis bovinus* took up more nitrogen than seedlings inoculated with the host generalist *Laccaria bicolor*. Indirect evidence for increased allocation of N by host specialist fungi compared to generalists is also supported by higher $^{15}N/^{14}N$ ratios among host specialist fungi (Taylor et al. 2003; Hobbie et al. 2005). Chu-Chou and Grace (1985) found that the pine specialists *Rhizopogon rubesens* and *R. luteolus* were more effective symbionts for *Pinus radiata* than the host generalists *Laccaria laccata* or *Hebeloma crustuliniforme*. *Rhizopogon vinicolor* provided greater drought tolerance to its specific host Douglas-fir than the generalists *Laccaria laccata* or *Pisolithus tinctorius* (Parke et al. 1983) and stimulated higher photosynthetic rates than *L. laccata* or *H. crustulinforme* (Dosskey et al. 1990). Further experimentation in all mycorrhizal systems is needed to explore the mechanisms for expression of mycorrhiza specificity, particularly the influence of resource allocation between the symbionts, and how this affects formation and function of CMNs.

In summary, the relevance of host specificity, ecological specificity, and host preference can be substantial in diverse ecosystems and influence the formation and function of CMNs. Host specificity and preferences affect the structure of plant and fungal communities and successional dynamics (see section on plant community dynamics). Several studies show that increased host diversity on the landscape is often accompanied by higher levels of host specificity and higher fungal diversity than in locations with low EM host diversity (Newton and Haigh 1998; Kernaghan et al. 2003; Kernaghan 2005; Debellis et al. 2006; Ishida et al. 2007; Dickie 2007; Tedersoo et al. 2012). Host specificity and preference create unique niches at the order of host roots, providing opportunities for multiple mycorrhizal fungi to persist and function, and also affect resource partitioning among sympatric hosts (Dickie 2007; Ishida et al. 2007; Peay et al. 2008; Tedersoo et al. 2008b; Horton et al. 2013). Host specificity and preference will also affect natural migration or exotic introductions of fungi, influencing their ability to form mycorrhiza with potential compatible hosts, and as such influence invasive potential (Karst et al. 2014; Vellinga et al. 2009; Wolfe and Pringle 2011). Finally, in an evolutionary context, potential expression of host specificity or preference may influence host shifts and thus contribute to fungal speciation (Kretzer et al. 1996; Halling 2001; den Bakker et al. 2004; Wilson et al. 2012).

1.4 Influence of Mycorrhiza Specificity on Plant Community Dynamics and Ecosystem Resiliency

Numerous reviews highlight that different plant species, often from different families, can be colonized by the same EM fungus when grown together in experimental bioassays (pot cultures) or naturally in the field (Table 1.2; Newman 1988; Simard and Durall 2004; Selosse et al. 2006). This potential is provided by the often abundant and dominant presence of fungal species with intermediate to broad host-ranges. Such overlap in host compatibilities and formation of functioning CMNs between diverse hosts can strongly influence plant community dynamics during primary and secondary plant succession, and overall ecosystem resiliency (Molina and Trappe 1982; Perry et al. 1989; Molina et al. 1992; Horton and van der Heijden 2008; van der Heijden and Horton 2009; Kennedy et al. 2012; Nara, Chap. 6, this volume). During plant succession many early seral plant species act as "legacy" or "refuge" plants wherein they establish (in primary succession; Nara 2006a, b, this volume) or maintain (in secondary succession; Horton et al. 1999) a diversity of EM fungi that will benefit later seral plants. Maintenance of EM fungal biodiversity and functional diversity creates a positive feedback beneficial to ecosystem recovery and resilience (Perry et al. 1989; Molina et al. 1992). Below we provide one robust example of legacy plant function and facilitation of plant community dynamics via potential ectomycorrhizal CMNs from our work with arbutoid mycorrhizal hosts in Western North America, and then briefly discuss other examples from diverse ecosystems worldwide (refer to Table 1.2 for additional details on the degree of overlap between hosts and experimental conditions of the exemplified studies).

While the examples below from different EM plant communities illustrate potential aspects of facilitation via CMNs, additional field studies are needed to add support for these ideas with clear empirical evidence. Attention is particularly needed on the functional differences between dominant generalist EM fungi and those fungi that show different levels of host preference or specificity.

1.4.1 Arbutoid Mycorrhizal Legacies in Secondary Succession

Research on arbutoid mycorrhizal host genera *Arbutus* and *Arctostaphylos* in the frequently disturbed Pinaceae forests of the western USA exemplifies the role mycorrhizal networks can play in plant community dynamics. Arbutoid hosts have a broad receptivity towards a diversity of EM fungi (Zak 1976a, b), leading Molina and Trappe (1982) to hypothesize that the plants maintain a reservoir of diverse EM fungi through disturbance events that support the establishment of later successional Pinaceae. Amaranthus and Perry (1989) and Borchers and Perry (1990) demonstrated a positive benefit in ectomycorrhiza formation and growth of *Pseudotsuga menziesii* (Douglas-fir) seedlings when inoculated with soil taken from beneath *Arbutus menziesii*. When seedlings of *A. menziesii* were grown in multispecies, pot cultures of Douglas-fir, grand fir (*Abies grandes*), and ponderosa pine (*Pinus ponderosa*) containing soil from a mature mixed evergreen forest, they developed seven EM morphotypes, and shared six of these with grand fir and ponderosa pine, and five with Douglas-fir (Massicotte et al. 1999). Notably, one of the fungi shared with *Arbutus* was a *Rhizopogon* sp., a Pinaceae specialist. Horton et al. (1999) were the first to investigate the facilitating nature of *Arctostaphylos glandulosa* associated EM networks for Douglas-fir seedlings. Douglas-fir established significantly better under *Arctostaphylos* compared to under the AM *Adenostoma fasciculatum* even though most environmental factors were conducive for Douglas-fir establishment under *Adenostoma*. They found that 17 of the 24 EM fungi colonizing Douglas-fir seedlings growing within the *Arctostaphylos* patches were also found on *Arctostaphylos*, and 49 % of the Douglas-fir EM root biomass associated with fungi also observed on *Arctostaphylos* roots in the same soil core. The authors hypothesized that Douglas-fir establishment in *Arctostaphylos* was likely to be facilitated via the EM fungi supported by *Arctostaphylos*.

Table 1.2 Context of ectomycorrhizal fungus colonization of multiple host species in diverse laboratory and field settings

Hosts	Overlap in EM Fungi	Methodology	Citation
Arbutus menziesii, *Arctostaphylos uva-ursi*, Pinaceae	30 species on arbutoid and Pinaceae	Pure culture synthesis	Zak (1976a, b) and Molina and Trappe (1982)
Arbutus menziesii, *Pseudotsuga menziesii*, *Abies grandis*, *Pinus ponderosa*	6 EMF on *Arbutus/Abies/Pinus* 5 EMF on *Arbutus/Pseudotsuga*	Soil bioassays/morphotype ID	Massicotte et al. (1994)
Arctostaphylos uva-ursi, *Pinus resinosa*	5 EMF on both hosts	Field EM roots/morphotype ID	Visser (1995)
Tsuga heterophylla, *Pseudotsuga menziesii*	11 EMF on both, one *Pseudotsuga* specialist observed on *Tsuga*	Soil bioassay/morphotype ID	Smith et al. (1995)
Pseudotsuga menziesii, *Betula papyrifera*	91 % of Betula and 56 % of *Pseudotsuga* EM morphotypes were on both hosts	Field EM roots/morphotype ID	Jones et al. (1997)
Pseudotsuga menziesii, *Betula papyrifera*	7 of 11 EMF on both hosts	Soil bioassay/morphotype ID	Simard et al. (1997a)
Pseudotsuga menziesii, *Pinus muricata*	12 of 16 EMF on both hosts	Field EM roots/molecular ID	Horton and Bruns (1998)
Arctostaphylos glandulsa, *Pseudotsuga menziesii*	17 of 24 EMF observed on *Pseudotsuga* also observed on *Arctostaphylos*	Field EM roots/molecular ID	Horton et al. (1999)
Lithocarpus densifolia, *Pseudotsuga menziesii*, *Abies grandis*, *Pinus ponderosa*	*Lithocarpus* showed 50 % overlap with Pinaceae EMF	Soil bioassay/morphotype ID	Massicotte et al. (1999)
Pinus contorta, *Picea glauca*, *Abies lasiocarpa*	74 EMF, 35 on all three hosts	Field EM roots/morphotype ID	Kranabetter et al. (1999)
Pinus contorta, *Picea engelmannii*	28 EMF, 21 on both hosts, 5 specific to *Picea*, none specific to *Pinus*	Field EM roots/molecular ID	Cullings et al. (2000)
Arctostaphylos uva-ursi, *Pseudotsuga menziesii*	17 morphotypes on *Pseudotsuga*, 14 morphotypes on *Arctostaphylos*, 10 EMF morphotypes (6 confirmed with RFLP typing) found on both	Field EM roots/morphotype and molecular ID	Hagerman et al. (2001)

(continued)

Table 1.2 (continued)

Hosts	Overlap in EM Fungi	Methodology	Citation
Lithocarpus densiflora, Pseudotsuga menziesii	56 EMF, 17 on both hosts	Field EM roots/molecular ID	Kennedy et al. (2003)
Helianthemum bicknellii, Quercus spp/Q.macrocarpa	8 EMF on *Helianthemum*, 7 of which also on *Quercus*	Field EM roots/molecular ID	Dickie et al. (2004)
Arbutus unedo, Quercus ilex	46 RFLP types on *A. unedo* 18 RFLP types also on *Quercus*	Field EM roots/molecular ID	Richard et al. (2005)
Tsuga heterophylla, Pseudotsuga menziesii	55 % overlap in early successional setting 14 % overlap in late successional setting	Field EM roots/molecular ID	Horton et al. (2005)
Cistus landanifer, Pinus pinaster	30 EMF with *Cistus*, many known *Pinus* associates	Fruitbody occurrence	Martin-Pinto et al. (2006)
Betula papyrifera, Pseudotsuga menziesii	105 EMF, 42 on both hosts, 23 only on *Pseudotsuga*, 40 only on *Betula*	Field EM roots/molecular ID	Twieg et al. (2007)
Betulaceae, Fagaceae, Pinaceae	14 EMF on only one host family, 37 EMF on both conifer and broadleaf host, 19 EMF on all three families, 6 EMF on Pinaceae and Betulaceae, 12 EMF on Pinaceae and Fagaceae 24 EMF on Betulaceae and Fagaceae	Field EM roots/molecular ID	Ishida et al. (2007)
Arbutus menziesii, Pseudotsuga menziesii, Pinus spp.	Study 1: 126 EMF on *Arbutus*, 17 also with Pseudotsuga or *Pinus* Study 2: 82 EMF, 25 on *Arbutus* and *Pseudotsoga*	Field EM roots/molecular ID	Kennedy et al. (2012)
Pterospora andromedea, Pinus strobus	Numerous EMF on *Pinus*, 1 *Rhizopogon* on both hosts	Field EM roots and bioassay/molecular ID	Hazard et al. (2012)
Pakaraimaea dipterocarpacea, Dicymbe jenmanii	16 of 52 OTUs shared, 13 of the 17 most common shared	Field EM roots/molecular ID	Smith et al. (2013)

(continued)

Table 1.2 (continued)

Hosts	Overlap in EM Fungi	Methodology	Citation
Castanea dentata, 8 other EM hosts in Fagaceae, Pinaceae and Betulaceae	71 RFLP types (EMF) with 41 found only on a single host, 22 of which were observed on a single root tip. *Castanea* was colonized by 24 EMF also associated with *Quercus rubra*, 7 EMF also associated with *Fagus grandifolia*, 6 EMF also with *Quercus alba, 4 EMF also with Betula lenta, 2 each with Ostrya virginiana, Tusga canadensis, and Pinus strobus.*	Field bioassay with *Castanea dentate* seed planted into a mixed deciduous forest. Field EM roots/molecular ID of plants and fungi from root tips.	Dulmer et al. (2014)
Arbutus unedo, *Cistus albidis*, *Quercus ilex*, *Q. coccifera*	151 total OTUs discovered, 3.3 % were found on all four host species, 14 % on three species, 27 % on two species, and 56 % on single hosts; multi-host fungi were the most frequent with 5 of the 8 found on all four hosts	Field EM roots/molecular ID	Taschen et al. (2015)

Arbutoid hosts in other regions also potentially facilitate successional dynamics of neighboring EM hosts. In Eastern Canada, Danielson (1984) synthesized in pure culture several EM fungi associated with *Pinus resinosa* onto *Arctostaphylos uva-ursi*, a common understory prostrate shrub that survives after timber removal. Visser (1995) and Hagerman et al. (2001) concluded that *A. uva-ursi* acts as an important refuge plant following timber harvest and natural disturbance, maintaining fungal diversity and inoculum important to later seral Pinaceae. Similarly, Richard et al. (2005) noted that of 46 rflp EM types found on *Arbutus unedo*, 18 types were shared with *Quercus ilex* in an old-growth Mediterranean forest dominated by *Q. ilex*; they hypothesized that *Arbutus unedo* may play an important role in the early succession dynamics of *Q. ilex*. In a follow-up study, Richard et al. (2009) demonstrated that *A. unedo* facilitated the establishment of *Q. ilex* in a shrub dominated community by enhancing seedling survival and EM colonization, mirroring results of Horton et al. (1999) for Douglas-fir seedling establishment under *Arctostaphylos* in a California chaparral.

More recently, Kennedy et al. (2012) used molecular tools in a field study of the EM fungal community of *Arbutus menziesii* in two sites with intermixed Pinaceae hosts in southwest Oregon. On one site they encountered 126 total fungal taxa on *Arbutus*, 17 of which also occurred with Pinaceae (*Pseudotsuga menziesii* and *Pinus* spp.); in the second site, of 82 total fungal taxa found, 25 colonized *Arbutus* and Douglas-fir, 13 of which were detected on both hosts in single soil cores. They also noted that the EM fungal community associated with *Arbutus menziesii* was phylogenetically similar in structure to that seen with Pinaceae and angiosperms in the genera *Quercus* and *Cercocarpus* in the region. In addition to the numerous “multi-host” fungi supported by *Arbutus*, they also noted that *Arbutus* plants were colonized by two *Rhizopogon* species that are well known Pinaceae specialists, similar to the results reported by Massicotte et al. (1999). Kennedy et al. (2012) describe *Arbutus menziesii* as a “hub” in the CMN in space and time, promoting ecosystem resiliency by maintaining EM fungal diversity, soil microbial processes, and facilitating the establishment of later seral trees. Overall, results from Horton et al. (1999), Richard et al. (2009) and Kennedy et al. (2012) strongly support the earlier hypotheses by Molina and Trappe (1982) and Perry et al. (1989) on the positive feedback provided by these pioneering shrubs and trees in these frequently disturbed plant communities.

1.4.2 Other Examples of Potential Facilitation in Plant Community Dynamics

Similar to arbutoid mycorrhizal plants, pioneering Cistaceae species may facilitate *Pinus* and *Quercus* species establishment in Mediterranean ecosystems (Martin-Pinto et al. 2006), and Quercus in oak savanaahs of the central USA (Dickie et al. 2004). Overlap in EM fungal associates can also shape successional patterns between dominant and subdominant EM forest trees. In Western North America, such facilitation may be involved among *Pinus contorta*, *Picea glauca*, and *Abies lasiocarpa* in British Columbia (Kranabetter et al. 1999), between pioneering lodgepole pine (*Pinus contorta*) and later seral Engelmann spruce (*Picea engelmannii*) in Wyoming, USA, and between Douglas-fir and western hemlock (*Tsuga heterophylla*) in the wet temperate forests of the Pacific Northwest, North America (Smith et al. 1995; Horton et al. 2005). The subdominant tree tanoak (*Lithocarpus densiflora*, Fagaceae), which also occurs as a shrub and stump sprouts following fire, may influence successional dynamics of neighboring Pinaceae, particularly Douglas-fir, through the CMNs of shared EM fungi (Massicotte et al. 1999; Kennedy et al. 2003). The extensive ectomycorrhiza research on the paper birch (*Betula pendula*)—Douglas-fir ecosystem in British Columbia confirmed not only the operlap in shared EM fungi between these EM hosts (Simard et al. 1997a; Jones et al. 1997; Twieg et al. 2007), but also demonstrated the transfer of isotopically labeled carbon from the pioneering birch to the later seral Douglas-fir via commonly shared EM fungi in the field (Simard et al. 1997b).

In one of the larger temperate forest molecular studies of overlap in shared EM fungi between codominant EM tree species, Ishida et al. (2007) examined 8 hosts belonging to 6 genera in three families (Betulaceae, Fagaceae, Pinaceae) in two mixed conifer-broadleaf forests of Japan. Although the EM fungal communities were similar among hosts, a significant portion of fungi showed host specificity, primarily at the family level, but also at the genus level. Some host generalists also showed statistically supported preference towards particular tree species. The authors emphasized that high diversity of EM hosts and the strong display of various host specificities and preferences by the fungi increased diversity among the total EM fungal community, i.e., high EM host diversity in a stand contributes to high EM fungal diversity, a pattern also recently revealed at the global level (Tedersoo et al. 2014). From a successional standpoint, they noted that the higher proportion of broad-host ranging fungi in the secondary forest stand compared to the old growth (primary) forest may benefit *Abies homolepis* (a late successional tree), as it is able to share several mycorrhizal fungi with the early seral broadleaf (angiosperm) species.

Many small stature, understory forest plants associate with diverse EM fungi, and may thus influence EM community dynamics of dominant overstory EM tree species. For example, several species of understory plants in the Pyroleae are widespread in northern temperate forests, survive in dense shade of the forest canopy, and form mycorrhiza with many EM fungi common to overstory EM trees (Tedersoo et al. 2007a; Zimmer et al. 2007; Massicotte et al. 2008; Hynson and Bruns 2009). Although many of the fungi recovered on the roots of Pyroleae in these studies were considered host generalists, Tedersoo et al. (2007a) noted some host preference for *Tricholoma* species and that some pyroloids hosted the Pinaceae specialist *Suillus variegatus*. Similarly, Zimmer et al. (2007) and Hynson and Bruns (2009) noted the Pinaceae specialists *Rhizopogon* spp. in roots of *Pyrola picta*. Such strong overlap between understory myxotrophic Pyroleae and overstory EM hosts for shared EM fungi raises interesting questions regarding the potential ecological interactions between these forest plants.

Potential facilitation via CMNs has also been proposed in tropical forests and the savanaahs of Africa. Alexander et al. (1992) found that seedlings of *Instia palembanica* (Caesalpinaceae) planted in proximity to *Shorea leprosula* (Dipterocarpaceae) more rapidly formed ectomycorrhiza than seedlings planted distant to *Shorea*, and emphasized the practical nature of maintaining *Shorea* trees on harvested forest sites to maintain fungal inoculum for seedlings of *Instia* or other EM hosts. Similarly, Onguene and Kuyper (2002) found survival and ectomycorrhiza formation of *Paraberlinia bifoliota* seedlings planted in contact with four adult EM hosts species were higher for seedlings under *Brachystegia cynometriodes* than under conspecific adults; they noted that this observation may influence a forester's choice in selecting and maintaining particular tree species after tree harvest. Tedersoo et al. (2011) examined EM fungi on roots of over 30 EM tree host species across 4 sites in wooded savannahs and rainforests of continental Africa and

Madagascar and found that their results support earlier hypotheses that "pioneer Phylanthaceae may facilitate the establishment of late-succession Fabaceae and potentially other EM hosts by providing compatible fungal inoculum in deforested and naturally disturbed ecosystems of tropical Africa." Diédhiou et al. (2010) demonstrated the EM fungus linkage potential between seedling and adult roots of five EM hosts from four genera in a tropical rain forest of Guinea: *Anthonotha*, *Cryptosepalum* and *Paramacrolobium* in Fabaceae, and *Uapaca* in Phyllanthaceae. They concluded that the adult hosts in the EM network "likely function as 'nurse trees' for conspecific and non-conspecific seedlings and therefore promote diversity and coexistence of species in this forest". In neotropical forests, Tedersoo et al. (2010b) and Smith et al. (2011) report similar results for EM fungi associated with diverse EM Fabaceae trees (*Dicymbe corymbosa*, *D. altosonii*, and *Aldina insignis*). They found that the dominance of these Fabaceae EM hosts in this ecosystem, the diversity of EM fungi they support, and the strong overlap in shared EM fungi among the trees may facilitate perpetuation of this EM guild of plants and fungi.

1.4.3 Primary Succession

Pioneering plants in primary succession are well known to facilitate many biological processes, often acting as foci for establishment of later seral plants, and presence of mycorrhizal fungi are key to their establishment (Nara 2008; Nara, Chap.6, this volume). In a series of experiments, Nara and colleagues have demonstrated the facilitative nature of pioneering dwarf willow (*Salix reinii*) on the volcanic desert areas of Mt. Fuji in Japan (Nara and Hogetsu 2004; Nara 2006a, b). Seedlings of *Betula* and *Larix* are commonly observed in close proximity to adult dwarf willow shrubs in this area and were shown to form ectomycorrhizae with several of the same generalist fungi found on willow. Overall, their studies lend strong support to the facilitative nature of pioneering dwarf willow on successional dynamics of later tree establishment as affected by CMNs of mutually compatible fungi. See Nara (Chap. 6, this volume) for more details on this set of elegant experiments.

1.4.4 Potential Exceptions to Facilitation

Although we cite above many potential examples of facilitation as suggested by the frequent ability of diverse host species to share multi-host EM fungi, there are probably widespread exceptions, because EM hosts and fungi can also display various degrees of specificity and preference. Perhaps the best example in this regard is displayed by the EM nature of the genus *Alnus*. As discussed previously, *Alnus* species worldwide associate with a relatively low diversity of EM fungi, many of which are host specialists with *Alnus* (Molina 1981; Tedersoo et al. 2009; Kennedy and Hill 2010; Kennedy et al. Chap. 8, this volume), thus lowering the potential for

CMNs with other nearby EM host species (Horton et al. 2013). In a recent study, Bent et al. (2011) examined shared EM fungal communities and the potential for successional facilitation of *Betula papyrifera*, *Picea glauca*, and *Populus tremuloides* seedlings growing near pioneer *Alnus viridis* shrubs in a recent fire-disturbed boreal ecosystem in interior Alaska. They found that the *A. viridis* EM fungal community contained several *Alnus* specialist fungi, and was distinct from the fungal communities noted on the adjacent three hosts. Although there was minimal overlap between alder and the other three hosts via minor ribotypes, the *Populus* and *Betula* seedlings showed a strong overlap in shared fungi and a moderate overlap with *Picea* seedlings. They concluded that a facilitative relationship between *Alnus* and the other three hosts was unlikely, but that the *Populus*, *Betula*, and *Picea* hosts had high potential for interacting with each other through CMNs. For more on *Alnus* mycorrhizal networks see Chap. 8 in this volume by Kennedy and colleagues.

Some distantly related EM host genera might display little overlap in compatible EM fungi when growing in proximity, limiting potential facilitation. For example, Smith et al. (2009) found different EM fungal community structure and strong host preference in a stand of mixed *Quercus* and *Pinus*; with the exception of a few "multi-host" fungal taxa, most of the dominant EM fungi on *Pinus* rarely associated with—or had low frequency on *Quercus*, and vice versa. They concluded that "multi-host" fungi may be less dominant in some EM systems than previously thought. However, high host preference in an EM system does not necessarily rule out abundant overlap in shared EM fungi and potential facilitation effects. For example, Morris et al. (2008) found contrasting EM fungal communities on two co-occurring oaks in California. However, 40 of the 140 fungal taxa identified occurred on both hosts, and 13 of the 16 most frequent fungi were shared between the two oaks. When Morris et al. (2009) examined two neighboring oak species in a tropical cloud forest in southern Mexico, of 154 EM species recovered, 62 (40 %) occurred only on *Quercus laurina*, 52 (34 %) only on *Q. crassifolia*, yet 40 (26 %) occurred on both. Similarly, Aponte et al. (2010) noted significant host preference among EM fungi for neighboring *Q. canariensis* and *Q. suber* in southern Spain, yet they still shared about 20 % of the total EM fungi recovered. Tedersoo et al. (2008a) also showed strong host preference in a Tasmanian wet sclerophyll forest among *Eucalyptus regnans*, *Ponaderrisa petala*, and *Nothofagus cunninghamii*; "two thirds of the most common EM fungi from several linages was significantly influenced by host species." Still, they note that the ability of *Nothofagus cunninghamii* and *Eucalyptus* to associate with the same EM fungi may facilitate late-successional *N. cunninghamii*.

1.4.5 Potential for Long-Term EM Legacies to Affect Plant Migration During Climate Change

EM hosts migrated during past episodes of climate change and will continue to do so under current climate change scenarios (Jacobson et al. 1987). The ability for EM fungal species to colonize diverse EM hosts will affect EM host migration

(Perry et al. 1990; Molina et al. 1992). The EM condition of *Dryas octopetala* in Ireland provides an interesting example of past and potential future legacy. Although primarily a widespread arctic and alpine shrub, *D. octopetala* occurs as a relict population in the lowland grassland-heath in northern Ireland (called the Burren); this area harbored *Pinus sylvestris* forests about 1500 years ago. Harrington (2003) and Harrington and Mitchel (2002, 2005a, b) conducted extensive studies of the EM fungi (sporocarps and EM root tip surveys) and found high diversity of EM fungi. Many of the species found fruiting and on roots are well known EM associates of European woodland forest trees and exhibit a broad host range (Harrington and Mitchell 2002). They hypothesized that *D. octopetala* has maintained the EM fungal community it shared in common with *Pinus* when the two hosts cohabitated the sites. Thus, if EM host trees migrate northward into this region as climate warms, the relict EM fungal community maintained by the *Dryas* legacy would be present to facilitate tree establishment. A similar case can be made for *Dryas* and other EM shrub hosts in Arctic and alpine communities. Ryberg et al. (2009) describe a rich EM fungal community on *D. octopetala* and *Salix reticulata* in an alpine cliff ecosystem; they identified about 70 potential EM fungi and noted low host specificity or preference between *Dryas* and *Salix*. They concluded that the hosts seem likely to facilitate succession of the alpine tundra to subalpine forest by serving as mycorrhizal partners for establishing pioneer trees. The rich EM fungal community on *Dryas* and *Salix* in alpine and arctic locations in Norway (Frederikke et al. 2010), Sweden (Ryberg et al. 2011) and North American (Fujimura and Egger 2012; Timling and Taylor 2012) may do likewise. Timling and Taylor (2012) note that 73 % of the EM fungal ITS OTUs they found on EM root tips of *Dryas integrifolia* and *Salix arctica* in Arctic habitats occur in regions outside the Arctic. These communities are ripe for manipulative experiments to test the hypothesis that ectomycorrhizal fungi associated with *Dryas* and *Salix* will facilitate establishment of migrating EM hosts during climate change.

Krpata et al. (2007) demonstrate similar potential for *Arctostaphylos uva-ursi* to act as legacy plants for afforestation of subalpine and alpine habitats. They observed a diverse group of fungi associated with *A. uva-ursi* based on sporocarp collections, mycorrhiza morphotypes and molecular identification from root tips. Although they considered the majority of EM fungi "generalists", they also discovered that several EM fungi considered specific to other EM hosts developed arbutoid mycorrhizae, e.g., *Lactarius deterrimus* (host specific to *Picea*), *Suillus plorens* and *S. grevillei* (host specific to *Larix*). Mühlmann and Göbl (2006) also discovered *L. deterrimus* forming arbutoid mycorrhizae with *A. uva-ursi* in the Swiss Alps. As noted above for arbutoid host legacies, Krpata et al. (2007) conclude that the ability of *A. uva-ursi* to maintain a high diversity of generalist and specialist EM fungi in these ecosystems, even though they have been treeless for 400 years, may facilitate future afforestation or natural migration of EM tree hosts.

1.4.6 Evolutionary Processes in Specificity Phenomena

This is an exciting time to be doing research on mycorrhizal symbioses with genomic approaches providing important new insights into the evolution of the symbioses and a better understanding of mycorrhiza ecology, including specificity phenomena. To date the genome of one species in Glomeromycota has been sequenced while over 25 ectomycorrhizal fungi have been sequenced from both Basidiomycota and Ascomycota. Glomeromycota are an ancient lineage (Berbee and Taylor 1993; Redecker et al. 2000; Remy et al. 1994) whose appearance in the fossil record over 400 million years before present coincides with the first land plants. Although thought to be entirely clonal, there is evidence for recombination in the group (Croll et al. 2008)) and meiosis specific proteins have now been found (Tisserant et al. 2012). To date, there are close to 250 Glomeromycota species identified by spore morphology, but analyses of environmental ribosomal sequences suggests the richness is closer to 350 or even 1600 operational taxonomic units (Kõljalg et al. 2013; Öpik et al. 2013; Schüssler 2015). Our ability to clarify the diversity of this enigmatic group is hampered because they are not free living, they are ceonocytic with few or no septa separating the nuclei into discrete cells, most have not been cultured (e.g., with bait plants or transformed carrot roots; Ohsowski et al. 2014) and individuals contain multiple genotypes making sequencing problematic. These same issues make our ability to analyze specificity phenomena difficult given the difficulty in identifying fungal species and even individuals in Glomeromycota.

While the Glomeromycota appear to have evolved the mycorrhizal habit only once (van der Heijden et al. 2015), ectomycorrhizal fungi may have independently evolved the symbiotic habit more than 78 times (Tedersoo and Smith 2013). The first ectomycorrhizal fungal species evolved from litter decay fungi about 200 million years ago with little evidence for any species evolving from pathogenic fungi (James et al. 2006; Plett and Martin 2011). Although ectomycorrhizal fungi evolved from saprotrophic lineages and maintain varying levels of saprotrophic capabilities (i.e., can grow on agar), they lack many genes involved in litter decomposition and to date, none are known to be free living in nature (Martin et al. 2008, 2010; van der Heijden et al. 2015).

Genomic evidence from all three phyla of mycorrhizal fungi (Glomeromycota, Ascomycota, and Basidiomycota) indicates that mycorrhizal fungi disrupt the host defense system. However, the molecular tools evolved independently with unique pathways as one may expect given their unique evolutionary histories. For instance, Laccaria bicolor produces a small secreted protein (MiSSP7) that disrupts the plant's jasmonic acid signaling pathway and suppresses the host defense system (Plett et al. 2014). Another small secreted protein, SP7, is produced by the AM fungus *Rhizophagus irregularis* (formally *Glomus intraradices*) and also disrupts the host immune response (Kloppholz et al. 2011). The general pattern may hold for other species as well, and with different proteins and through interactions with different parts of the host defense system. Specificity phenomena are likely governed by similar host-fungus genetic cross talk, but much work is needed on the genetics of specificity interactions. The number of complete ectomycorrhizal fungus genomes available is now at 27 species (van der Heijden et al. 2015) and many more are on the way (Grigoriev et al. 2014). Similar advances are on the way in Glomeromycota, but this group is proving difficult because of its unique genetic system. With these genomic data, our understanding of the evolution of mycorrhiza specificity will rapidly improve.

1.5 Summary and Conclusions

The review of host specificity phenomena by Molina et al. (1992) was shaped largely by evidence from fruitbody occurrence in field settings and direct observations of compatibility based on bioassays and pure culture synthesis experiments. Although these same approaches remain useful for understanding mycorrhiza specificity, the revolution in molecular approaches over the last 20 years has enhanced our ability to examine mycorrhiza specificity at the root tip scale in both greenhouse and field settings (Gardes and Bruns 1993; Bruns et al. 1998; Horton and Bruns 2001). Genome sequences and modern phylogenetic analyses provide considerable new data to examine evolutionary patterns of mycorrhizal symbioses (Kohler et al. 2015; Plett and Martin 2011) with implications for understanding specificity relationships between diverse taxa of plants and fungi. Most importantly, molecular technology has spread to labs around the world, providing data and enhancing our understanding of EM fungi from diverse ecosystems and plant communities.

The following points summarize our general conclusions.

1. Given the complexity of specificity patterns discussed and the dependency of results on the context of the study, we urge caution in overgeneralizing on global patterns of specificity phenomena among diverse fungal and plant taxa.
2. Mycorrhizal networks are directly impacted by specificity phenomena. Evidence strongly suggests that CMNs play a role in plant community dynamics but additional field-based empirical evidence for this role is needed. This knowledge provides tools for resource managers to maintain resilient ecosystems in the face of growing resource extraction and climate change (e.g., maintaining legacy host plants, plant diversity, and avoiding oversimplification of ecosystems).
3. Varying degrees of specificity occur at all levels in the taxonomic hierarchy, although specificity at the species level appears rare. It is important for authors to clearly define terms when discussing host specificity. For instance, we suggest labeling an EM fungus a *Quercus* specialist or Fagaceae specialist depending on the phylogenetic breadth of known hosts.
4. Virtually all ecological communities are dominated by a few species with many others being rare, and ectomycorrhizal fungus communities are no different. Most species that are frequently sampled tend to be found on multiple hosts. In contrast, rarely or infrequently sampled species tend not to be found on multiple

hosts. Therefore, the number samples included in a study impacts the view of host preference and even specificity. Increased sampling efforts and statistical tools for assessing specificity phenomena in the field are enhancing our ability to assess host specificity and preference patterns (Hoeksema, Chap. 9, this Volume).
5. Global understanding of specificity phenomena requires sampling of fungi in multiple communities and in multiple ecological settings—most fungal species remain undersampled.
6. Genomic studies are providing and will continue to provide insights into specificity phenomena.

Acknowledgements We thank Dr. James Trappe and Dr. Jeremy Hayward for their thoughtful discussions on specificity phenomena and comments on earlier drafts of this chapter. Financial support was provided to T.R.H. by and the US Forest Service, PNW Research Station, National Research Initiative award no. 99-35107-7843 from the USDA Cooperative State Research Education and Extension Service, and National Science Foundation awards, DEB 0614384 and 949175.

References

Alexander I, Ahad N, See LS (1992) The role of mycorrhiza in the regeneration of some Malaysian forest trees. Phil Trans R Soc Lond B 335:379–388
Allen TR, Millar T, Berch SM, Berbee ML (2003) Culturing and direct DNA extraction find different fungi from the same ericoid mycorrhizal roots. New Phytol 160:255–272
Amaranthus MP, Perry DA (1989) Interaction effects of vegetation type and Pacific madrone soil inocula on survival, growth, and mycorrhiza formation of Douglas-fir. Can J For Res 19:550–556
Aponte C, García LV, Marañón Gardes M (2010) Indirect host effect on ectomycorrhizal fungi: leaf fall and litter quality explain changes in fungal communities on the roots of co-occurring Mediterranean oaks. Soil Biol Biochem 42:788–796
Baar J, de Vries FW (1995) Effects of manipulation of litter and humus layers on ectomycorrhizal colonization potential in Scots pine stands of different age. Mycorrhiza 5:267–272
Bahram M, Kõljalg Kohout P, Mirshahvaladi S, Tederso L (2013) Ectomycorrhizal fungi of exotic pine plantations in relation to native host trees in Iran: evidence of host range expansion by local symbionts to distantly related host taxa. Mycorrhiza 23:11–19
Bent E, Kiekel P, Brenton R, Taylor DL (2011) Root-associated ectomycorrhizal fungi shared by various boreal forest seedling naturally regenerating after a fire in interior Alaska and correlation of different fungi with host growth responses. Appl Environ Microb 77:3351–3359
Berbee ML, Taylor JW (1993) Dating of the evolutionary radiations of the true fungi. Can J Bot 71:1114–1127
Bergero R, Perotto S, Girlanda M, Vidano G, Luppi AM (2000) Ericoid mycorrhizal fungi are common root associates of a Mediterranean ectomycorrhizal plant (*Quercus ilex*). Mol Ecol 9:1639–1649
Bever JD, Richardson SC, Lawrence BM, Holmes J, Watson M (2009) Preferential allocation to beneficial symbionts with spatial structure maintains mycorrhizal mutualism. Ecol Lett 12:13–21
Bidartondo I (2005) The evolutionary ecology of myco-heterotrophy. New Phytol 167:335–352
Bidartondo BI, Burghardt B, Gebauer G, Bruns TD, Read DJ (2004) Changing partners in the dark: isotopic and molecular evidence of ectomycorrhizal liaisons between forest orchids and trees. Proc R Soc Lond B 271

Borchers SL, Perry DA (1990) Growth and ectomycorrhiza formation of Douglas-fir seedlings grown in soils collected at different distances from pioneering hardwoods in southwest Oregon clear-cuts. Can J For Res 20:712–721

Brundrett M (2004) Diverstiy and classification of mycorrhizal associations. Biol Rev 79:473–495

Bruns TD, Szaro TM, Gardes M, Cullings KW, Pan JJ, Taylor DL, Horton TR, Kretzer A, Garbelotto M, Li Y (1998) A sequence database for the identification of ectomycorrhizal Basidiomycetes by phylogenetic analysis. Mol Ecol 7:257–272

Cairney JWG, Rees BJ, Allaway WG, Ashford AE (1994) A basidiomycete isolated from a *Pisonia* mycorrhiza forms sheathing mycorrhizas with transfer cells on *Pisonia grandis* R. Br. New Phytol 126:91–98

Cavender-Bares J, Izzo A, Robinson R, Lovelock CE (2009) Changes in ectomycorrhizal community structure on two containerized oak hosts across an experimental hydrologic gradient. Mycorrhiza 19:133–142

Cázares E, Smith JE (1995) Occurrence of vesicular-arbuscular mycorrhizae in *Pseudotsuga menziesii* and *Tsuga heterophylla* seedlings grown in Oregon Coast Range soils. Mycorrhiza 6:65–67

Chambers SM, Hitchcock CJ, Cairney JW (2005) Ectomycorrhizal mycobionts of *Pisonia grandis* on coral cays I the Capricorn-Bunker group Great Barrier Reef, Australia. Mycol Res 109:1105–1111

Chambers SM, Curlevski NJA, Cairney JWG (2008) Ericoid mycorrhizal fungi are common root inhabitants of non-Ericaceae plants in a south-eastern Australian sclerophyll forest. FEMS Microbiol Ecol 65:263–270

Chu-Chou M, Grace LJ (1985) Comparative efficiency of the mycorrhizal fungi *Laccaria laccata, Hebeloma crustuliniforme* and *Rhizopogon* species on growth of radiata pine seedlings. New Zeal J Bot 23:417–424

Conn C, Dighton J (2000) Litter quality influences on decomposition, ectomycorrhizal community structure and mycorrhizal root surface acid phosphatase activity. Soil Biol Biochem 32:489–496

Croll D, Wille L, Gamper HA, Mathimaran N, Lammers PJ, Corradi N, Sanders IR (2008) Genetic diversity and host plant preferences revealed by simple sequence repeat and mitochondrial markers in a population of the arbuscular mycorrhizal fungus *Glomus intraradices*. New Phytol 178:672–687

Cullings KW, Vogler DR, Parker VT, Finley SK (2000) Ectomycorrhizal specificity patterns in a mixed *Pinus contorta* and *Picea englmannii* forest in Yellowstone National Park. Appl Environ Microbiol 66:4988–4991

Cullings KW, New MH, Makihija S, Parker VT (2003) Effects of litter addition on ectomycorrhizal associates of a lodgepole pine (*Pinus contorta*) stand in Yellowstone National Park. Appl Environ Microbiol 69:3772–3776

Danielson RM (1984) Ectomycorrhizal association in jack pine stands in northeastern Alberta. Can J Bot 62:932–939

Davidson J, Öpik M, Daniell TJ, Moora M, Zobel M (2011) Arbuscular mycorrhizal fungal communities in plant roots are not random assemblages. FEMS Microbiol Ecol 78:102–115

DeBellis T, Kernaghan G, Bradley R, Widden P (2006) Relationships between stand composition and ectomycorerhizal community structure in boreal mixed-wood forests. Microbiol Ecol 52:114–126

den Bakker HC, Zucarello GC, Kuyper ThW, Noordeloos ME (2004) Evolution and host specificity in the ectomycorrhizal genus *Leccinum*. New Phytol 163:201–215

Dickie IA (2007) Host preference, niches and fungal diversity. New Phytol 174:230–233

Dickie IA, Moyerson B (2008) Towards a global view of ectomycorrhizal ecology. New Phytol 180:263–265

Dickie IA, Guza RC, Krazewski SE, Reich PB (2004) Shared ectomycorrhizal fungi between a herbaceous perennial (*Helianthemum bicknellii*) and oak (*Quercus*) seedlings. New Phytol 164:375–382

Dickie IA, Bolstridge N, Cooper JA, Peltzer DA (2010) Co-invasion by *Pinus* and its mycorrhizal fungi. New Phytol 187:475–484
Diédhiou AG, Selosse MA, Galiana A, Biabaté Dreyfus B, Bâ AM, de Faria SM, Béna G (2010) Multi-hist ectomycorrhizal fungi are predominant in a Guinean tropical rainforest and shared between canopy trees and seedlings. Environ Microbiol 12:2219–2232
Dighton J, Bonilla ASM, Jiminez-Nunez RA, Martinez N (2000) Determinants of leaf litter patchiness in mixed species New Jersey pine barrens forest and it possible influence on soil and soil biota. Biol Fert Soils 31:288–293
Dosskey MG, Linderman RG, Boersma L (1990) Carbon-sink stimulation of photosynthesis in Douglas fir seedlings by some ectomycorrhizas. New Phytol 115:269–274
Douhan GW, Huryn KL, Douhan LI (2007) Significant diversity and potential problems associated with inferring population structure within the *Cenococcum geophilum* species complex. Mycologia 99:812–819
Dulmer KM, LeDuc SD, Horton TR (2014) Ectomycorrhizal inoculum potential of northeastern US forest soils for American chestnut restoration: results from field and laboratory bioassays. Mycorrhiza 24:65–74
Frederikke M, Bjorbaekmo M, Carlsen T, Brysting A, Vralstad T, Hoiland K, Ugland KI, Geml J, Schmacher T, Kauserud H (2010) High diversity of root associated fungi in both alpine and arctic *Dryas octopetala*. BMC Plant Biol 10:244–255
Fujimura KE, Egger KN (2012) Host plant and environment influence community assembly of high arctic root-associated fungal communities. Fungal Ecol 5:409–418
Gardes M, Bruns TD (1993) ITS primers with enhanced specificity for basidiomycetes—application to the identification of mycorrhizae and rusts. Mol Ecol 2:113–118
Geml J, Tulloss RE, Laursen GA, Sazanova NA, Taylor DL (2008) Evidence for strong inter-and intracontinental phylogeographic structure in *Amanita muscaria*, a wind-dispersed ectomycorrhizal basidiomycete. Mol Phylogenet Evol 48:694–701
Gianinazzi-Pearson V, Gianinazzi S (1983) The physiology of vesicular-arbuscular mycorrhizal roots. Plant Soil 71:197–209
Gorissen A, Kuyper TW (2000) Fungal species-specific responses of ectomycorrhizal Scots pine (*Pinus sylvestris*) to elevated [CO_2]. New Phytol 146:163–168
Grelet GA, Johnson D, Paterson E, Anderson IC, Alexander IJ (2009) Reciprocal carbon and nitrogen transfer between an ericaceous dwarf shrub and fungi isolated from *Piceirhiza bicolorata* ectomycorrhizas. New Phytol 182:359–366
Grelet GA, Johnson D, Vrålstad T, Alexander IJ, Anderson IC (2010) New insights into the mycorrhizal *Rhizoscyphus ericae* aggregate: spatial structure and co-colonization of ectomycorrhizal and ericoid roots. New Phytol 188:210–222
Grigoriev IV, Nikitin R, Haridas S, Kuo A, Ohm R, Otillar R, Riley R, Salamov A, Zhao X, Korzeniewski F, Smirnova T, Nordberg H, Dubchak I, Shabalov I. (2014) MycoCosm portal: gearing up for 1000 fungal genomes. Nucleic Acids Res. 42(1):D699–704
Hagerman SM, Sakakibara SM, Durall DM (2001) The potential for wood understory plants to provide refuge for ectomycorrhizal inoculum at an interior Douglas-fir forest after clear-cut logging. Can J For Res 31:711–721
Halling RE (2001) Ectomycorrhizae: co-evolution, significance, and biogeography. Ann Missouri Bot Gard 88:5–13
Harley JL, Smith SE (1983) Mycorrhizal symbiosis. Academic Press, London
Harrington TJ (2003) Relationships between macrofungi and vegetation in the Burren. Biol Environ Proc Roy Irish Acad 103B:147–159
Harrington TJ, Mitchel DT (2002) Characterization of *Dryas octopetala* ectomycorrhizas form limestone karst vegetation western Ireland. Can J Bot 80:970–982
Harrington TJ, Mitchel DT (2005a) Ectomycorrhizas associated with a relic population of *Dryas octopetala* in the Burren, western Ireland I. Distribution of ectomycorrhizas in relation to vegetation and soil characteristics. Mycorrhiza 15:425–433

Harrington TJ, Mitchel DT (2005b) Ectomycorrhizas associated with a relic population of *Dryas octopetala* in the Burren, western Ireland II. Composition, structure and temporal variation in the ectomycorrhizal community. Mycorrhiza 15:435–445
Hart MM, Klironomos JN (2002) Diversity of arbuscular mycorrhizal fungi and ecosystem functioning. In: van der Heijden MGA, Sanders I (eds) Mycorrhizal ecology. Ecological studies, vol 157. Springer, Berlin, pp 225–242
Hart MM, Reader RJ, Klironomos JN (2003) Plant coexistence mediated by arbuscular mycorrhizal fungi. TREE 18:418–423
Hausmann NT, Hawkes CV (2009) Plant neighborhood control of arbuscular mycorrhizal community composition. New Phytol 183:1188–1200
Hausmann NT, Hawkes CV (2010) Order of plant host establishment alters the composition of arbuscular mycorrhizal communities. Ecology 91:2333–2343
Hayward JA, Horton TR (2012) Edaphic factors do not govern the ectomycorrhizal specificity of *Pisonia grandis* (Nyctaginaceae). Mycorrhiza 22:647–652
Hayward J, Horton TR, Nuñez M (2015a) Ectomycorrhizal communities coinvading with Pinaceae species in Argentina: Gringos bajo el bosque. New Phytol. doi:10.1111/nph.13453
Hayward J, Horton TR, Pauchard A, Nuñez MA (2015b) A single ectomycorrhizal fungal species can enable a *Pinus* invasion. Ecology 96:1438–1444
Hazard C, Lilleskov EA, Horton TR (2012) Is rarity of pinedrops (*Pterospora andromeda*) in eastern North America linked to rarity of its unique mycorrhizal host? Mycorrhiza 22:393–402
Helgason T, Merryweather JW, Denison J, Wilson P, Young JPW, Fitter AH (2002) Selectivity and functional diversity in arbuscular mycorrhizas of co-occurring fungi and plants from a temperate deciduous woodland. J Ecol 90:371–384
Helgason T, Merryweather JW, Young JPW, Fitter AH (2007) Specificity and resilience in the arbuscular mycorrhizal fungi of a natural woodland community. J Ecol 95:623–630
Hibbett DS, Matheny PB (2009) The relative ages of ectomycorrhizal mushrooms and their plant hosts estimated using Bayesian relaxed molecular clock analyses. BMC Biol 7:13
Hobbie EA, Jumpponen A, Trappe J (2005) Foliar and fungal ^{15}N:^{14}N ratios reflect development of mycorrhizae and nitrogen supply during primary succession: testing analytical models. Oecologia 146:258–268
Horton TR, Bruns TD (1998) Multiple-host fungi are the most frequent and abundant ectomycorrhizal types in a mixed stand of Douglas fir (*Pseudotsuga menziesii*) and bishop pine (*Pinus muricata*). New Phytol 139:331–339
Horton TR, Bruns TD (2001) The molecular revolution in ectomycorrhizal ecology: peeking into the black-box. Mol Ecol 10:1855–1871
Horton TR, van der Heijden MGA (2008) The role of symbiosis in seedling establishment and survival. In: Leck M, Parker VT, Simpson B (eds) Seedling ecology and evolution. Cambridge University Press, Cambridge, pp 150–171
Horton TR, Bruns TD, Parker VT (1999) Ectomycorrhizal fungi associated with *Arctostaphylos* contribute to *Pseudotsuga menziesii* establishment. Can J Bot 77:93–102
Horton TR, Molina R, Hood K (2005) Douglas-fir ectomycorrhizae in 40- and 400-year-old stands: mycobiont availability to late successional western hemlock. Mycorrhiza 15:393–403
Horton TR, Hayward J, Tourtellot SG, Taylor DL (2013) Uncommon ectomycorrhizal networks: richness and distribution of *Alnus*-associating ectomycorrhizal fungal communities. New Phytol 198(4):978–980
Hynson NA, Bruns TD (2009) Evidence of a myco-heterotroph in the plant family *Ericaceae* that lacks mycorrhizal specificity. Proc R Soc Lond B Biol Sci 276:4053–4059
Hynson NA, Merckx VSFT, Perry BA, Treseder KK (2013) Identities and distributions of the co-invading ectomycorrhizal fungal symbionts of exotic pines in the Hawaiian Islands. Biol Invasions 15:2373–2385
Ishida TA, Nara K, Hogetsu T (2007) Host effects on ectomycorrhizal fungal communities: insight from eight host species in mixed conifer-broadleaf forests. New Phytol 174:430–440

Jacobson Jr GL, Webb III T, Grimm EC (1987) Patterns and rates of vegetation change during the deglaciation of eastern North America. The geology of North America. Geological Society of America. Boulder Co, USA, pp 277–288

James TY, Kauff F, Schoch CL, Matheny PB, Hofstetter V, Cox CJ, et al. (2006) Reconstructing the early evolution of fungi using a six-gene phylogeny. Nature 433:818–822

Johnson NC, Graham JH, Smith FA (1997) Functioning of mycorrhizal associations along the mutualism-parasitism continuum. New Phytol 135:575–586

Jones MD, Durall DM, Harniman SMK, Classen DC, Simard SW (1997) Ectomycorrhizal diversity on *Betula papyrifera* and *Pseudotsuga menziesii* seedlings grown in the greenhouse or outplanted in single-species and mixed plots in southern British Columbia. Can J For Res 27:1872–1889

Karst J, Morgan RJ, Gehring CA (2014) Consequences for ectomycorrhizal fungi of the selective loss or gain of pine across landscapes. Botany 92:855–865

Kennedy PG, Hill LT (2010) A molecular and phylogenetic analysis of the structure and specificity of *Alnus rubra* ectomycorrhizal assemblages. Fungal Ecol 3:195–204

Kennedy PG, Izzo AD, Bruns TD (2003) There is high potential for the formation of common mycorrhizal networks between understorey and canopy trees in a mixed evergreen forest. J Ecol 91:1071–1080

Kennedy PG, Smith DP, Horton TR, Molina RJ (2012) *Arbutus menziesii* (Ericaceae) facilitates regeneration dynamics in mixed evergreen forest by promoting mycorrhizal fungal diversity and host connectivity. Amer J Bot 99:1–11

Kernaghan G (2005) Mycorrhizal diversity: cause and effect. Pedobiologia 49:511–520

Kernaghan G, Widden P, Bergeron Y, Légaré S, Paré D (2003) Biotic and abiotic factors affecting ectomycorrhizal diversity in boreal mixed-woods. Oikos 102:497–504

Kiers ET, van der Heiden MGA (2006) Mutualistic stability in the arbuscular mycorrhizal symbiosis: exploring hypotheses of evolutionary stability. Ecology 87:1627–1636

Kiers ET, Duhamel M, Beesetty Y, Mensah JA, Franken O, Verbruggen E, Fellbaum CR, Kowalchuk GA, Hart MM, Bago A, Palmer TM, West SA, Vandenkornhuyse P, Jansa J, Bücking H (2011) Reciprocal rewards stabilize cooperation in the mycorrhizal symbiosis. Science 333:880–882

Kloppholz S, Kuhn H, Requena N (2011) A secreted fungal effector of Glomus intraradices promotes symbiotic biotrophy. Curr Biol 21:1204–1209

Kohler A, Kuo A, Nagy LG et al (2015) Convergent losses of decay mechanisms and rapid turnover of symbiosis genes in mycorrhizal mutualists. Nat Genet doi:10.1038/ng.3223

Kohout P, Sykorová Bahram, Hadincová V, Albrechtová J, Tedersoo L, Vohník M (2011) Ericaceous dwarf shrubs affect ectomycorrhizal fungal community of the invasive *Pinus strobus* and native *Pinus sylvestres* in a pot experiment. Mycorrhiza 21:403–412

Koide RT, Suomi L, Stevens CM, McCormick L (1998) Interactions between needles of *Pinus resinosa* and ectomycorrhizal fungi. New Phytol 140:539–547

Kõljalg U, Nilsson RH, Abarenkov K, Tedersoo L, Taylor AF, Bahram M, Bates ST, Bruns TD, Bengtsson-Palme J, Callaghan TM (2013) Towards a unified paradigm for sequence-based identification of fungi. Mol Ecol 22:5271–5277

Koske RE, Gemma JN, Englander L (1990) Vesicular-arbuscular mycorrhizae in Hawaiian Ericales. Amer J Bot 77:64–68

Kranabetter JM, Hayden S, Wright EF (1999) A comparison of ectomycorrhiza communities from three conifer species planted on forest gap edges. Can J Bot 77:1193–1198

Kretzer A, Li Y, Szaro T, Bruns TD (1996) Internal transcribed spacer sequences from 38 recognized species of *Suillus* sensu lato: phylogenetic and taxonomic implications. Mycologia 88:776–785

Krpata D, Mühlmann O, Kuhnert R, Ladurner H, Friederike Göbl, Peintner U (2007) High diversity of ectomycorrhizal fungi associated with *Arctostaphlos uva-ursi* in subalpine and alpine zones: potential inoculum for afforestation. For Ecol Manag 250:167–175

Martin F, Díez J, Dell B, Delaruelle C (2002) Phylogeography of the ectomycorrhizal *Pisolithus* species as inferred from nuclear ribosomal DNA ITS sequences. New Phytol 153:345–357

Martin F, Aerts A, Ahren D, Brun A, Danchin EGJ, Duchaussoy F, Gibon J, Kohler A, Lindquist E, Pereda V et al (2008) The genome of *Laccaria bicolor* provides insights into mycorrhizal symbiosis. Nature 452:88–92

Martin F, Kohler A, Murat C, Balestrini R, Coutinho PM, Jaillon O, Montanini B, Morin E, Noel B, Percudani R et al (2010) Perigord black truffle genome uncovers evolutionary origins and mechanisms of symbiosis. Nature 464(7291):1033–1038

Martin-Pinto P, Vaquerizo H, Peñalver F, Olaizola J, Oria-de-Rueda JA (2006) Early effects of wildfire on the diversity and production of fungal communities in Mediterranean vegetation types dominated by *Cistus ladanifer* and *Pinus pinaster* in Spain. For Ecol Manag 225:296–305

Massicotte HB, Molina R, Luoma DL, Smith JE (1994) Biology of the ectomycorrhizal genus *Rhizopogon*. II. Patterns of host-fungus specificity following spore inoculation of diverse hosts grown in mono- and dual cultures. New Phytol 126:677–690

Massicotte HB, Molina R, Tackaberry LE, Smith JE, Amaranthus MP (1999) Diversity and host specificity of ectomycorrhizal fungi retrieved from three adjacent forest sites by five host species. Can J Bot 77:1053–1076

Massicotte HB, Melville LH, Tackaberry LE, Rl Peterson (2008) A comparative study of mycorrhiza in several genera of Pyroleae (Ericaceae) from western Canada. Botany 86:610–622

Melin E (1922) Untersuchungen über die *Larix* mycorrhiza I. Synthese der mycorrhiza in reinkulture. Svensk Bot Tidskrift 17:161–196

Melin E (1923) Experimentelle Untersuchungen über die Birkin un Espenmycorrhizen und ihre Pilzsymbionten. Svensk Bot Tidskrift 17:479–520

Molina R (1981) Ectomycorrhizal specificity in the genus *Alnus*. Can J Bot 59:325–334

Molina R, Trappe JM (1982) Lack of mycorrhizal specificity by the ericaceous host *Arbutus menziesii* and *Artoctaphylos uva-ursi*. New Phytol 90:495–509

Molina R, Massicotte H, Trappe JM (1992) Specificity phenomena in mycorrhizal symbioses: community-ecological consequences and practical implications. In: Allen MF (ed) Mycorrhizal functioning: an integrative plant-fungal process. Chapman and Hall, New York, pp 357–423

Molina R, Smith JE, Mckay D, Melville LH (1997) Biology of the ectomycorrhizal genus *Rhizopogon*. III. Influence of co-cultured species on mycorrhizal specificity with the arbutoid host *Arctostaphylos uva-ursi* and *Arbutus menziesii*. New Phytol 137:519–528

Molina R, Trappe JM, Grubisha LC, Spatafora JW (1999) *Rhizopogon*. In: Cairney JWG, Chambers SM (eds) Ectomycorrhizal fungi: key genera in profile. Springer, Berlin, pp 129–161

Morris MH, Smith ME, Rizzo DM, Rejmánek M, Bledsoe CS (2008) Contrasting ectomycorrhizal fungal communities on the roots of co-occurring oaks (*Quercus* spp.) in a California woodland. New Phytol 178:167–176

Morris MH, Pérez-Pérez A, Smith ME, Bledsoe CA (2009) Influence of host species on ectomycorrhizal communities associated with two co-occurring oaks (*Quercus* spp.) in a tropical cloud forest. FEMS Microb Ecol 69:274–287

Mühlmann O, Göbl F (2006) Mycorrhiza of the host-specific *Lactarius deterrimus* on the roots of *Picea abies* and *Arctostaphylos uva-ursi*. Mycorrhiza 16:245–250

Nara K (2006a) Ectomycorrhizal networks and seedling establishment during early primary succession. New Phytol 169:169–178

Nara K (2006b) Pioneer dwarf willow may facilitate tree succession by providing late colonizers with compatible ectomycorrhizal fungi in a primary successional volcanic desert. New Phytol 171:187–197

Nara K (2008) Community developmental patterns and ecological functions of ectomycorrhizal fungi: implications from primary succession. In: Varma A (ed) Mycorrhiza: state of the art, genetics and molecular biology, eco-function, biotechnology, eco-physiology, structure and systematics. Springer, Berlin, pp 581–599

Nara K, Hogetsu T (2004) Ectomycorrhizal fungi on established shrubs facilitate subsequent seedling establishment of successional plant species. Ecology 85:1700–1707

Newman EI (1988) Mycorrhizal links between plants: their functioning and ecological significance. Adv Ecol Res 18:243–270
Newton AC (1991) Mineral nutrition and mycorrhizal infection of seedling oak and birch. III. Epidemiological aspects of ectomycorrhizal infection, and the relationship to seedling growth. New Phytol 117:53–60
Newton AC, Haigh JM (1998) Diversity of ectomycorrhizal fungi in Britain: a test of the species-area relationship, and the role of host specificity. New Phytol 138:619–627
Nuñez MTA, Horton TR, Simberloff D (2009) Lack of belowground mutualisms hinders Pinaceae invasions. Ecology 90:2352–2359
Ohsowski BM, Zaitsoff PD, Öpik M, Hart MM (2014) Where the wild things are: looking for uncultured Glomeromycota. New Phytol 204:171–179
Onguene NA, Kuyper TW (2002) Importance of the ectomycorrhizal network for seedling survival and ectomycorrhiza formation in rain forest of south Cameroon. Mycorrhiza 12:13–17
Öpik M, Zobel M, Cantero JJ, Davison J, Facelli JM, Hiiesalu I, Jairus T, Kalwij JM, Koorem K, Leal ME (2013) Global sampling of plant roots expands the described molecular diversity of arbuscular mycorrhizal fungi. Mycorrhiza 23:411–430
Parke JL, Linderman RG, Black CH (1983) The role of ectomycorrhizas in drought tolerance of Douglas-fir seedlings. New Phytol 95:83–95
Peay KG, Kennedy PG, Bruns TD (2008) Fungal community ecology: a hybrid beast with a molecular master. Bioscience 58:799–810
Perry DA, Amaranthus MP, Borchers JG, Borchers SL, Brainerd RE (1989) Bootstrapping in ecosystems. Bioscience 39:230–237
Perry DA, Borchers JG, Borchers SL, Amaranthus MP (1990) Species migrations and ecosystem stability during climate change: the belowground connection. Cons Biol 4:266–274
Peterson RL, Massicotte HB, Melville LH (2004) Mycorrhizas: anatomy and cell biology. NRC Research Press, Canada
Plett JM, Martin F (2011) Blurred boundaries: lifestyle lessons from ectomycorrhizal fungal genomes. Trends Genet 27:14–22
Plett JM, Daguerre Y, Wittulsky S, et al (2014) Effector MiSSP7 of the mutualistic fungus Laccaria bicolor stabilizes the Populus JAZ6 protein and represses jasmonic acid (JA) responsive genes. PNAS 111:8299–8304. doi:10.1073/pnas.1322671111
Polme S, Bahram M, Yamanaka T, Nara K, Dai YC, Grebenc T, Kraigher H, Toivonen M, Wang P-H, Matsuda Y, Naadel T, Kennedy PG, Kõljalg U, Tedersoo L (2013) Biogeography of ectomycorrhizal fungi associated with alders (*Alnus* spp.) in relation to biotic and abiotic variables at the global scale. New Phytol 198:1239–1249
Redecker D, Kodner R, Graham LE (2000) Glomalean fungi from the Ordovician. Science 289:1920–1921
Remy W, Taylor TN, Hass H, Kerp H (1994) Four hundred-million-year-old vesicular arbuscular mycorrhizae. PNAS 9:11841–11843
Richard F, Millot S, Gardes M, Selosse MA (2005) Diversity and specificity of ectomycorrhizal fungi retrieved from an old-growth Mediterranean forest dominated by *Quercus ilex*. New Phytol 166:1011–1023
Richard F, Selosse MA, Gardes M (2009) Facilitated establishment of *Quercus ilex* in shrub-dominated communities within a Mediterranean ecosystem: do mycorrhizal partners matter? FEMS Microbio Ecol 68:14–24
Roy M, Rochet J, Manzi S, Jargeat P, Gryta H, Moreau P-A, Gardes M (2013) What determines *Alnus*-associated ectomycorrhizal community diversity and specificity? A comparison of host and habitat effects at a regional scale. New Phytol 198:1228–1238
Ryberg M, Larsson E, Molau U (2009) Ectomycorrhizal diversity on *Dryas octopetala* and *Salix reticulata* in an alpine cliff ecosystem. Arct Antarct Alp Res 41:506–514
Ryberg M, Andreasen M, Björk RG (2011) Weak habitat specificity I ectomycorrhizal communities associated with *Salix herbacea* and *Salix polaris* in alpine tundra. Mycorrhiza 21:289–296

Schüssler A (2015) Glomeromycota: species list (WWW document) http://schuessler.userweb.mwn.de/amphylo
Selosse MA, Bauer R, Moyersoen B (2002a) Basal hymenomycetes belonging to the Sebacinaceae are ectomycorrhizal on temperate deciduous trees. New Phytol 155:183–195
Selosse MA, Weiss M, Jany JL, Tillier A (2002b) Communities and populations of sebacinoid basidiomycetes associated with the achlorophylos orchid *Neottia nidus-avis* (L.) L.C.M. Rich and neighboring tree ectomycorrhizae. Mol Ecol 11:1831–1844
Selosse M-A, Richard R, He X, Simard SW (2006) Mycorrhizal networks: des liaisons dangereuses. TREE 21:621–628
Setaro S, Weiss M, Oberwinkler F, Kottke I (2006) Sebacinales form ectendomycorrhizas with *Cavendishia nobilis*, a member of the Andean clade of Ericaceae, in the mountain rain forest of southern Ecuador. New Phytol 169:355–365
Simard SW, Austin ME (2010) The role of mycorrhizas in forest soil stability with climate change. Clim Change Variability 275–302
Simard SW, Durall DM (2004) Mycorrhizal networks: a review of their extent, function, and importance. Can J Bot 82:1140–1165
Simard SW, Molina R, Smith JE, Perry DA, Jones MD (1997a) Shared compatibility of ectomycorrhizae on *Pseudotsuga menziesii* and *Betula papyrifera* seedlings grown in mixture in soils from southern British Columbia. Can J For Res 27:331–342
Simard SW, Perry DA, Jones MD, Myrold D, Durral DM, Molina R (1997b) Net transfer of carbon between ectomycorrhizal tree species in the field. Nature 388:579–582
Simard SW, Durall D, Jones M (2002) Carbon and nutrient fluxes within and between mycorrhizal plants. In: van der Heijden MGA, Sanders IR (eds) Mycorrhizal ecology. Springer, Berlin, pp 33–74
Simard SW, Beiler KJ, Bingham MA, Deslippe JR, Philip LJ, Teste FP (2012) Mycorrhizal networks: mechanisms, ecology and modeling. Fungal Biol 26:39–60
Smith SE, Read D (2008) Mycorrhizal symbiosis. Academic Press, London
Smith JE, Molina R, Perry DA (1995) Occurrence of ectomycorrhizas on ericaceous and coniferous seedlings grown in soils for the Oregon Coast Range. New Phytol 129:73–81
Smith ME, Douhan GW, Fremier AK, Rizzo DM (2009) Are true multihost fungi the exception or the rule? Dominant ectomycorrhizal fungi on *Pinus sabiniana* differ from those on co-occurring *Quercus* species. New Phytol 182:295–299
Smith ME, Henkel TW, Aime MC, Fremier AK, Vigalys R (2011) Ectomycorrhizal fungal diversity and community structure on three co-occurring leguminous canopy tree species in a Neotropical rainforest. New Phytol 192:699–712
Smith ME, Henkel TW, Uehling JK, Fremier AK, Clarke HD, Vilgalys R (2013) The ectomycorrhizal fungal community in a neotropical forest dominated by the endemic dipterocarp *Pakarimaea dipterocarpacea*. PLoS 1(8):e55160
Stanley MR, Koide RT, Shumway DL (1993) Mycorrhizal symbiosis increases growth, reproduction and recruitment of *Abutilon theophrasti* Medic. in the field. Oecologia 94:30–35
Suvi T, Tedersoo L, Abarenkov K, Beaver K, Gerlach J, Köljalg U (2010) Mycorrhizal symbionts of *Pisonia grandis* and *P. sechellarum* in Seychelles: identification of mycorrhizal fungi and description of new *Tomentella* species. Mycologia 102:522–533
Tansley AG (1935) The use and abuse of vegetational concepts and terms. Ecology 16:284–307
Taschen E, Sauve M, Taudiere A, Parlarde J, Selosse MA, Richard F (2015) Whose truffle is this? Distribution patterns of ectomycorrhizal fungal diversity in *Tuber meloanosporum* brûlés developed in multi-host Mediterranean plant communities. Environ Microb doi:10.1111/1462-2920.12741
Taylor AFS (2002) Fungal diversity in ectomycorrhizal communities: sampling effort and species detection. Plant Soil 244:19–28
Taylor DL, Bruns DD, Leake JR, Read DJ (2002) Mycorrhizal specificity and function in myco-heterotrophic plants. In: van der Heijden MGA, Sanders I (eds) Mycorrhizal ecology Ecological studies, vol 157. Springer, Berlin, pp 375–413

Taylor AFS, Fransson PM, Högberg P, Högberg MN, Plamboeck Ah (2003) Species level patterns in C-13 and N-15 abundance of ectomycorrhizal and saprotrophic fungal sporocarps. New Phytol 159:757–774
Tedersoo L, Pellet P, Kõljalg U, Selosse MA (2007a) Parallel evolutionary paths to mycoheterotrophy in understorey Ericaceae and Orchidaceae: ecological evidence for mixotrophy in Pyroleae. Oecologia 151:206–217
Tedersoo L, Suvi T, Beaver K, Kõljalg U (2007b) Ectomycorrhizal fungi of the Seychelles: diversity patterns and host shifts from the native *Vateriopsis seychellarum* (Dipterocarpaceae) and *Intsia bijuga* (Caesalpinaiaceae) to the introduced *Eucalyptus robusta* (Myrtaceae), but not to *Pinus caribea* (Pinaceae). New Phytol 175:321–333
Tedersoo L, Jairus T, Horton BM, Abarenkov K, Suv T, Saar I, Kõljalg U (2008a) Strong host preference of ectomycorrhizal fungi in a Tasmanian wet sclerophyll forest as revealed by DNA barcoding and taxon-specific primers. New Phytol 180:479–490
Tedersoo L, Suvi T, Jairus T, Kõljalg U (2008b) Forest microsite effects on community composition of ectomycorrhizal fungi on seedlings of *Picea abies* and *Betula pendula*. Environ Microbiol 10:1189–1201
Tedersoo L, Suvi T, Jairus T, Ostonen I, Põlme S (2009) Revisiting ectomycorrhizal fungi of the genus *Alnus*: differential host specificity, diversity and determinants of the fungal community. New Phytol 182:727–735
Tedersoo L, May TW, Smith ME (2010a) Ectomycorrhizal lifestyle in fungi: global diversity, distribution, and evolution of phylogenetic lineages. Mycorrhiza 20:217–263
Tedersoo L, Sadam A, Zambrano M, Valencia R, Bahram M (2010b) Low diversity and high host preference of ectomycorrhizal fungi in Western Amazonia, a neotropical biodiversity hotspot. ISME J 4:465–471
Tedersoo L, Bahram M, Jairus T, Bechem E, Chinoya S, Mpumba SR, Leal M, Randrianjohany E, Razafimandimbison S, Sadam A, Naadel T, Kõljalg U (2011) Spatial structure and the effects of host and soil environments on communities of ectomycorrhizal fungi in wooded savannas and rain forest of Continental Africa and Madagascar. Mol Ecol 20:3071–3080
Tedersoo L, Bahram M, Toots M, Diédhiou AG, Henkel TW, KjollerKjøller Morris MH, Nara K, Nouhra E, Peay KG, Põlme S, Rybert M, Smith ME, Kõljalg U (2012) Towards global patterns in the diversity and community structure of ectomycorrhizal fungi. Mol Ecol 21:4160–4170
Tedersoo L, Smith ME (2013) Lineages of ectomycorrhizal fungi revisited: foraging strategies and novel lineages revealed by sequences from belowground. Fungal Biol Rev 27:83–99. doi:10.1016/j.fbr.2013.09.001
Tedersoo L, Bahram M, Põlme S et al (2014) Global diversity and geography of soil fungi. Science 346 (6213). doi:10.1126/science.1256688
Timling I, Taylor DL (2012) Peeking through a frosty window: molecular insights into the ecology of Arctic soil fungi. Fungal Ecol 5:419–429
Tisserant E, Kohler A, Dozolme-Seddas P, Balestrini R, Benabdellah K, Colard A, Croll D, Da Silva C, Gomez SK, Koul R et al (2012) The transcriptome of the arbuscular mycorrhizal fungus *Glomus intraradices* (DAOM 197198) reveals functional tradeoffs in an obligate symbiont. New Phytol 193(3):755–769
Trappe JM (1962) Fungus associates of ectotrophic mycorrhizae. Bot Rev 28:538–606
Trappe JM (1977) Selection of fungi for ectomycorrhizal inoculation in nurseries. Ann Rev Phytopath 15:203–222
Trappe JM (2005) AB Frank and mycorrhizae: the challenge to evolutionary and ecological theory. Mycorrhiza 15:277–281
Twieg BD, Durall DM, Simard SW (2007) Ectomycorrhizal fungal succession in mixed temperate forests. New Phytol 176:437–447
Upson R, Read DJ, Newsham KK (2007) Widespread association between the ericoid mycorrhizal fungus *Rhizoscyphus ericae* and a leafy liverwort in the maritime and sub-Antartic. New Phytol 176:460–471
Urban A, WEIß M, Bauer R (2003) Ectomycorrhizas involving sebacinoid mycobionts. Mycol Res 107:3–14

van der Heijden MGA, Horton TR (2009) Socialism in soil? The importance of mycorrhizal fungal networks for facilitation in natural ecosystems. J Ecol 97:1139–1150
van der Heijden MGA, Martin FM, Selosse M-A, Sanders IR (2015) Mycorrhizal ecology and evolution: the past, the present, and the future. New Phytol 205:1406–1423
Vellinga EC, Wolfe BE, Pringle A (2009) Global patterns of ectomycorrhizal introductions. New Phytol 181:960–973
Villarreal-Ruiz L, Anderson IC, Alexander IJ (2004) Interaction between an isolate from the *Hymenoscyphus ericae* aggregate and roots of *Pinus* and *Vaccinium*. New Phytol 164:183–192
Visser S (1995) Ectomycorrhizal fungal succession in jack pine stands following wildfire. New Phytol 129:389–401
Vrålstad T (2004) Are ericoid and ectomycorrhizal fungi part of a common guild? New Phytol 164:7–10
Wilson AW, Binder M, Hibbet DS (2012) Diversity and evolution of ectomycorerhizal host associations in the Sclerodermatineae (Boletales, Basidiomycota). New Phytol 104:951–961
Wolfe BE, Pringle A (2011) Geographically structured host specificity is caused by the range expansions and host shifts of a symbiotic fungus. ISME J 6:745–755
Zak B (1976a) Pure culture synthesis of bearberry mycorrhizae. Can J Bot 54:1297–1305
Zak B (1976b) Pure culture synthesis of Pacific madrone ectendomycorrhizae. Mycologia 68:362–369
Zimmer K, Hynson NA, Gebauer G, Allen EB, Allen MF, Read DJ (2007) Wide geographical and ecological distribution of nitrogen and carbon gains from fungi in pyroloids and monotoropoids (Ericaceae) and in orchids. New Phytol 175:166–175

Chapter 2
Functional Significance of Anastomosis in Arbuscular Mycorrhizal Networks

Manuela Giovannetti, Luciano Avio and Cristiana Sbrana

Abstract Arbuscular mycorrhizal fungi (AMF) are obligate biotrophs (Glomeromycota), which live symbiotically in the roots of most land plants and facilitate mineral nutrition of their hosts. Their spores are able to germinate in the absence of host-derived signals, but are unable to complete the life cycle without establishing a functional symbiosis with a host plant. Such behaviour did not represent a selective disadvantage, as a result of diverse survival strategies allowing them to compensate for the lack of host-regulated germination and to overcome their obligate biotrophic state. The ability to form hyphal fusions (anastomoses) between compatible mycelia may represent an important mechanism evolved by AMF to increase their chances of survival, since fungal germlings can plug into pre-existing extraradical mycelial networks, thus gaining immediate access to plant-derived carbon before asymbiotic growth arrest. In fusions between hyphae of the same or different individual germlings of the same isolate, perfect anastomoses occur with the highest frequency and are characterized by protoplasm continuity and complete fusion of hyphal walls. High anastomosis frequencies are also detected between extraradical mycelial networks produced by the same isolate, spreading from plants of different species, genera and families. Pre- and post-fusion incompatibility are often observed in hyphal interactions between asymbiotic and symbiotic mycelium and between genetically different germlings belonging to the same isolate, while pre-fusion incompatible responses, hindering hyphal fusions, occur between germlings of geographically different isolates. The analysis of vegetative compatibility/incompatibility during hyphal fusions represents a

M. Giovannetti (✉)
Department of Agriculture, Food and Environment, University of Pisa, Pisa, Italy
e-mail: manuela.giovannetti@unipi.it

L. Avio
Institute of Agricultural Biology and Biotechnology, CNR, Milan, Italy
e-mail: avio@ibba.cnr.it

C. Sbrana
IBBA-CNR UOS Pisa c/o Dipartimento di Scienze Agrarie,
Alimentari e Agroambientali, Via del Borghetto 80, 56124 Pisa, Italy
e-mail: sbrana@ibba.cnr.it

T.R. Horton (ed.), *Mycorrhizal Networks*,
Ecological Studies 224, DOI 10.1007/978-94-017-7395-9_2

valuable tool for genetic studies of AMF, which are recalcitrant to axenic cultivation. Molecular analyses of the progeny of mycelium derived from nonself vegetative fusions of genetically different germlings of *R. irregularis* showed that genetic exchange occurs, despite low anastomosis frequencies and post-fusion incompatible responses, suggesting that anastomosis between genetically different mycelia may represent a recombination mechanism in the absence of an evident sexual cycle.

Keywords Mycorrhizal networks • Network nutrient transfer • Hyphal anastomosing ability • Hyphal recognition • Non-self hyphal compatibility

2.1 Introduction

The majority of land plants establish mutualistic symbioses with arbuscular mycorrhizal (AM) fungi (AMF), a group of beneficial soil microorganisms fundamental for plant nutrition and ecosystem biodiversity and productivity, affecting the composition of plant communities in terms of survival, competition and diversity of plants (van der Heijden et al. 1998; Wardle et al. 2004; Smith and Read 2008). AMF belong to the Glomeromycota, are obligate biotrophs and colonise the roots of their host plants obtaining sugars, which they are not able to synthesize. In exchange, the host plants receive mineral nutrients, absorbed and translocated through a fine network of extraradical hyphae extending from the roots to the surrounding soil. Such belowground networks represent the key structure for soil nutrient uptake and transfer to the roots, and are thought to have played a fundamental role in land colonisation by early terrestrial plants, which, lacking an extended root system, could have been facilitated by the AMF symbionts functioning as an auxiliary absorbing structure (Pirozynski and Malloch 1975). Fossil records and molecular data have supported such a view, considering AMF as evolutionarily successful living fossils, having survived 460 to possibly 980 million years (Simon et al. 1993; Remy et al. 1994; Phipps and Taylor 1996; Redecker et al. 2000; Blair 2009).

Mycorrhizal networks spreading from colonised roots into the surrounding soil represent the structure where the flow of nutrients is realized. Such a flow consists of a bidirectional flux of mineral nutrients, mainly P, N, Cu, Fe, K, Zn (Smith and Read 2008), from the soil to the host plant, and of sugars acquired by intraradical hyphae and transferred to other fungal structures in the soil, i.e. mycelium and spores. Moreover, mycorrhizal networks are of fundamental importance for plants, since they can grow indefinitely in every direction, foraging for soil nutrients far from the roots with high efficiency, given the very fine dimensions of hyphae (5–10 μm diameter).

Data on the mechanisms of absorption of mineral nutrients, in particular phosphate, confirmed the key role played by mycorrhizal networks in plant nutrition. Phosphorus can be absorbed in the soil-plant interface by both root hair and epidermal cells and in the soil-fungus interface by fungal hyphae that transfer P to root cells in the root cell-fungus interface (Karandashov and Bucher 2005). Molecular studies show that genes for phosphate uptake are differentially expressed in extraradical hyphae (Harrison and van Buuren 1995; Maldonado-Mendoza et al. 2001; Casieri et al. 2013), and that the mycorrhizal network is structurally and functionally able to capture high quantities of phosphate from the soil (Smith et al. 2003).

A reverse flow of sugars occurs from host plants to fungal symbionts. The amount of C, obtained from the host plant and transformed by the fungal symbiont into trehalose and other polyols may reach 20 % of total photosynthate, depending on different plant-fungus combinations (Jakobsen and Rosendahl 1990).

Since AMF have a wide host range, mycorrhizal networks can simultaneously colonise diverse root systems, interconnecting plants belonging to the same and different species, genera and families (Eason et al. 1991; Lerat et al. 2002; Giovannetti et al. 2004). Thus, common mycorrhizal networks (CMNs) represent the physical structures through which carbon, mineral nutrients and water can move from plant to plant (Johansen and Jensen 1996; Egerton-Warburton et al. 2007; Simard et al., Chap. 5, this Vol.), allowing plants to share ecosystem resources which may modify and/or facilitate plant coexistence.

The occurrence of AMF mediated interplant C transfer, requiring a net flux of C from the fungal symbiont to the host, was reported in some mycoheterotrophic plants (Bidartondo et al. 2002), while C allocation from one green plant to another through AMF mycelial networks is much more controversial. Early findings suggesting C transfer between plants through AMF hyphae (Hirrel and Gerdemann 1979; Francis and Read 1984; Grime et al. 1987; Martins 1993) were followed by other reports that showed the occurrence of interplant C flow, but pointed out that transferred C remained in fungal root tissues without moving into the shoots (Watkins et al. 1996; Graves et al. 1997; Fitter et al. 1998). Some findings supported the view of an exchange of C between plants, at least in particular conditions (Lerat et al. 2002; Carey et al. 2004), while others, utilising in vitro mycorrhizal root organ cultures or plants, further confirmed that C originating from a donor plant was retained in fungal cells (Pfeffer et al. 2004; Voets et al. 2008; Lekberg et al. 2010).

It has been suggested that N and P can move from a plant to another through mycorrhizal networks (Whittingham and Read 1982; Haystead et al. 1988). Studies on N fixing plants utilizing ^{15}N showed that AMF mediated N transfer may occur (Frey and Schüepp 1992; Martins and Cruz 1998), although also indirect pathways may be significant (Ikram et al. 1994; Rogers et al. 2001). However, a laboratory experiment that utilized two plant compartments linked only by AMF hyphae and separated by an air gap, confirmed N transfer together with transfer of analogs of P and K (Meding and Zasoski 2008). Direct interplant P transfer through hyphal connections, suggested by early field and laboratory studies (Chiariello et al. 1982; Francis et al. 1986), was not confirmed by other experiments, suggesting that the observed flow could result from the release of P from donor roots into the soil or to the mobilization of nutrients from a dying donor plant (Newman and Ritz 1986; Newman and Eason 1989, 1993; Johansen and Jensen 1996). An elegant experimentation using

^{32}P as a tracer confirmed belowground P transfer from donor to receiver plants mediated by interconnected mycorrhizal networks (Mikkelsen et al. 2008). Recently, a differential allocation of P and N to plant hosts either belonging to diverse species or with high/low C source strength was demonstrated for CMNs formed by different AMF isolates (Fellbaum et al. 2014; Walder et al. 2015).

Mycorrhizal networks interconnecting different plants may function as plant-plant underground communication ways, allowing signals transfer among plants and activating defence pathways before pathogen attacks. For example, tomato plants connected by *Funneliformis mosseae* (formally *Glomus mosseae*) CMNs showed increased expression of defence-related genes and higher levels of disease resistance enzymes in healthy "receiver" plants after inoculation of 'donor' plants with the pathogen *Alternaria solani* (Song et al. 2010). In addition, aphid-free *Vicia faba* connected to aphid-infested plants via a CMN showed aphid repellence and aphid enemy attraction due to systemic changes in the production of volatiles (Babikova et al. 2013). CMNs are also able to widen the action of allelochemicals in natural environments and to affect interactions within plant communities (see Jakobsen and Hammer, Chap. 4, this Vol.), as plant-derived allelopathic substances and herbicides supplied to mycorrhizal plants may be transferred to the target plant, leading to their accumulation at levels that cannot be reached by soil diffusion (Barto et al. 2011; Achatz et al. 2013). In addition, CMNs may allow water flux between interconnected plants, facilitating plant survival during drought (Egerton-Warburton et al. 2007).

Further studies aimed at detecting and quantifying mineral nutrient and carbon transfer in the fungal network could improve our understanding of its functional significance and of the role played by AMF in the distribution of resources in plant communities. Moreover, since mycorrhizal networks may also represent a channeling system for a wide exchange of information molecules between plants, they appear to play a fundamental role in the dynamics and evolution of the complex network of interactions regulating ecosystem functioning.

In this chapter we will review the developments which contributed to give a picture of mycorrhizal networks as one of previously unimagined dynamism. We will discuss the structure of AMF networks, the cellular events leading to anastomosis formation, and the phenomenon of self/nonself recognition and nonself incompatibility between hyphae belonging to the same and to genetically different AMF.

2.2 Structure of Mycorrhizal Networks

The structure, viability, extent and interconnectedness of AMF mycelium are of critical importance for the establishment and maintenance of the flow of nutrients from soil to plant roots and were investigated by many authors. Using a destructive approach, the extent of AMF networks was estimated to range from 2.7 to 20.5 m/g of soil, depending on the fungal species (Giovannetti and Avio 2002; Mikkelsen et al. 2008).

Some non destructive observations of AMF extraradical mycelium (ERM), carried out by using root observation chambers (Friese and Allen 1991) and in vitro dual systems (Bago et al. 1998a), provided qualitative information on its architecture and development before and after symbiosis establishment. A non-destructive in vivo bi-dimensional experimental system (sandwich system), allowed the visualization and quantification of the whole intact AMF network produced by the AM fungus *F. mosseae* living in symbiosis with three different plant species: *Allium porrum*, *Thymus vulgaris* and *Prunus cerasifera* (Fig. 2.1). After 7 days' growth the length of ERM spreading out from root-based hyphae into the surrounding environment ranged from 5169 mm in *T. vulgaris* to 7471 mm in *A. porrum*, corresponding to 10 and 40 mm mm^{-1} root length, respectively. The mean growth rate was 738–1067 mm d^{-1}, depending on the host plant (Giovannetti et al. 2001). By contrast, in a tri-dimensional soil system lower values were detected, ranging from 3.1 to 3.8 mm d^{-1} for *F. mosseae* and *F. caledonius* ERM spreading from *Trifolium subterraneum* mycorrhizal roots (Mikkelsen et al. 2008).

Besides ERM extent, fungal biomass is important for the role played by mycorrhizal networks in the transfer of C from atmosphere to soil (Bago et al. 2000; Treseder and Allen 2000), since they can supposedly sequester large quantities of organic C in their walls in the form of recalcitrant compounds, such as chitin and

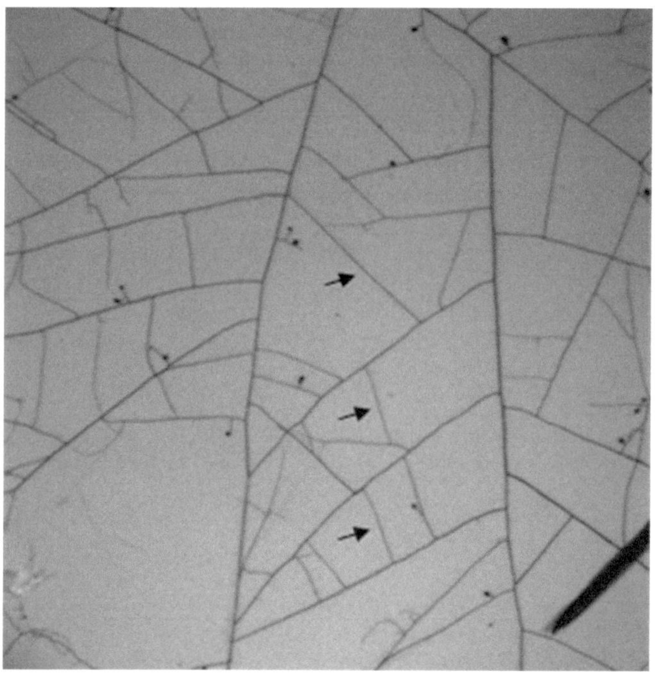

Fig. 2.1 Visualisation of intact extraradical mycelium of *Funneliformis mosseae* spreading from colonised roots of *Allium porrum* showing the network structure realised by means of frequent anastomoses interconnecting nearby hyphae (*arrows*)

chitosan (Gooday 1990). Specific fungal weights, assessed using different AMF species and experimental systems, ranged from 3.85 to 7.84 μg/m of mycelium (Bethlenfalvay and Ames 1987; Frey et al. 1994; Miller et al. 1995; Fortuna et al. 2012). The extraradical network of the AM fungus *F. mosseae* IMA 1, visualised by means of a bi-dimensional experimental system, appeared highly branched (0.86–0.97 branches mm^{-1}), while its viability, determined by the localization of formazan salts depositions (SDH activity), was 100 %, after 7 days' growth (Fig. 2.2) (Giovannetti et al. 2001).

The interconnectedness of mycorrhizal networks is the result of fusions (anastomoses) between contacting hyphae. The number of anastomoses in extraradical mycelium ranged between 0.75 per 100 cm of hypha in *Gigaspora margarita* to 0.51 per mm of hypha in *F. mosseae*, whereas fusion frequencies ranged between 0 in *Ambispora leptoticha*, *Gigaspora albida*, *Gigaspora gigantea* and *Dentiscutata heterogama*, to 64 % in *F. mosseae* (Table 2.1). Fusion frequencies showed the highest values in in vivo systems, both in bi-dimensional (up to 64 %) and in tri-dimensional models (up to 37.4 %) (Table 2.1). In some isolates of the AMF species *Rhizoglomus clarus*, anastomosis frequencies recorded in extraradical (symbiotic) mycelium were different from those observed in hyphae originating from germinating spores (asymbiotic) mycelium (Purin and Morton 2013), whereas slight differences were reported for *F. mosseae* (Giovannetti et al. 1999, 2004). In an in vitro root organ culture system, characterised by high soluble nutrient levels, a low number of anastomoses was recorded, with a maximum of 17 fusions per meter of *Rhizoglomus proliferus* hyphae, although 100 % of such fusions were beteen different hyphae in *Rhizoglomus intraradices*, *R. proliferus* and *Glomus hoi* (de la Providencia et al. 2005; Voets et al. 2006). Networks formed by Gigasporaceae showed no fusions in vivo (Purin and Morton 2011), whereas a low number of anastomoses was produced in root-organ cultures: interestingly, about 95 % of

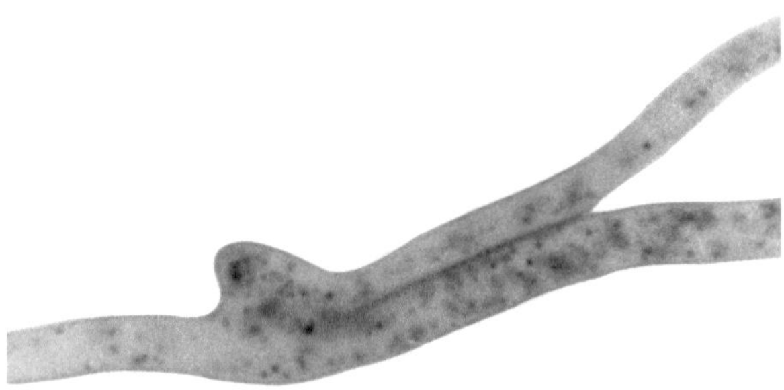

Fig. 2.2 Visualisation of succinate dehydrogenase activity (SDH) evidenced by formazan salt depositions, showing complete fusions of hyphal walls and the establishment of protoplasmic continuity in anastomosing extraradical hyphae of *Funneliformis mosseae*

Table 2.1 Summary of the different patterns of anastomosis formation across different lineages of arbuscular mycorrhizal fungi during both the symbiotic and asymbiotic stage (a spore germling without a plant host connection)

Within isolate	Asymbiotic mycelium				Symbiotic mycelium				References
	PF	PrFI	PFI	NI	PF	PrFI	PFI	NI	
Acaulospora scrobiculata A38	42–72.9	0	0	27.1–58					Barreto de Novais et al. (2013)
Acaulospora spinosa 95-UFLA	14–32.8	0	0	67.2–86					Barreto de Novais et al. (2013)
Ambispora leptoticha CR312					0	10.7	-	89.3	Purin and Morton (2011)
Claroideoglomus (*Glomus*) *etunicatum* SCT101A	63.8–75.4	0	0	24.6–36.2					Barreto de Novais et al. (2013)
Funneliformis (*Glomus*) *caledonius* BEG 20	34–55	–	–	–					Giovannetti et al. (1999)
Funneliformis (*Glomus*) *mosseae*									
IMA1	40–60.4	0	0	39.6	44–64	0–8.9	0	36–56	Giovannetti et al. (1999, 2003, 2004), Sbrana et al. (2011)
IN101C	75.8	0	0	24.2					Giovannetti et al. (2003)
BEG25	76.7	0	0	23.3					Giovannetti et al. (2003)
AZ225C	85.1	0	0	14.9					Giovannetti et al. (2003)
BEG69	72.0	0	0	28					Giovannetti et al. (2003)
Glomus formosanum A20	79.8–91.4	0	0	8.6–20.2					Barreto de Novais et al. (2013)
Glomus hoi					5.7/100 cm (100 % bh)				de la Providencia et al. (2005)
Paraglomus occultum WY112A					0	0	-	100	Purin and Morton (2011)
Rhizoglomus (*Glomus*) *proliferus*					6.6/100 cm (100 % bh)				de la Providencia et al. (2005)
Rhizoglomus (*Glomus*) *proliferus* MUCL 41827					0.172/cm				Voets et al. (2006)
Rhizoglomus (*Glomus*) *clarus*									
MUCL46238	64.2	–	–	–					Cárdenas-Flores et al. (2011)

(continued)

Table 2.1 (continued)

Within isolate	Asymbiotic mycelium				Symbiotic mycelium				References
	PF	PrFI	PFI	NI	PF	PrFI	PFI	NI	
351-UFLA	47.9	0	0	52.1					Barreto de Novais et al. (2013)
WV101					6.7	32.1	-	61.2	Purin and Morton (2013)
1UnC#7	6.3–8	25*	0*		15.6–37.4	5*	0*		Purin and Morton (2013)
6AmA#2	6.6–19.7				10.3–24.5				Purin and Morton (2013)
CRwest#8	17.4–37.5				11.7–32.4				Purin and Morton (2013)
WV123A#6	13.4–30.3				2.4–5.4				Purin and Morton (2013)
WV123A#7	11.4–18.1				4.7–11.1				Purin and Morton (2013)
WV310#5	20.6–35.7				6.8–17.8				Purin and Morton (2013)
Rhizoglomus (*Glomus*) *intraradices*									
LPA 8 (BEG 141)	59	–	–	–					Giovannetti et al. (1999)
IMA 5	69–90	–	–	–					Giovannetti et al. (1999)
MUCL43204					5.4/100 cm (100 % bh)				de la Providencia et al. (2005)
ON201B					13.9	7.5	-	78.6	Purin and Morton (2011)
Rhizoglomus (*Glomus*) *irregularis*									
DAOM197198	60	0	0	40					de la Providencia et al. (2013)
DAOM240415	38	0	0	62					de la Providencia et al. (2013)
DAOM234328	55	0	0	45					de la Providencia et al. (2013)
Lineages of MUCL43194	41–78	0	0	-					Cárdenas-Flores et al. (2010)
MUCL41833	89	0	0	-					Cárdenas-Flores et al. (2010)

(continued)

Table 2.1 (continued)

Within isolate	Asymbiotic mycelium				Symbiotic mycelium				References
	PF	PrFI	PFI	NI	PF	PrFI	PFI	NI	
MUCL43194					7.6/100 cm (100 % bh)				de la Providencia et al. (2005)
MUCL43194					0.086/cm				Voets et al. (2006)
A4	48.1	0	0	51.9					Croll et al. (2009)
C2	51	0	0	49					Croll et al. (2009)
C3	47.5	0	0	52.5					Croll et al. (2009)
D1	45.8	0	0	54.2					Croll et al. (2009)
B3	49.4	0	0	50.6					Croll et al. (2009)
Rhizoglomus (*Glomus*) *manihotis* A83	9.3–24.3	0	0	75.7–90.7					Barreto de Novais et al. (2013)
Dentiscutata (*Scutellospora*) *heterogama* SN722					0	13.2	-	86.8	Purin and Morton (2011)
Dentiscutata (*Scutellospora*) *reticulata*									
EMBRAPA CNPAB1					0.02/cm (94.7 % wh)				Voets et al. (2006)
EMBRAPA CNPAB11					0.79/100 cm (5.2 % bh, 94.8 % wh)				de la Providencia et al. (2005)
EMBRAPA CNPAB11					<1				de Souza and Declerck (2003)
Gigaspora albida URM-FMA 01	0	0	0	100					Barreto de Novais et al. (2013)
Gigaspora gigantea MN414D					0	15.4	-	84.6	Purin and Morton (2011)
Gigaspora margarita BEG 34					0.75/100 cm				de la Providencia et al. (2005)

(continued)

Table 2.1 (continued)

Within isolate	Asymbiotic mycelium				Symbiotic mycelium				References
	PF	PrFI	PFI	NI	PF	PrFI	PFI	NI	
					(9.8 % bh, 90.2 % wh)				
Gigaspora margarita BEG 34					0.009/cm (95.5 % wh)				Voets et al. (2006)
Gigaspora rosea									
BEG 9	0	–	–	–					Giovannetti et al. (1999)
A35	0	0	0	100					Barreto de Novais et al. (2013)
BEG 9					1.2/100 cm (4.2 % bh, 95.8 % wh)				de la Providencia et al. (2005)
Racocetra (Scutellospora) castanea BEG 1	0	–	–	–					Giovannetti et al. (1999)
Scutellospora calospora A80	0	0	0	100					Barreto de Novais et al. (2013)

Between isolates		Asymbiotic mycelium				Symbiotic mycelium	References
		PF	PrFI	PFI	NI	PF	
Funneliformis (Glomus) mosseae	IMA1 × BEG25	0	51	0	49		Giovannetti et al. (2003)
	AZ225C × IN101C	0	49	0	51		
	AZ225C × BEG25	0	46	0	54		
	IMA1 × AZ225C	0	43	0	57		
	AZ225C × BEG69	0	43	0	57		
	IN101C × BEG25	0	36	0	64		
	IMA1 × IN101C	0	33	0	67		
	SY710 × IN101C	0	32	0	68		

(continued)

Table 2.1 (continued)

Between isolates		Asymbiotic mycelium				Symbiotic mycelium	References
		PF	PrFI	PFI	NI	PF	
Rhizoglomus (*Glomus*) *irregularis*	DAOM240415 × 234328	1.2	1.3	1.3	96		de la Providencia et al. (2013)
	DAOM197198 × 240415	0	2.3	0.7	97		
	DAOM197198 × 234328	0	0	2	98		
	A4 × C2	4.6	1.9	13.9	79.6		Croll et al. (2009)
	A4 × C3	10.3	4.1	11.3	74.2		
	A4 × D1	0	7.3	13.4	79.3		
	A4 × B3	1.9	8.5	8.5	81.1		
	C2 × C3	5.4	0	19.4	77.4		
	C2 × D1	1.9	8.5	20.8	70.8		
	C2 × B3	1.9	6.6	12.3	79.2		
	C3 × D1	1.1	4.2	15.8	82.1		
	C3 × B3	1	5.8	10.7	82.5		
	D1 × B3	4.6	3.4	9.2	87.4		
Rhizoglomus (*Glomus*) *clarus*	6AmA#2 3 1UnC#7	0				0	Purin and Morton (2013)
	6AmA#2 3 CRwest#8	0				0	
	1UnC#7 3 CRwest#8	0				0	
	1UnC#5 3 1UnC#7	5.8				0	
	1UnC#7 3 WV310#5	0				0	
	1UnC#7 3 WV123A#6	0				0	
	1UnC#7 3 WV123A#7	0				0	
	WV310#5 3 WV123A#6	0.9				0	
	WV123A#7 3 WV123A#6	1.6				0	
	WV123A#7 3 WV310#5	2.2				0	

(continued)

Table 2.1 (continued)

Between isolates		Asymbiotic mycelium				Symbiotic mycelium	References
		PF	PrFI	PFI	NI	PF	
Between lineages							
Rhizoglomus (Glomus) irregularis	MUCL 43194	41–77	0	0	–		Cárdenas-Flores et al. (2010)

Between life cycle phases		PF	PrFI	PFI	NI	Reference
Funneliformis (Glomus) mosseae IMA1	Asymbiotic × symbiotic hyphae	4.9–23.9	5.2–17.6	1.2–14.9	46.6–88.6	Sbrana et al. (2011)

*Indicate mean data for all the isolates.

Data are given as frequencies (percent) of perfect fusions (PF), pre-fusion incompatibility (PrFI), post-fusion incompatibility (PFI), no interactions (NI) or as number of perfect fusions per hyphal length. In the latter case, frequencies of fusions within the same hypha (wh) and between different hyphae (bh) are reported.

fusions detected in these two genera occurred within the same hypha (de la Providencia et al. 2005; Voets et al. 2006).

It is interesting to note that, using data obtained from studies of ERM development in soil and hyphal fusion frequencies in tri-dimensional agar or soil systems, even the apparently low anastomosis rate results in the production of 100–410 interhyphal connections per gram of soil (Giovannetti and Avio 2002; Voets et al. 2006; Mikkelsen et al. 2008).

Other studies, carried out on mycorrhizal networks produced by geographically different isolates of the globally distributed AMF species *F. mosseae* and *R. intraradices* (two isolates for each species) living in symbiosis with *Medicago sativa*, revealed that the structure of the network significantly differed among AMF isolates, since the hyphal length ranged from 4 to 21 mm per mm of root length and the number of anastomoses per hyphal contact varied between 30 and 67; Avio et al. 2006). Such a high interconnectedness was shown to play an important role in the translocation and flow of nutrients from soil to host plants, affecting plant growth and nutrition (Avio et al. 2006).

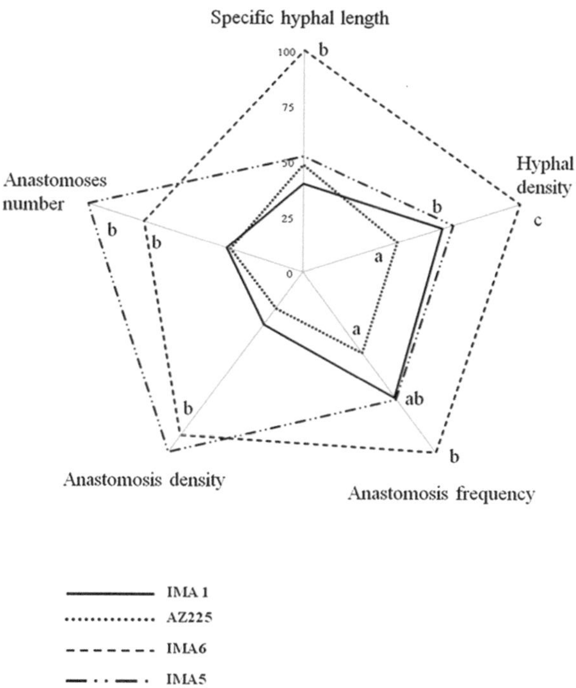

Fig. 2.3 Radar chart showing mycorrhizal network variables characterizing different isolates of *Funneliformis mosseae*, IMA1 and AZ225, and *Rhizoglomus intraradices*, IMA5 and IMA6, expressed as percentages of the highest obtained value. Fungal variables are measured as: specific hyphal length, mm mm^{-1} colonised root length; hyphal density, mm mm^{-2}; anastomosis number mm^{-1} hyphal length; anastomosis density mm^{-2}; anastomosis frequency, percentage of hyphal contacts leading to hyphal fusions

Although the sandwich system is bi-dimensional and may affect hyphal anastomosis and growth rate, its use extended the knowledge of the mechanisms underlying plant interconnectedness, revealing an unexpected and remarkable outcome: the root systems of plants belonging to different species, genera and families and colonised by the same fungal symbiont could be interconnected by means of linkages between contiguous mycorrhizal networks (Giovannetti et al. 2004). The extraradical hyphae spreading from *Allium porrum* root system were able to establish connections with those originating from *Daucus carota*, *Gossypium hirsutum*, *Lactuca sativa*, *Solanum melongena*, colonized by the same strain of the AM symbiont *F. mosseae* (Fig. 2.4). The percentages of hyphal contacts leading to anastomosis between extraradical networks originating from different plant species, ranging from 44 % in the pairing *A. porrum-S. melongena* to 49 % in *A. porrum-G. hirsutum*, were significantly lower than those detected between hyphae belonging to the same plant, which ranged from 46 % in *D. carota* to 64 % in *L. sativa* and showed also a host plant effect.

According to such data, connections between different plants are not exclusively established through hyphae spreading from mycorrhizal roots and colonising contiguous host plants (Newman 1988; Graves et al. 1997; Van der Heijden et al.

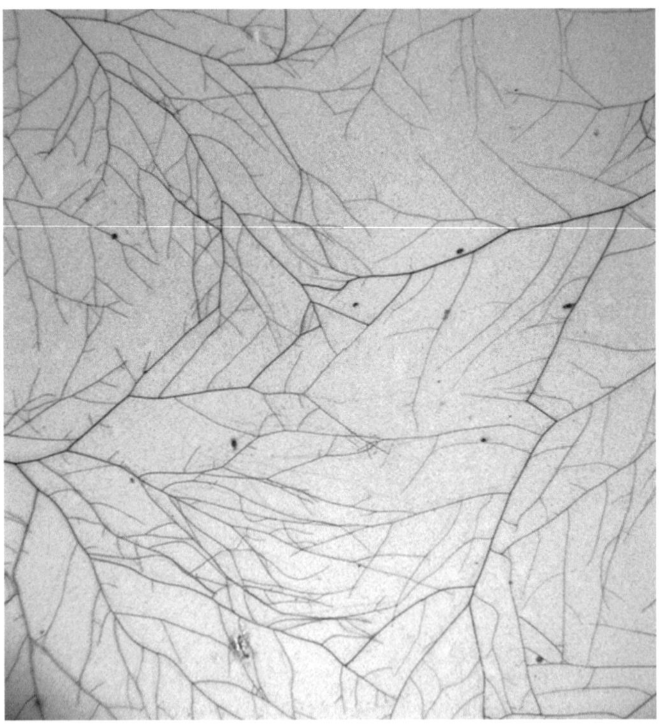

Fig. 2.4 Hyphal connections established between extraradical mycorrhizal networks originating from *Allium porrum* (*left*) and *Solanum melongena* (*right*) colonised by the same *Funneliformis mosseae* isolate IMA1

1998), but also by means of fusions between mycorrhizal networks originating from different plants, which could potentially create indefinitely large numbers of linkages through which nutrients can be transported over long distances. As the visualisation of such linkages in soil is not possible, because every kind of sampling would destroy the structure of the fungal network, an indirect approach confirmed the occurrence of anastomosis between contiguous ERM in a soil experimental system (Mikkelsen et al. 2008).

In our laboratory we recently demonstrated the ability of hyphae originating from individual germinated spores to fuse and incorporate into hyphae of the mycorrhizal network produced by plants colonised by the same fungal strain, and to establish vital connections with nuclei flowing in anastomosis bridges (Sbrana et al. 2011; Table 2.1). This phenomenon represents an important mechanism evolved by AMF to increase their chances of survival. Indeed, although AMF are obligate biotrophs, their spores can germinate in the absence of host-derived chemical signals, giving rise to coenocytic colonies where an active bidirectional flow of nuclei, mitochondria, fat droplets, vacuoles and organelles is easily detectable (Bago et al. 1998b; Logi et al. 1998) and whose extent may range from 18 to 50 mm (Bécard and Piché 1989; Gianinazzi-Pearson et al. 1989; Logi et al. 1998). Such asymbiotic hyphae, being unable to establish a symbiosis, rapidly undergo a programmed growth arrest accompanied by protoplasm withdrawal and resource reallocation towards mother spores (Mosse 1959; Hepper 1983; Bécard and Piché 1989; Logi et al. 1998). This energy-saving behaviour, though important to allow long-term maintenance of spore viability and host-infection ability (Beilby and Kidby 1980; Koske 1981; Tommerup 1984; Logi et al. 1998), could have represented an evolutionary selective disadvantage. The ability of fungal germlings to plug into pre-existing extraradical mycelium may increase their probability of survival, allowing them to gain access to plant-derived carbon circulating in the network before asymbiotic growth arrest.

2.3 Cytology of Anastomosis Formation

The word anastomosis derives from Greek and originally referred to an opening or junction through a mouth as of one body of water with another. In human anatomy, it commonly refers to a connection that is created between tubular structures, such as blood vessels, involving the concept of fluid flow. In mycology, anastomoses (vegetative hyphal fusions), first described in 1933 (Buller 1933), occur between hyphae of Ascomycota and Basidiomycota (Gregory 1984; Ainsworth and Rayner 1986; Leslie 1993). Anastomoses were formerly believed to be lacking or rare in Zygomycetes (Gregory 1984; Carlile 1995) but some authors mentioned their occurrence without giving any quantitative data on their frequency or the cytological events involved (Godfrey 1957; Mosse 1959; Tommerup 1988; Giovannetti et al. 1993).

Anastomoses between living hyphae of AMF were first studied and monitored in asymbiotic mycelium originating from individually germinated spores (Giovannetti et al. 1999). By using a combination of time-lapse and video-enhanced light microscopy, image analysis, and epifluorescence microscopy the dynamics of anastomosis formation was monitored, cytoplasmic flow and nuclear exchange were visualised, and the occurrence and frequency of anastomosis between hyphae of germlings belonging to the same and to different isolates, species and genera were assessed. Anastomoses formed in hyphae belonging to the same germling or to different germlings of the same strain were characterized by cellular compatibility, consisting in complete fusion of hyphal walls, cytoplasmic flow and migration of organelles and nuclei through hyphal bridges (Table 2.1). The histochemical localization of formazan salts in hyphal fusions, evidencing SDH-succinate dehydrogenase activity, allowed the detection of successful anastomoses, characterised by viable hyphal connections and protoplasmic continuity (Giovannetti et al. 1999).

The morphological types of hyphal fusions were mainly tip-to-side, and rare tip-to-tip fusions were observed. During pre-contact interactions, approaching hyphal tips were actively attracted towards the nearby hyphae and showed growth orientation, while the approached hyphae initiated new hyphal tips, suggesting the existence of an interhyphal remote signalling. When a tip contacted a trunk hypha, it stopped growing and in some cases appeared swollen, but more often the walls fused without any apparent tip swelling, while a protoplasmic flow was established through the fusion pore. The cascade of cellular and biochemical events, including cell wall degradation and synthesis, leading to the formation of a hyphal bridge connecting the two previously separated hyphae remains to be unravelled. Further investigations should be performed to answer the question as to whether the complex process of anastomosis formation starts with a physiological switch making hyphae growing nearby fusion-competent as a result of remote chemical signals controlling pre-fusion events, similarly to what happens during the sexual phase of other fungi (Bistis 1981; Snetselaar et al. 1996).

The complete formation of hyphal fusions in living hyphae of AMF was accomplished in 35 min, after hyphal contact in *F. mosseae* and *F. caledonius* (Giovannetti et al. 1999), and in 4 h after a hyphal tip showed directed growth towards another hypha in *R. irregularis* (Sbrana, unpublished results). In hyphal fusions, the intense protoplasmic flow subsequent to anastomosis was visualised by the bidirectional movement of particles—vacuoles, mitochondria, nuclei, and fat droplets—migrating at the speed of 1.8–0.26 $\mu m\ s^{-1}$ in *F. caledonius*, *F. mosseae* and *R. irregularis* (Giovannetti et al. 1999; Sbrana, unpublished results). Nuclear occurrence in hyphal bridges, evidenced by DAPI staining and epifluorescence microscopy was detected between hyphae belonging to the same germling and to different germlings of the same AMF isolate, in *F. caledonius*, *R. intraradices*, *F. mosseae*, showing the complete compatibility and interconnectedness of the mycelia.

Nuclear migration in AMF hyphal fusion bridges was confirmed by the visualisation—by immunofluorescence microscopy—of nuclei closely associated to cytoplasmic microtubules, which are believed to mediate nuclear division and migration processes in fungi (Morris and Enos 1992; Ästrom et al. 1994). In fungi three types of cytoplasmic microtubule (cMT)-dependent nuclear movements have been observed using live cell imaging: short-range oscillations (up to 4.5 μm/min), rotations (up to 180° in 30 s), and long-range nuclear bypassing (up to 9 μm/min) (Lang et al. 2010). In *Ashbya gossypii* long-range nuclear movements were regulated by cytoplasmic microtubule cytoskeleton emanating from each nucleus and by dynein, and nuclear pulling was due to cytoplasmic microtubule cytoskeleton cortical sliding (Grava et al. 2011).

The migration and intermingling of nuclei in hyphal bridges indicate that anastomoses in AMF play a fundamental role not only in the establishment of "mycelial interconnectedness", allowing intrahyphal communication and homeostasis, as proposed by Rayner (1996), but also in the information flow leading to a physiological and genetic integration among vegetatively compatible individual germlings. Cytological observations of *C. etunicatum* mycelium showed that nuclear mobility contributed to mix different lineages of nuclei within the coenocytic hyphae, and that the occurrence of asynchronous nuclear replication allowed changes in relative rates of such nuclear lineages. Moreover, a selective elimination of compromised nuclei, through a programmed death process, was observed during spore development, suggesting that also a nuclear-level selection operates in Glomeromycota (Jany and Pawlowska 2010).

Anastomosis frequency ranged from 35 to 69 % between contacting hyphae of the same germling and from 6 to 90 % between hyphae of different germlings of the same strain (Table 2.1) in different experimental systems. However, no information is available on the factors controlling anastomosis frequency, involving either the extracellular environment or the intrahyphal microenvironment, possibly differentiating hyphae into fusion-competent regions, as observed in other fungal species (Hickey et al. 2002).

No hyphal fusions over 220 and 460 contacts were detected in *Gigaspora rosea* and *Racocetra castanea* mycelium, revealing an additional character differentiating the family Glomeraceae from the Gigasporaceae (Giovannetti et al. 1999). Such a difference was confirmed by in vitro experiments carried out using RiT-DNA transformed carrot roots, which reported very low values of anastomosis formation between different hyphae in AMF species belonging to Gigasporaceae (de Souza and Declerck 2003; de la Providencia et al. 2005; Table 2.1). Interestingly, the fusions observed were likened to a healing process (Gerdemann 1955; de Souza and Declerck 2003), which could be functional to the restriction of damages as a result of ageing, lytic events or physical lesions. In some species, short-length hyphal sections were able to undergo septa formation rapidly to shelter from the external environment and new hyphal tips growing from detached sections formed anastomoses among them. A differential behaviour was observed between *Dentiscutata reticulata*, where only a recovery of hyphal integrity was achieved, and *R. clarus* where the healing mechanism led to hyphal recovery and to a new growth into the surrounding medium (de la Providencia et al. 2007). Differences in hyphal fusion regulating mechanisms between these two species, mostly still unknown, could explain such different behaviour, supporting the view that Glomeraceae and Gigasporaceae have developed different survival strategies.

Anastomosis between vegetative hyphae may represent the first step of the parasexual cycle, allowing the formation of a heterokaryotic coenocytic mycelium where distinct nuclear genotypes are maintained for an indefinite/definite period of time (Pontecorvo 1956). However, no evidence of parasexual hybridization by means of hyphal fusions has so far been reported in AMF, as described in other fungi (Schardl et al. 1994). Though, in *R. intraradices* high-mobility domains containing transcriptional factors, with significant similarity to genes controlling mating type in *Phycomyces blakesleeanus* (Idnurm et al. 2008) and transcripts encoding for the meiotic recombination machinery, as well as meiosis-specific proteins (Tisserant et al. 2012), were detected. Moreover, 51 genes showing homology to those required for the proper completion of meiosis in *Saccharomyces cerevisiae* were identified in *Glomus* spp. (Halary et al. 2011), indicating the possibility of sexual reproduction in AMF. Indeed, findings consistent with recombination were reported for different AMF species, suggesting the occurrence of gene exchange, which could be realised by means of intermingling of nuclei during anastomosis (Vandenkoornhuyse et al. 2001; Croll et al. 2009; den Bakker et al. 2010; de la Providencia et al., 2013; Beaudet et al. 2015; Boon et al. 2015; Weichert and Fleißner 2015). Further research is needed to understand how fusions between genetically different lineages may alter the genetic structure and the reproductive success of AMF populations.

2.4 Vegetative Compatibility and Incompatibility in Anastomosing Hyphae

Studies on fungal somatic fusions revealed beneficial outcomes of frequent self-anastomoses, which increase the absorbing surface and the foraging ability of hyphal colonies (Aanen et al. 2008; Richard et al. 2012). Although enhanced mycelial fitness has been reported also after nonself fusions, their frequency is low, as vegetative compatibility is under the control of het or vic (heterokaryon or vegetative incompatibility genes) genes (Glass and Kuldau 1992; Leslie 1993; Glass and Kaneko 2003). The occurrence of incompatible het/vic alleles in fusing hyphae triggers incompatible responses, including programmed cellular compartmentalization and death (Glass and Dementhon 2006). Such incompatibility systems have probably evolved to limit mycelial damages resulting from genetic conflicts, due to DNA exchange, and from the transfer of pathogenic elements (viruses, deleterious mitochondria and plasmids) (Biella et al. 2002; Malik and Vilgalys 1999).

In AMF, experiments carried out on hyphae of germlings belonging to different genera and species, and to geographic isolates of the same species, revealed their ability to discriminate self from nonself. Hyphae belonging to different species or genera do not form anastomoses and, during interspecific and intergeneric interactions, do not show any contact interference. For example, no hyphal fusions were detected on a total of 90, 140, 232 and 98 hyphal contacts between hyphal germlings of *F. mosseae* and *F. caledonius*, *F. mosseae* and *Gigaspora rosea*, *F. caledonius* and *G. rosea*, *G. rosea* and *R. castanea*, respectively.

No hyphal compatibility between germlings belonging to geographically different isolates of *F. mosseae* was observed, even if pre-contact tropism, directional growth and branching of approaching hyphae occurred. In the interaction between *F. mosseae* isolates IN101, BEG25 and AZ225C (Giovannetti et al. 2003), approaching hyphae showed directed growth, branching and initiation of tips contacting the receiving hyphae, which were able to sense the presence of approaching hyphae and produced new lateral tips growing towards them. Interestingly, either prior to or after physical contact between hyphae, clear pre-fusion incompatible responses (rejection responses), were evidenced, such as apical swellings, wall thickening and cell wall depositions in the contacting hypha, followed by protoplasm withdrawal from hyphal tips, vacuolization and septa formation (Fig. 2.5).

The different ranges of events leading to the development of hyphal bridges and to the formation of anastomoses suggested the existence of a highly regulated system of self-recognition, leading to compatibility between hyphae belonging to the same network and between germlings and mycelia originated from spores produced by the same isolate. Such events are mirrored by nonself discrimination mechanisms, leading to nonself incompatibility between hyphae of AMF belonging to different genera, species, and geographic isolates of the same species. Though,

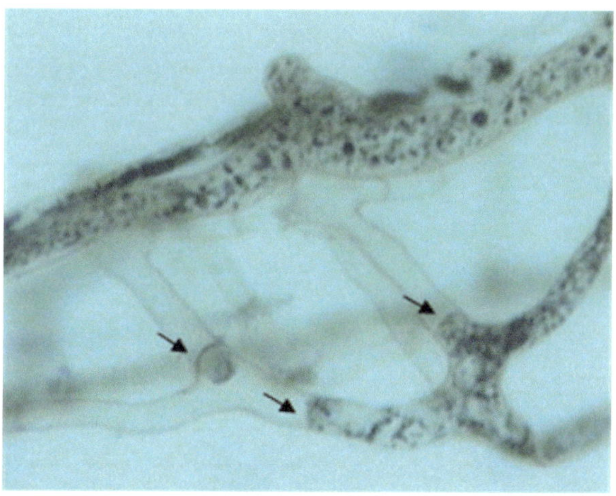

Fig. 2.5 Pre-fusion incompatible interactions between hyphae belonging to two geographically different isolates of the AMF species *Funneliformis mosseae*, IMA1 and AZ225, after SDH staining. Note the retraction septa developed by an approaching hypha after protoplasm withdrawal (*arrows*)

the specific chemical signals triggering interhyphal attraction and regulating vegetative compatibility/incompatibility, and leading to self recognition and nonself discrimination, remain poorly understood.

Post-fusion incompatible interactions, showing protoplasm withdrawal and cross wall formation in fused hyphae, were demonstrated in germinating spores and vegetative hyphae of Ascomycota, where incompatibility results from heterodimers of het or vic proteins (Glass et al. 2000; Saupe 2000; Glass et al. 2004). In AMF, nonself vegetative fusions (Fig. 2.6) were detected between genetically different single spore isolates (clonal lineages) of *R. irregularis*, which established vital connections, thereby creating the possibility for genetic exchange (Table 2.1; Fig. 2.7). Molecular analyses of the progeny of the mycelium derived from such nonself vegetative fusions evidenced the transmission of specific genetic markers, showing that genetic exchange had indeed occurred, despite the low anastomosis frequencies (Croll et al. 2009). Recent findings confirmed the occurrence of nonself anastomoses in *R. clarus* and the possibility of genetic exchange and heteroplasmy as a result of either perfect fusions or post-fusion incompatible interactions in *Rhizoglomus* isolates (Purin and Morton 2011, 2013; de la Providencia et al. 2013; Beaudet et al. 2013; Lin et al. 2014).

In conclusion, AMF hyphae are capable of recognition and fusion, thus producing large mycorrhizal networks where important nutritional, genetic and information flows are active. Such property is crucial for the survival of AMF populations, because it can directly affect their fitness, viability and reproductive success. The visualisation of AMF networks and of their structure unravelled a high level of interconnectedness, fundamental for facilitating the interchange of mineral nutrients, water and sugars flowing from soil to plants and from plants to soil.

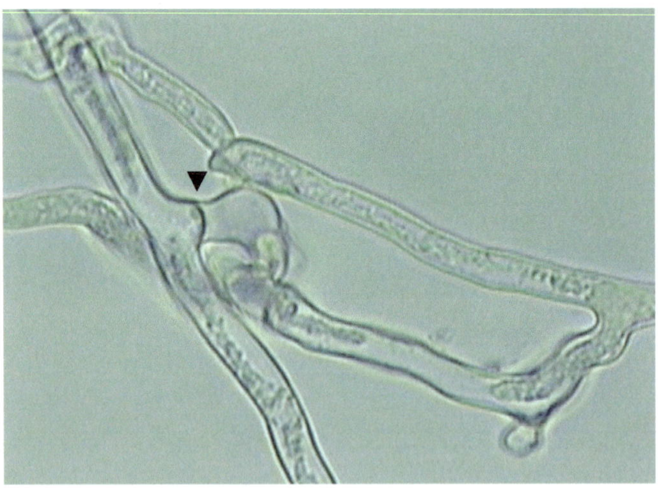

Fig. 2.6 Post-fusion incompatible interactions between hyphae belonging to two genetically different isolates of *Rhizoglomus irregularis*. Note the retraction septum and protoplasm withdrawal developed after fusion (*arrow*)

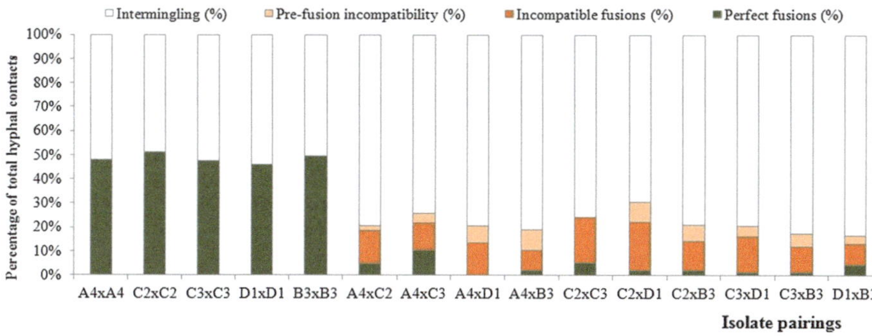

Fig. 2.7 Frequencies of the different types of interaction between hyphae of the same and genetically different lineages of *Rhizoglomus irregularis* (isolates A4, B3, C2, C3 and D1 originated from different nearby fields; isolates C2 and C3 originated from the same field. Modified from Croll et al. 2009)

In addition, the ability of self recognition and nonself discrimination of AMF hyphae suggests that the mycorrhizal network is also a site of information flow. The capacity of extraradical hyphae of fusing by means of anastomosis, interconnecting many different plants in the community, confirms that mycorrhizal networks can contribute to the formation of indefinitely large potential functional guilds (see Simard et al., Chap. 5, this Vol.), playing a key role in the complex web of interactions that regulates the functioning of natural and agricultural ecosystems.

Acknowledgments The authors acknowledge the support by the University of Pisa (Fondi di Ateneo) and by CNR.

References

Aanen DK, Debets AJ, de Visser JA, Hoekstra RF (2008) The social evolution of somatic fusion. BioEssays 30:1193–1203

Achatz M, Morris EK, Müller F, Hilker M, Rillig MC (2013) Soil hypha mediated movement of allelochemicals: arbuscular mycorrhizae extend the bioactive zone of juglone. Funct Ecol. doi:10.1111/1365-2435.12208

Ainsworth AM, Rayner ADM (1986) Responses of living hyphae associated with self and non-self fusions in the basidiomycete *Phanerochaete velutina*. J Gen Microbiol 132:191–201

Äström H, Giovannetti M, Raudaskoski M (1994) Cytoskeletal components in the arbuscular mycorrhizal fungus *Glomus mosseae*. Mol Plant Microbe Interact 7:309–312

Avio L, Pellegrino E, Bonari E, Giovannetti M (2006) Functional diversity of arbuscular mycorrhizal fungal isolates in relation to extraradical mycelial networks. New Phytol 172:347–357

Babikova Z, Gilbert L, Bruce TJ, Birkett M, Caulfield JC, Woodcock C, Pickett JA, Johnson D (2013) Underground signals carried through common mycelial networks warn neighbouring plants of aphid attack. Ecol Lett 16:835–843

Bago B, Azcon-Aguilar C, Piché Y (1998a) Architecture and developmental dynamics of the external mycelium of the arbuscular mycorrhizal fungus *Glomus intraradices* grown under monoxenic conditions. Mycologia 90:52–62

Bago B, Zipfel W, Williams RM, Chamberland H, Lafontaine JG, Webb WW, Piché Y (1998b) In vivo studies on the nuclear behavior of the arbuscular mycorrhizal fungus *Gigaspora rosea* grown under axenic conditions. Protoplasma 203:1–15

Bago B, Pfeffer PE, Shachar-Hill Y (2000) Carbon metabolism and transport in arbuscular mycorrhizas. Plant Physiol 124:949–958

Barreto de Novais C, Sbrana C, Saggin OJ Jr, Siqueira JO, Giovannetti M (2013) Vegetative compatibility and anastomosis formation within and among individual germlings of tropical isolates of arbuscular mycorrhizal fungi (Glomeromycota). Mycorrhiza 23:325–331

Barto EK, Hilker M, Muller F, Mohney BK, Weidenhamer JD, Rillig MC (2011) The fungal fast lane: common mycorrhizal networks extend bioactive zones of allelochemicals in soils. PLoS ONE 6:e27195

Beaudet D, Nadimi M, Iffis B, Hijri M (2013) Rapid mitochondrial genome evolution through invasion of mobile elements in two closely related species of arbuscular mycorrhizal fungi. PLoS ONE 8:e60768

Beaudet D, de la Providencia IE, Labridy M, Roy-Bolduc A, Daubois L, Hijri M (2015) Intra-isolate mitochondrial genetic polymorphism and gene variants coexpression in arbuscular mycorrhizal fungi. Genome Biol Evol 7:218–227

Bécard G, Piché Y (1989) Fungal growth stimulation by CO_2 and root exudates in vesicular-arbuscular mycorrhizal symbiosis. Appl Environ Microbiol 55:2320–2325

Beilby JP, Kidby DK (1980) Biochemistry of ungerminated and germinated spores of the vesicular-arbuscular mycorrhizal fungus, *Glomus caledonius*. Lipids 15:375–378

Bethlenfalvay GJ, Ames RN (1987) Comparison of two methods for quantifying extraradical mycelium of vesicular-arbuscular mycorrhizal fungi. Soil Sci Soc Am J 51:834–837

Bidartondo MI, Redecker D, Hijri I, Wiemken A, Bruns TD, Dominguez L, Sersic A, Leake JR, Read DJ (2002) Epiparasitic plants specialized on arbuscular mycorrhizal fungi. Nature 419:389–392

Biella S, Smith ML, Aist JR, Cortesi P, Milgroom MG (2002) Programmed cell death correlates with virus transmission in a filamentous fungus. Proc Roy Soc Lond B 269:2269–2276

Bistis GN (1981) Chemotropic interactions between trichogynes and conidia of opposite mating-type in *Neurospora crassa*. Mycologia 73:959–975

Blair JE (2009) Fungi. In: Hedges SB, Kumar S (eds) The timetree of life. Oxford University of Press, Oxford, pp 215–219

Boon E, Halary S, Bapteste E, Hijri M (2015) Studying genome heterogeneity within the arbuscular mycorrhizal fungal cytoplasm. Genome Biol Evol 7:505–521

Buller AHR (1933) Researches on fungi. Longmans, Green & Co., London, pp 33–49

Carey EV, Marler MJ, Callaway RM (2004) Mycorrhiza transfer carbon from a native grass to an invasive weed: evidence from stable isotopes and physiology. Plant Ecol 172:133–141

Cárdenas-Flores A, Draye X, Bivort C, Cranenbrouck S, Declerck S (2010) Impact of multispores in vitro subcultivation of Glomus sp. MUCL 43194 (DAOM 197198) on vegetative compatibility and genetic diversity detected by AFLP. Mycorrhiza 20:415–425

Cárdenas-Flores A, Cranenbrouck S, Draye X, Guillet A, Govaerts B, Declerck S (2011) The sterol biosynthesis inhibitor molecule fenhexamid impacts the vegetative compatibility of *Glomus clarum*. Mycorrhiza 21:443–449

Carlile MJ (1995) The success of the hypha and mycelium. In: Gow NAR, Gadd GM (eds) The growing fungus. Chapman & Hall, London

Casieri L, Lahmidi NA, Doidy J, Veneault-Fourrey C, Migeon A, Bonneau L, Courty P-E, Garcia K, Charbonnier M, Delteil A, Brun A, Zimmermann S, Plassard C, Wipf D (2013) Biotrophic transportome in mutualistic plant–fungal interactions. Mycorrhiza 23:597–625

Chiariello N, Hickman JC, Mooney HA (1982) Endomycorrhizal role for interspecific transfer of phosphorus in a community of annual plants. Science 217:941–943

Croll D, Giovannetti M, Koch AM, Sbrana C, Ehinger M, Lammers PJ, Sanders IR (2009) Nonself vegetative fusion and genetic exchange in the arbuscular mycorrhizal fungus *Glomus intraradices*. New Phytol 181:924–937

de la Providencia IE, Fernandez F, Declerck S (2007) Hyphal healing mechanism in the arbuscular mycorrhizal fungi *Scutellospora reticulata* and *Glomus clarum* differs in response to severe physical stress. FEMS Microbiol Lett 268:120–125

de la Providencia IE, de Souza FA, Fernandez F, Séjalon-Delmas N, Declerck S (2005) Arbuscular mycorrhizal fungi exhibit distinct pattern of anastomoses formation and hyphal healing mechanism between different phylogenic groups. New Phytol 165:261–271

de la Providencia IE, Nadimi M, Beaudet D, Rodriguez Morales G, Hijri M (2013) Detection of a transient mitochondrial DNA heteroplasmy in the progeny of crossed genetically divergent isolates of arbuscular mycorrhizal fungi. New Phytol 200:211–221

de Souza FA, Decleck S (2003) Mycelium development and architecture, and spore production of *Scutellospora reticulata* in monoxenic culture with Ri T-DNA transformed carrot roots. Mycologia 95:1004–1012

den Bakker HC, Van Kuren NW, Morton JB, Pawlowska TE (2010) Clonality and recombination in the life history of an asexual arbuscular mycorrhizal fungus. Mol Biol Evol 27:2474–2486

Eason WR, Newman EI, Chuba PN (1991) Specificity of interplant cycling of phosphorus: the role of mycorrhizas. Plant Soil 137:267–274

Egerton-Warburton LM, Querejeta JI, Allen MF (2007) Common mycorrhizal networks provide a potential pathway for the transfer of hydraulically lifted water between plants. J Exp Bot 58:1473–1483

Fellbaum CR, Mensah JA, Cloos AJ, Strahan GE, Pfeffer PE, Kiers ET, Bücking H (2014) Fungal nutrient allocation in common mycorrhizal networks is regulated by the carbon source strength of individual host plants. New Phytol 203:646–656

Fitter AH, Graves JD, Watkins NK, Robinson D, Scrimgeour C (1998) Carbon transfer between plants and its control in networks of arbuscular mycorrhizas. Funct Ecol 12:406–412

Fortuna P, Avio L, Morini S, Giovannetti M (2012) Fungal biomass production in response to elevated atmospheric CO_2 in a *Glomus mosseae–Prunus cerasifera* model system. Mycol Progress 11:17–26

Francis R, Finlay RD, Read DJ (1986) Vesicular-arbuscular mycorrhiza in natural vegetation systems IV. Transfer of nutrients in inter- and intra-specific combinations of host plants. New Phytol 102:103–111

Francis R, Read DJ (1984) Direct transfer of carbon between plants connected by vesicular-arbuscular mycorrhizal mycelium. Nature 307:53–56

Frey B, Vilarino A, Schuepp H, Arines J (1994) Chitin and ergosterol content of extraradical and intraradical mycelium of the vesicular-arbuscular mycorrhizal fungus *Glomus intraradices*. Soil Biol Biochem 26:711–717

Frey B, Schüepp H (1992) Transfer of symbiotically fixed nitrogen from berseem (*Trifolium alexandrinum* L.) to maize via vesicular-arbuscular mycorrhizal hyphae. New Phytol 122:447–454

Friese C, Allen MF (1991) The spread of VA mycorrhizal fungal hyphae in the soil: inoculum types and external hyphal architecture. Mycologia 83:409–418

Gerdemann JW (1955) Wound-healing of hyphae in a phycomycetous mycorrhizal fungus. Mycologia 47:916–918

Gianinazzi-Pearson V, Branzanti B, Gianinazzi S (1989) In vitro enhancement of spore germination and early hyphal growth of a vesicular-arbuscular mycorrhizal fungus by host root exudates and plant flavonoids. Symbiosis 7:243–255

Giovannetti M, Avio L, Sbrana C, Citernesi AS (1993) Factors affecting appressorium development in the vesicular-arbuscular mycorrhizal fungus *Glomus mosseae* (Nicol. & Gerd.) Gerd. & Trappe. New Phytol 123:115–122

Giovannetti M, Avio L (2002) Biotechnology of arbuscular mycorrhizas. In: Khachatourians GG, Arora DK (eds) Applied mycology and biotechnology: agriculture and food production, vol 2. Elsevier, Amsterdam, pp 275–310

Giovannetti M, Azzolini D, Citernesi AS (1999) Anastomosis and nuclear and protoplasmic exchange in arbuscular mycorrhizal fungi. Appl Environ Microbiol 65:5571–5575

Giovannetti M, Fortuna P, Citernesi AS, Morini S, Nuti MP (2001) The occurrence of anastomosis formation and nuclear exchange in intact arbuscular mycorrhizal networks. New Phytol 151:717–724

Giovannetti M, Sbrana C, Avio L, Strani P (2004) Patterns of below-ground plant interconnections established by means of arbuscular mycorrhizal mycorrhizal networks. New Phytol 164:175–181

Giovannetti M, Sbrana C, Strani P, Agnolucci M, Rinaudo V, Avio L (2003) Genetic diversity of isolates of *Glomus mosseae* from different geographic areas detected by vegetative compatibility testing and biochemical and molecular analysis. Appl Environ Microbiol 69:616–624

Glass NL, Dementhon K (2006) Non-self recognition and programmed cell death in filamentous fungi. Curr Opin Microbiol 9:553–558

Glass NL, Jacobson DJ, Shiu PKT (2000) The genetics of hyphal fusion and vegetative incompatibility in filamentous ascomycete fungi. Annu Rev Genetics 34:165–186

Glass NL, Rasmussen C, Roca MG, Read ND (2004) Hyphal homing, fusion and mycelial interconnectedness. Trends Microbiol 12:135–141

Glass L, Kaneko I (2003) Fatal attraction: nonself recognition and heterokaryon incompatibility in filamentous fungi. Eukaryot Cell 2:1–8

Glass NL, Kuldau GA (1992) Mating type and vegetative incompatibility in filamentous ascomycetes. Annu Rev Phytopathol 30:201–224

Godfrey RM (1957) Studies on British species of endogone. III. Germination of spores. Trans Br Mycol Soc 40:203–210

Gooday GW (1990) The ecology of chitin degradation. Adv Microb Ecol 11:387–430

Grava S, Keller M, Voegeli S, Seger S, Lang C, Philippsen P (2011) Clustering of nuclei in multinucleated hyphae is prevented by dynein-driven bidirectional nuclear movements and microtubule growth control in *Ashbya gossypii*. Eukaryot Cell 10:902–915

Graves JD, Watkins NK, Fitter AH, Robinson D, Scrimgeour C (1997) Intraspecific transfer of carbon between plants linked by a common mycorrhizal network. Plant Soil 192:153–159

Gregory PH (1984) The fungal mycelium: an historical perspective. Trans Br Mycol Soc 82:1–11

Grime JP, Mackey JML, Hillier SH, Read DJ (1987) Floristic diversity in a model system using experimental microcosm. Nature 328:420–422

Halary S, Malik S-B, Lildhar L, Slamovits CH, Hijri M, Corradi N (2011) Conserved meiotic machinery in *Glomus* spp., a putatively ancient asexual fungal lineage. Genome Biol Evol 3:950–958

Harrison MJ, van Buuren ML (1995) A phosphate transporter from the mycorrhizal fungus *Glomus versiforme*. Nature 378:626–629

Haystead A, Malajczuk N, Grove TS (1988) Underground transfer of nitrogen between pasture plants infected with vesicular arbuscular mycorrhizal fungi. New Phytol 108:417–423

Hepper CM (1983) Limited independent growth of a vesicular-arbuscular mycorrhizal fungus in vitro. New Phytol 93:537–542

Hickey PC, Jacobson DJ, Read ND, Glass NL (2002) Live-cell imaging of vegetative hyphal fusion in *Neurospora crassa*. Fungal Genet Biol 37:109–119

Hirrel MC, Gerdemann JW (1979) Enhanced carbon transfer between onions infected with a vesicular–arbuscular mycorrhizal fungus. New Phytol 83:731–738

Idnurm A, Walton FJ, Floyd A, Heitman J (2008) Identification of the sex genes in an early diverged fungus. Nature 451:193–196

Ikram A, Jensen ES, Jakobsen I (1994) No significant transfer of N and P from *Pueraria Phaseoloides* to *Hevea Brasiliensis* via hyphal links of arbuscular mycorrhiza. Soil Biol Biochem 26:1541–1547

Jakobsen I, Rosendahl S (1990) Carbon flow into soil and external hyphae from roots of mycorrhizal cucumber plants. New Phytol 115:77–83

Jany JL, Pawlowska TE (2010) Multinucleate spores contribute to evolutionary longevity of asexual Glomeromycota. Am Nat 175:424–435

Johansen A, Jensen ES (1996) Transfer of N and P from intact or decomposing roots of pea to barley interconnected by an arbuscular mycorrhizal fungus. Soil Biol Biochem 28:73–81

Karandashov V, Bucher M (2005) Symbiotic phosphate transport in arbuscular mycorrhizas. Trends Plant Sci 10:22–29

Koske RE (1981) Multiple germination by spores of Gigaspora gigantea. Trans Br Mycol Soc 76:320–330

Lang C, Grava S, van den Hoorn T, Trimble R, Philippsen P, Jaspersen SL (2010) Mobility, microtubule nucleation and structure of microtubule-organizing centers in multinucleated hyphae of *Ashbya gossypii*. Mol Biol Cell 21:18–28

Lekberg Y, Hammer EC, Olsson PA (2010) Plants as resource islands and storage units: adopting the mycocentric view of arbuscular mycorrhizal networks. FEMS Microbiol Ecol 74:336–345

Lerat S, Gauci R, Catford JG, Vierheilig H, Piché Y, Lapointe L (2002) C-14 transfer between the spring ephemeral *Erythronium americanum* and sugar maple saplings via arbuscular mycorrhizal fungi in natural stands. Oecologia 132:181–187

Leslie JF (1993) Fungal vegetative compatibility. Annu Rev Phytopathol 31:27–150

Lin K, Limpens E, Zhang Z, Ivanov S, Saunders DGO, Mu D, Pang E, Cao H, Cha H, Lin T, Zhou Q, Shang Y, Li Y, Sharma T, van Velzen R, de Ruijter N, Aanen DK, Win J, Kamoun S, Bisseling T, Geurts R, Huang S (2014) Single nucleus genome sequencing reveals high similarity among nuclei of an endomycorrhizal fungus. PLoS Genet 10:e1004078

Logi C, Sbrana C, Giovannetti M (1998) Cellular events involved in survival of individual arbuscular mycorrhizal symbionts growing in the absence of the host. Appl Environ Microbiol 64:3473–3479

Maldonado-Mendoza IE, Dewbre GR, Harrison MJ (2001) A phosphate transporter gene from the extraradical mycelilum of an arbuscular mycorrhizal fungus *Glomus intraradices* is regulated in response to phosphate in the environment. Mol Plant Microbe Interact 14:1140–1148

Malik M, Vilgalys R (1999) Somatic incomapatibility in fungi. In: Worrall JJ (ed) Structure and dynamics of fungal populations. Kluwer Academic Publishers, Dodrecht

Martins MA (1993) The role of the external mycelium of arbuscular mycorrhizal fungi in the carbon transfer process between plants. Mycol Res 97:807–810

Martins MA, Cruz AF (1998) The role of the external mycelial network of arbuscular mycorrhizal fungi: III. A study of nitrogen transfer between plants interconnected by common mycelium. Microbiol Rev 29:289–294

Meding SM, Zasoski RJ (2008) Hyphal-mediated transfer of nitrate, arsenic, cesium, rubidium, and strontium between arbuscular mycorrhizal forbs and grasses from a California oak woodland. Soil Biol Biochem 40:126–134

Mikkelsen BL, Rosendahl S, Jakobsen I (2008) Underground resource allocation between individual networks of mycorrhizal fungi. New Phytol 180:890–898

Miller RM, Reinhardt DR, Jastrow JD (1995) External hyphal production of vesicular arbuscular mycorrhizal fungi in pasture and tallgrass prairie communities. Oecologia 103:17–23

Mosse B (1959) The regular germination of resting spores and some observations on the growth requirements of an *Endogone* sp. causing vesicular arbuscular mycorrhiza. Trans Br Mycol Soc 42:273–286

Morris NR, Enos AP (1992) Mitotic gold in a mold: Aspergillus genetics and the biology of mitosis. Trends Genet 8:32–37

Newman EI (1988) Mycorrhizal links between plants: their functioning and ecological significance. Adv Ecol Res 18:243–270

Newman EI, Eason WR (1989) Cycling of nutrients from dying roots to living plants, including the role of mycorrhizas. Plant Soil 115:211–215

Newman EI, Eason WR (1993) Rates of phosphorus transfer within and between ryegrass (*Lolium perenne*) plants. Funct Ecol 7:242–248

Newman EI, Ritz K (1986) Evidence on the pathways of phosphorus transfer between vesicular–arbuscular mycorrhizal plants. New Phytol 104:77–87

Pfeffer PE, Douds DD Jr, Bucking H, Schwartz DP, Shachar-Hill Y (2004) The fungus does not transfer carbon to or between roots in an arbuscular mycorrhizal symbiosis. New Phytol 163:617–627

Phipps CJ, Taylor TN (1996) Mixed arbuscular mycorrhizae from the Triassic of Antarctica. Mycologia 88:707–714

Pirozynski KA, Malloch DW (1975) The origin of land plants: a matter of mycotrophism. BioSystems 6:153–164

Pontecorvo G (1956) The parasexual cycle in fungi. Annu Rev Microbiol 10:393–400

Purin S, Morton JB (2011) In situ analysis of anastomosis in representative genera of arbuscular mycorrhizal fungi. Mycorrhiza 21:505–514

Purin S, Morton JB (2013) Anastomosis behaviour differs between asymbiotic and symbiotic hyphae of *Rhizophagus clarus*. Mycologia 12:589–602

Rayner ADM (1996) Interconnectedness and individualism in fungal mycelia. In: Sutton BC (ed) A century of mycology. University of Cambridge Press, Cambridge, pp 193–232

Redecker D, Kodner R, Graham LE (2000) Glomalean fungi from the Ordovician. Science 289:1920–1921

Remy W, Taylor TN, Hass H, Kerp H (1994) Four hundred-million-year-old vesicular arbuscular mycorrhizae. Proc Natl Acad Sci USA 91:11841–11843

Richard F, Glass NL, Pringle A (2012) Cooperation among germinating spores facilitates the growth of the fungus *Neurospora crassa*. Biol Lett 8:419–422

Rogers JB, Laidlaw AS, Christie P (2001) The role of arbuscular mycorrhizal fungi in the transfer of nutrients between white clover and perennial ryegrass. Chemosphere 42:153–159

Saupe SJ (2000) Molecular genetics of heterokaryon incompatibility in filamentous ascomycetes. Microbiol Mol Biol Rev 64:489–502

Sbrana C, Fortuna P, Giovannetti M (2011) Plugging into the network: belowground connections between germlings and extraradical mycelium of arbuscular mycorrhizal fungi. Mycologia 103:307–316

Schardl CL, Leuchtmann A, Tsai HF, Collett MA, Watt DM, Scott DB (1994) Origin of a fungal symbiont of perennial ryegrass by interspecific hybridization of a mutualist with the ryegrass choke pathogen, *Epichloë typhina*. Genetics 136:1307–1317

Simon L, Bousquet J, Levesque RC, Lalonde M (1993) Origin and diversification of endomycorrhizal fungi and coincidence with vascular land plants. Nature 363:67–69

Smith SE, Read DJ (2008) Mycorrhizal symbiosis, 3rd edn. Academic Press, San Diego

Smith SE, Smith FA, Jakobsen I (2003) Mycorrhizal fungi can dominate phosphate supply to plants irrespective of growth responses. Plant Physiol 133:16–20

Snetselaar KM, Boelker M, Kahmann R (1996) *Ustilago maydis* mating hyphae orient their growth toward pheromone sources. Fungal Genet Biol 20:299–312

Song YY, Zeng RS, Xu JF, Li J, Shen X, Woldemariam Gebrehiwot Y (2010) Interplant communication of tomato plants through underground common mycorrhizal networks. PLoS ONE 5:e13324

Tisserant E et al (2012) The transcriptome of the arbuscular mycorrhizal fungus *Glomus intraradices* (DAOM 197198) reveals functional tradeoffs in an obligate symbiont. New Phytol 193:755–769

Tommerup IC (1984) Persistence of infectivity by germinated spores of vesicular-arbuscular mycorrhizal fungi in soil. Trans Br Mycol Soc 82:275–282

Tommerup IC (1988) The vesicular-arbuscular mycorrhizas. Adv Plant Pathol 6:81–91

Treseder KK, Allen MF (2000) Mycorrhizal fungi have a potential role in soil carbon storage under elevated CO_2 and nitrogen deposition. New Phytol 147:189–200

van der Heijden MGA, Klironomos JN, Ursic M, Moutoglis P, Streitwolf-Engel R, Boller T, Wiemken A, Sanders IR (1998) Mycorrhizal fungal diversity determines plant biodiversity, ecosystem variability and productivity. Nature 396:69–72

Vandenkoornhuyse P, Leyval C, Bonnin I (2001) High genetic diversity in arbuscular mycorrhizal fungi: evidence for recombination events. Heredity (Edinb) 87:243–253

Voets L, de la Providencia IE, Declerck S (2006) Glomeraceae and Gigasporaceae differ in their ability to form mycelium networks. New Phytol 172:185–188

Voets L, Goubau I, Olsson PA, Merckx R, Declerck S (2008) Absence of carbon transfer between *Medicago truncatula* plants linked by a mycorrhizal network, demonstrated in an experimental microcosm. FEMS Microbiol Ecol 65:350–360

Walder F, Brulé D, Koegel S, Wiemken A, Boller T, Courty P-E (2015) Plant phosphorus acquisition in a common mycorrhizal network: regulation of phosphate transporter genes of the Pht1 family in sorghum and flax. New Phytol 205:1632–1645

Wardle DA, Bardgett RD, Klironomos JN, Setälä H, van der Putten WH, Wall DH (2004) How belowground biota drives the aboveground subsystem. Ecological linkages between aboveground and belowground biota. Science 304:1629–1633

Watkins NK, Fitter AH, Graves JD, Robinson D (1996) Carbon transfer between C3 and C4 plants linked by a common mycorrhizal network, quantified using stable carbon isotopes. Soil Biol Biochem 28:471–477

Weichert M, Fleißner A (2015) Anastomosis and heterokaryon formation. In: Genetic transformation systems in fungi, vol 2. Springer, Berlin, pp 3–21

Whittingham J, Read DJ (1982) Vesicular arbuscular mycorrhiza in natural vegetation systems. II. Nutrient transfer between plants with mycorrhizal interconnections. New Phytol 90:277–284

Chapter 3
The Importance of Ectomycorrhizal Networks for Nutrient Retention and Carbon Sequestration in Forest Ecosystems

Håkan Wallander and Alf Ekblad

Abstract Extramatrical mycelium (EMM) of mycorrhizal fungi have a fundamental role in carbon (C) cycling in forest ecosystems. This carbon is used for building extensive mycelial networks in the soil as well as for metabolic activity related to nutrient uptake. Here we discuss the factors that regulate the production and turnover of EMM and its role in soil C dynamics and nitrogen retention. C availability seems to be the key factor determining EMM production and possibly its standing biomass in forests but direct effects of mineral nutrient availability on the EMM can also be important. There is great uncertainty about the rate of turnover of EMM, and the increasing evidence that residues of EM fungi play a major role in the formation of stable N and C in soil organic matter highlights the need to include mycorrhizal effects in models of global soil C stores.

Keywords Carbon · Nitrogen · Nutrient retention · Extramatrical mycorrhizal mycelium · Ingrowth mesh bags

3.1 Introduction

Mycorrhizal fungi form extensive mycelial networks in the soils of boreal and temperate forests (Smith and Read 2008). Most of the trees form symbioses with ectomycorrhizal fungi (EMF), while shrubs and herbaceous plants are colonized by arbuscular mycorrhizal fungi, ericoid or orchid mycorrhizal fungi. Here we will

H. Wallander (✉)
Department of Biology, Microbial Ecology Group, Lund University, SE-223 62, Lund, Sweden
e-mail: hakan.wallander@biol.lu.se

A. Ekblad
School of Science and Technology, Örebro University, SE-701 82, Örebro, Sweden
e-mail: alf.ekblad@oru.se

T.R. Horton (ed.), *Mycorrhizal Networks*,
Ecological Studies 224, DOI 10.1007/978-94-017-7395-9_3

focus on the EMF mycelial networks formed by many forest trees, especially in Pinaceae, Fagaceae and Betulaceae.

Trees invest large amounts of carbon (C) to facilitate nutrient uptake by the EMF networks, especially in nutrient-poor sites, and the growth of these networks is regulated by the C flux from the trees (Smith and Read 2008). Network functionality diminishes shortly after termination of the current photosynthate flux from the trees, as has been shown both in microcosms after severing the connection between the fungus and the host (Söderström and Read 1987), and in field experiments when the belowground C flux has been terminated by girdling the trees (Högberg et al. 2001). When uptake is terminated, leaching of nutrients can follow. This is commonly seen after clear-cutting forests, especially in nutrient-rich sites, but nutrient retention usually recovers after a few years when field layer vegetation has been established (Futter et al. 2010). Nitrogen (N) can also leach from standing forests when N input has been continuously high and the forests are subjected to N saturation (Emmett 2007). This leaching is possibly an effect of impaired EMF function (Aber et al. 1998; Högberg et al. 2011).

C sequestration in forest soils is dependent on N availability since the C:N ratio of stable soil organic matter (SOM) in deeper soil layers is rather constant around 10–15. This value is similar to the ratio of EMF biomass (Wallander et al. 2003) and it has been suggested that the SOM in boreal forests to a large extent is composed of EMF residues (Högberg et al. 2011; Clemmensen et al. 2013; Fernandez et al. 2013; Fernandez and Kennedy 2015). However, the enhanced C sequestration that occurs after N fertilization (Franklin et al. 2003; Hyvönen et al. 2007) is difficult to attribute to EMF since many of these fungi decline under elevated N conditions (Wallenda and Kottke 1998; Nilsson and Wallander 2003; Högberg et al. 2011; Bahr et al. 2013). This makes assessing the role of EMF networks in C sequestration complicated. The enhanced tree growth and litter production, and the reduced decomposition of SOM (Nohrstedt et al. 1989; Franklin et al. 2003), usually found after N fertilization are possible reasons for enhanced C sequestration. But it is also possible that changes in ectomycorrhizal (EM) community composition that occurs after N fertilization plays a significant role since different EM species decompose at different rates (Langley and Hungate 2003; Koide and Malcolm 2009; Koide et al. 2011; Fernandez et al. 2013). Clemmensen et al. (2013) found larger C sequestration in old compared to young successional stages of boreal forests in northern Sweden and attributed this, at least in part, to different mycorrhizal communities (Clemmensen et al. 2015). Furthermore, certain species of EMF may reduce SOM accumulation by degrading recalcitrant compounds to obtain N that is delivered to the host trees (Talbot et al. 2008). Thus EMF appears to have a central role in C sequestration, although the overall effect is difficult to predict because of these opposing processes.

In this chapter we will discuss (1) how the production of EMF networks can be measured under field conditions, (2) how the production is regulated, and (3) to what extent EMF networks are important for ecosystem processes such as nutrient retention and C sequestration.

3.2 Methods to Study Production of EMF in the Field

3.2.1 Observational Methods

One of the problems with quantifying standing biomass or production of EM networks in the field is difficulties in separating mycorrhizal mycelia from saprotrophic mycelia. There is no biochemical or DNA-based marker to distinguish extramatrical mycorrhizal mycelia (EMM) from the complex fungal soil community since EM fungi do not form a monophyletic clade (Tedersoo et al. 2010), but there are various ways to overcome this problem. Production of EMM can be studied in the field through direct observation in root windows or minirhizotrons (Coutts and Nicoll 1990; Treseder et al. 2005; Pritchard et al. 2008), although the resolution is too low to observe individual hyphae. Coutts and Nicoll (1990) followed the growth rate of the advancing hyphal front of two EM species observed through observation windows installed on 2 m large tubes placed outside. The estimated growth rate of *Thelephora terrestris* was 1–3 mm day^{-1} during the plant growing season which is similar to what has been found for EMF in laboratory microcosms (Read 1992) as well as under axenic growth (Gafur et al. 2004). Furthermore, the mycelium continued to grow over the winter, although at a slower rate (0.3 mm day^{-1}). *Laccaria proxima* on the other hand grew slower and the mycelium disappeared during the autumn. In this way observation methods are useful for determining longevity of EMM and Pritchard et al. (2008) found that EM rhizomorphs observed in minirhizotrons lived much longer than mycorrhizal root tips (mean longevity 532 and 104 days respectively, Fig. 3.1). Observation methods are useful to study seasonal dynamics of EMF networks in the field but it is more difficult to quantify the production in terms of biomass.

3.2.2 Mesh Bags

The most common approach to quantify EMM production is the use of ingrowth mesh bags (Wallander et al. 2001; Fig. 3.2) or in-growth cores (Godbold et al. 2006; Hendricks et al. 2006). Such techniques have so far been used to estimate EMM production at ~140 different sites (Ekblad et al. 2013). The mesh bags or cores are usually filled with sand, free of fungal material, and incubated in the soil for various periods of time. The amount of fungal biomass detected at harvest is used as an estimate of EMM production. However, colonization by saprophytic fungi can also occur, although to a smaller extent, and to account for this, trenched plots can be used to measure production of the saprophytic mycelium only (Wallander et al. 2001). It should be noted that trenching by forcing down tubes into the soil will only last a limited amount of time since EMM may enter the tubes from below. In studies in Sweden such trenched plots were free of EMM for one growing season but EM fungi entered some of the tubes after two growing seasons

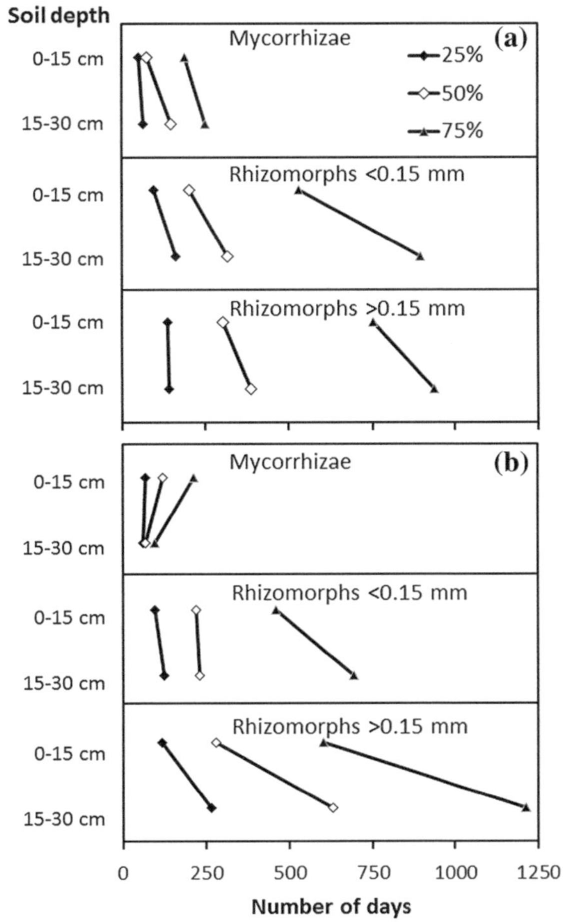

Fig. 3.1 Number of days until 25, 50 or 75 % mortality of mycorrhizal tips and of rhizomorphs of two diameter classes; **a** ambient CO_2, **b** elevated CO_2. Data from minirhizotrons installed in a loblolly pine forest in North Carolina (from data in Pritchard et al. 2008)

(Wallander et al. 2001, 2011). Molecular analysis of the fungal communities of ingrowth mesh bags has revealed that between 70 and 90 % of the sequences obtained from the bags originate from fungi known to form ectomycorrhizal symbiosis (Parrent and Vilgalys 2007; Wallander et al. 2010; Berner et al. 2012). Another approach to check for saprophytic ingrowth is to analyse the C isotopic signature of mycelia extracted from mesh bags (Wallander et al. 2001). Fruitbodies of wood—and litter decomposing fungi usually have a $\delta^{13}C$ value that is 2–3 ‰ higher than values for EMF fruitbodies (Högberg et al. 1999; Taylor et al. 2003; Hobbie et al. 2012) and the $\delta^{13}C$ of mycelia in mesh bags usually resemble values for EMF fruitbodies (Wallander et al. 2001; Hagerberg et al. 2003; Mikusinska et al. 2013). However, with time, it is possible that fungi that decompose the EMF mycelium will establish in the bags; whether the mycelium formed by these fungi differs in $\delta^{13}C$ from EMF is unknown.

Fig. 3.2 A harvested mesh bag after 2 years incubation in a Norway spruce forest in Sweden has been opened in the lab. Abundant mycelia and rhizomorphs of EM fungi are clearly visible. Photo: Adam Bahr

Studies using natural soil in mesh bags or cores have shown higher production rates of EMM than studies using sand (Hendricks et al. 2006). The reason for this is probably that soil is a more natural substrate for the fungi, but problems arise when using soil since it contains background fungal material that needs to be subtracted before EMM production can be calculated. This seems to work when the soil has low SOM content (Hendricks et al. 2006; Sims et al. 2007) but under other circumstances fungal biomass background values are too high to make ingrowth measurements reliable (Wallander personal observation). An alternative approach to estimate EMM production in mesh bags or cores filled with soil is to use the different C isotopic signatures of C_4 and C_3 plant material (Godbold et al. 2006; Wallander et al. 2011). In this case mesh bags or cores are filled with ^{13}C-enriched C_4 material (soil or plant material) and the change in isotopic composition that occurs when the bags/cores are colonized by ^{13}C depleted EMM is followed and used to calculate C flux into the bags. This approach was used by Godbold et al. (2006) who estimated a C flux to EMF of around 1000 g C m^{-2} during a period of 2.5 years in a poplar plantation in Italy. Wallander et al. (2011) used a mixture of sand and compost made of maize leaves (a C_4 plant) in mesh bags and estimated a C flux of around 100 g C m^{-2} over a three year period in Norway spruce forests in Sweden.

One of the problems of using ingrowth mesh bags or cores to quantify EMM is that the fungal community that colonizes the bags/cores may not be representative of the soil community. The reason for this is the use of artificial substrate (sand) and the fact that fungal-free bags or cores select for fast-growing EMF species. Fungi that proliferate in the mineral soil may be overrepresented in the bags when a sandy substrate is used. One way to overcome some of these problems may be to incubate mesh bags for longer time periods and analyse annual production and turnover of EMM in the mesh bags after colonization by fast-growing EMF species has terminated. However, it is not known whether the contribution of saprotrophic mycelium increases over time, and this needs to be tested. Production of specific species can probably be measured by qPCR or other molecular techniques, and the C flux into the bags or cores over time can be followed by isotopic techniques as discussed above. The mesh bag method is best suited for relative comparisons between different forest management practices or treatments and for comparing how different mesh bag amendments influence EMM production. Estimations of absolute amounts produced must be interpreted with caution. For a more detailed review on this subject see Wallander et al. (2013).

3.2.3 Exploration Types

EMF communities can be extremely diverse (Dahlberg 2001) and the composition of the EM community is probably of large importance for ecosystem processes such as nutrient retention and C sequestration. One approach to handle this diversity of EMF in functional terms has been to classify the species into exploration types based on the amount of hyphae emanating from the root tips and the presence and differentiation of rhizomorphs (Agerer 2001, 2006). The different C demand among the explorations types will most likely have profound effects on their ecological roles in terms of nutrient uptake/retention and their potential to sequester C. When more physiological data have been collected about the different exploration types it might be possible to incorporate them into ecosystem models with the aim to increase predictions of key ecological processes such as nutrient uptake, leaching of nutrients and SOM cycling. Work along this line has been started by Weigt et al. (2011, 2012a, b) who have quantified the amount of mycelium produced by representatives of a few different exploration types in laboratory experiments. If these values are applied to EM communities that have been identified on root tips in the field, it might be possible to extrapolate potential EMM production in field sites from EM community composition estimated from analysis of root tips.

3.3 Regulation of EM Growth by C Supplied from the Host Trees

3.3.1 Tree Growth

Since the EMM depends on C delivered from the host trees, higher photosynthetic rates can potentially result in higher EMM production. Support for this view was found by Korkama et al. (2007) who studied EMM production related to fast- and

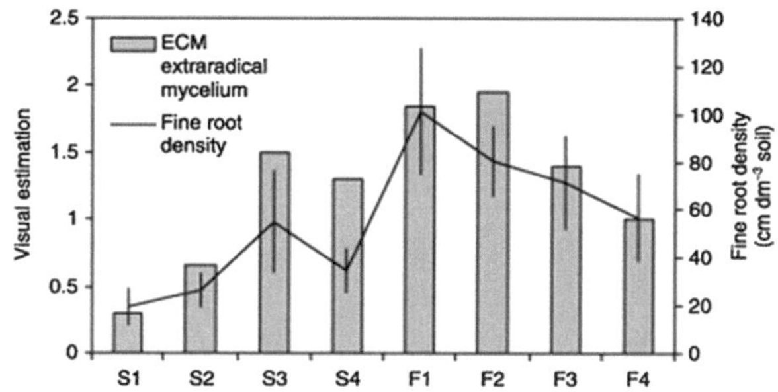

Fig. 3.3 The biomass of extraradical mycelia developing in mesh bags (bars) is related to the fine root density (lines) under slow (S1–S4) and fast-growing Norway spruce (*Picea abies*) clones (F1–F4). Fungal biomass was estimated visually under a dissecting microscope and according to aggregation of the sand. It was classified into four categories: (0) no mycelial strands and sand aggregation; (1) a few mycelial strands but no sand aggregation; (2) moderate number of mycelial strands and some sand aggregation; (3) considerable extraradical mycelium and sand aggregation (from Korkama et al. 2007)

slow-growing spruce clones. Significantly higher EMM production was found in the fast-growing compared to slow-growing ones (Fig. 3.3). Furthermore, EMM production was correlated to fine root biomass suggesting that enhanced belowground allocation of C was necessary to sustain the better growth of fast-growing clones. In a larger data set of Scandinavian Norway spruce forests (data from Ekblad et al. 2013), EMM growth and spruce productivity were positively correlated (Fig. 3.4), but other factors such as nutrient availability are important for the relative allocation above and belowground (see below) which complicates the picture.

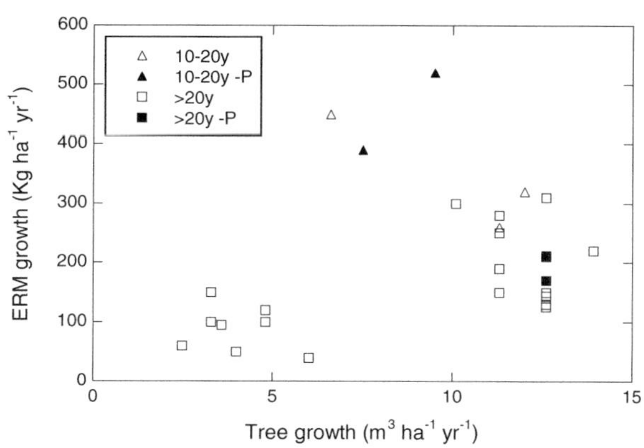

Fig. 3.4 Relationship between growth of extramatrical mycelium (EMM; Kg ha^{-1} year^{-1}) in the top 10 cm and wood production (m^3 ha^{-1} year^{-1}) in the Norway spruce stands from Table 2 in Ekblad et al. (2013). *Open triangles* (young stands 10–20 year), *closed triangles* (young stands with P deficiency, needle P < 1.3 mg P g^{-1}), *open squares* (stands older than 20 year), *closed squares* (older stands with P deficiency)

3.3.2 Tree Age

Trees usually peak in nutrient uptake during canopy closure when nutrient demand is highest. When the trees mature more nutrients are supplied through internal cycling (Kimmins 2004). EMM growth in the soils shows similar pattern (Kalliokoski et al. 2010; Wallander et al. 2010), and its peak in production seems to be close to that of the usual canopy closure stage of coniferous forests in southern Scandinavia (25–40 years, Schmalholz and Hylander 2009).

3.3.3 Seasonality

Winter is obviously a season with poor EMM growth in boreal and boreo-nemoral regions but might be a period of active growth in warmer climates. Thus, in pine forests of northeastern Spain the living EMM biomass, quantified by specific primers and qPCR, peaked in February for *Boletus edulis* and in December for *Lactarius deliciosus* (De la Varga et al. 2013). Also in cooler temperate regions some species are able to continue growing at a low rate during winter, at least in the study in UK by Coutts and Nicoll (1990). But EMM growth is probably not directly related to temperature in the same way as growth of saprotrophs in soil, since EMM growth depends on C allocated from the trees. In northern temperate and boreal regions maximal growth can be expected in the second half of the growing season when the above ground C sink in terms of tree growth has ceased. Support for this view was found by Nilsson et al. (2007) who found better EMM growth in oak forests during the colder fall period compared to the warmer summer period. Another factor that might be confounding in these studies is soil humidity which usually is higher in the fall than in the summer.

3.3.4 Elevated CO_2

Elevated CO_2 could potentially increase the photosynthetic rate and thereby increase EMM growth. However this will also depend on other factors since nutrient availability may limit photosynthesis resulting in less or no increased growth rate after elevated CO_2. Several studies in the laboratory have demonstrated increased EMM growth after elevated CO_2 (e.g. Rouhier and Read 1998), but although an increased rhizomorph production and longevity was observed in a *Pinus taeda* plantation (Pritchard et al. 2008), these results have been difficult to repeat in the field (Godbold et al. 2006; Parrent and Vilgalys 2007; Dawes et al. 2013). The effect of elevated CO_2 on EMF was recently reviewed by Fransson (2012).

3.3.5 Herbivory

Herbivory of tree foliage could potentially result in reduced EMM growth if photosynthetic capacity declines, but few studies have tested this. However, herbivory from scale insects reduced colonization of pinyon pines by EMF (Del Vecchio et al. 1993) and suilloid fungi have shown reduced growth more than other species after artificial herbivory (Kuikka et al. 2003). These species appear to have higher C demands when tested in the laboratory (Fransson et al. 2007) which may be one explanation for their decline after herbivory.

3.4 How Nutrient Availability Influence C Allocation and EMM Growth

Nutrient availability strongly affects C allocation pattern in the host trees. Work by Ericsson (1995; Fig. 3.5) demonstrated that allocation belowground increased when N, P, S and Fe was limiting growth, while limitation of K, Mg, and Mn resulted in reduced belowground allocation. This was explained by impaired photosynthesis at K, Mg and Mn limitation resulting in lower carbohydrate production. In contrast limitation of N, P, S and Fe resulted in impaired growth but lower direct effects on photosynthesis, leading to accumulation of carbohydrates in the leaves. These carbohydrates could be loaded into the phloem and transported belowground (Ericsson 1995). Much less is known about how EMM is influenced by nutrient

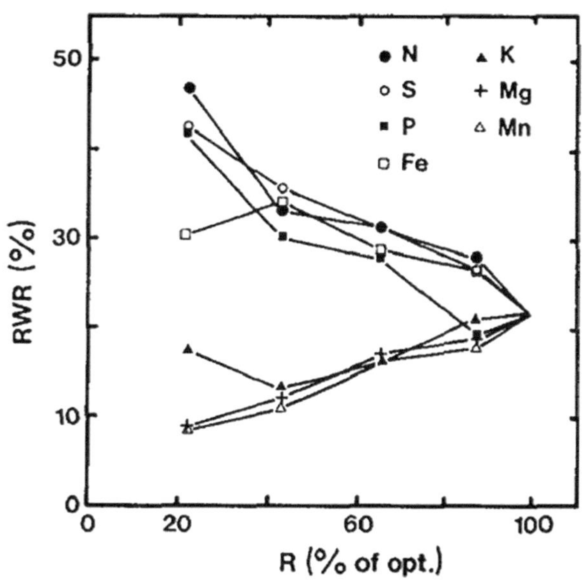

Fig. 3.5 Root weight ratio (RWR) in *Betula pendula* seedlings grown at different nutrient regimes. The indicated nutrients were added to give plant growth rates between 20 and 100 % of optimum. Reduced availability of N, S, P and Fe resulted in increased RWR while reduced availability of K, Mg and Mn resulted in increased RWR (data from Ericsson 1995)

limitation and excessive amounts of nutrients, but as demonstrated in the following section, the current knowledge appear in many respects to follow the results found by Ericsson (1995) for roots.

3.4.1 Nitrogen

The large production of EMM in boreal forests is attributed to N commonly being the limiting nutrient in these ecosystems. EMF are well adapted to cope with N limitation by producing large mycelia that can take up and store N during periods of high N availability (Mikusinska et al. 2013). In addition, many species can mobilize and take up organic forms of N (Perez-Moreno and Read 2001). Thus it is not surprising that the activity of EMF is reduced when N input to the system increases. This can be seen both in terms of fruitbody formation and in the number of mycorrhizal root tips (Wallenda and Kottke 1998; Peter et al. 2001; Lilleskov et al. 2002; Lilleskov et al. 2011). EMF species more efficient in taking up organic N are usually the ones that become less frequent in response to inorganic N loads (Taylor et al. 2000; Lilleskov et al. 2011).

Growth of EMM is strongly reduced after N addition in laboratory-grown seedlings (Wallander and Nylund 1992; Arnebrant 1994) and recent work using ingrowth mesh bags have confirmed this also for EMM growth in natural forests (Nilsson and Wallander 2003; Hendricks et al. 2006; Parrent and Vilgalys 2007; Kjøller et al. 2012). It should however be noted that this negative effect is reduced when N is balanced by other nutrients (Wallander et al. 2011). The number of mycorrhizal root tips or the fungal biomass on the root tips may remain similar after several years of annual N addition (Kåren and Nylund 1997), while the growth of the EMM was severely reduced in the same forest (Nilsson and Wallander 2003). This suggest that C demanding fungi (e.g. suilloid spp., Fransson et al. 2007), decline after N input. This was also found along a N deposition gradient in Alaska (Lilleskov et al. 2002) where the contact type *Lactarius theiogalus* (presumably low C demand) dominated (68.5 % of colonized root tips) in the most N-polluted site, while it decreased to only 7.4 % of root tips in the least N-polluted site. In contrast, medium-distance types like *Amphinema byssoides* and *Piloderma byssinum* (presumably high C demand) became more abundant in the least N-polluted site (40 %), while they were totally absent at higher N input sites. Along another short-distance N deposition gradient in a Norway spruce forest in Denmark, Kjøller et al. (2012) found that *Lactarius quietus* (contact type) dominated the root community at the forest edge (56 %) with the highest inorganic N deposition (43 kg N ha^{-1}) while short distance types (89 %, mainly *Tylospora fibrillosa* and *Cenococcum geophilum*) dominated further into the forest (N deposition 27 kg N ha^{-1}). Very few medium and long-distance types were formed in this forest (<5 % in the forest, 0 % at the edge), probably because of the rather high deposition of inorganic N. EM communities dominated by contact types, with less well developed mycorrhizal networks, may result in vulnerability to N leaching. This aspect will be further elaborated below.

3.4.2 Phosphorus

P deficiency is less common in boreal and temperate regions than in tropical regions, but mass balance calculations suggest that intensive harvesting of forest residues in combination with high N deposition will lead to P deficiencies in many forests in temperate and boreal regions (Akselsson et al. 2006). In laboratory grown seedlings, P deficiency resulted in strong enhancement of EMM production (Wallander and Nylund 1992; Ekblad et al. 1995). Phosphorus effects on the EMM has not been much studied in the field but Wallander and Thelin (2008) found that EMM ingrowth into sand-filled bags amended with apatite was higher than in bags filled with sand only, but this effect disappeared when the forests were fertilized with P and K. The magnitude of the EMM ingrowth response to apatite was negatively correlated to needle P status which supports the view that P availability in the soil is of great importance for the regulation of EMM production (Wallander and Thelin 2008). Since P is rarely limiting tree growth in temperate and boreal forests, the needle P status is probably not low enough to stimulate growth of EMM under field situations. However, Bahr et al. (2013) found a negative correlation between EMM growth and needle P status also when needle P levels were above 1.3 mg P g^{-1}, which was the threshold value where apatite stimulated EMM growth in the study by Wallander and Thelin (2008). Furthermore, Blum et al. (2002) found apatite to be an important calcium source for ectomycorrhizal trees in base-poor forest ecosystems in the US.

3.4.3 Other Nutrients

Under laboratory conditions, both K (Ekblad et al. 1995) and Mg (Wallander and Nylund 1992) deficiency has resulted in reduced EMM growth, supporting the finding by Ericsson (1995) of reduced belowground C allocations during such conditions (Fig. 3.5). These findings have however not been confirmed under field conditions, since Hagerberg et al. (2003) found no difference in EMM growth in Norway spruce forests with varying K availability. Some indications that Mg deficiency may impair EMM growth was found in a study in the Czech Republic by Berner (2013). EMM was much lower in Norway spruce forest growing on Mg-poor granite soil compared to similar forests growing on more Mg-rich amphibolite or serpentinite bedrock. Furthermore, a positive correlation between needle Mg concentration and EMM growth was found. It should however be noted that many other factors also varied among these sites and a causal relationship between Mg availability and EMM growth could not be established in this study. Mg deficiency can have severe effects on belowground C allocation since carbohydrate loading of the phloem can be impaired (Cakmak and Kirkby 2008). The forest die-back that occurred in central Europe during 1980–1990 was suggested to be caused by Mg deficiency as a result of acid rain (Schulze 1989). The Mg deficiency was proposed to be the result of dysfunctional mycorrhizal associations when belowground C allocation was impaired (Mejstrik 1989).

3.5 Ecological Consequences of Altered EMM Production

3.5.1 Nitrogen Leaching

Boreal forest soils have large retention capacity for N, especially when the C:N ratio of the organic layer (O horizon) is above 30. Nitrate leaching can be induced when ratios drop below 25, especially if N deposition exceeds 10 kg N ha^{-1} (Gundersen et al. 1998). Increased N retention was correlated to enhanced fungal proportion of the microbial biomass in a gradient of C:N ratios of Norway spruce forests in southern Scandinavia (Nilsson et al. 2012) and this may be attributed to the high capacity of EMF networks to assimilate N (Read et al. 2004). Wallander et al. (2004) estimated that EMM contained between 100 and 200 kg N ha^{-1} in Norway spruce and mixed oak/Norway spruce forests in southern Sweden. This amount is higher than what is found in the tree stems in the same forests (Thelin et al. 2002). The high N retention capacity of EMF is an effect of the large flux of C to the EMM in forest ecosystems, resulting in a well-developed EMM network in the soil that efficiently captures available N. Some of the N that is taken up is allocated to the host trees but significant amounts are also retained in the EMF biomass in the soil (Näsholm et al. 2013). Aber et al. (1998) suggested that N was exuded from EMF as enzymes that formed stable complexes with humus material in the soil, while Högberg et al. (2011) and Fernandez et al. (2013) proposed that the EMF mycelia itself could be a precursor for stable N. This view is supported by the fact that recalcitrant SOM deeper in the soil becomes more and more similar to EMF in terms of C:N ratio and N isotopic signatures (Boström et al. 2007; Lindahl et al. 2007; Högberg et al. 2011; Clemmensen et al. 2013). Furthermore, Näsholm et al. (2013) concluded that EMF immobilize large amounts of N in boreal forests to restrict establishment of species that are more N demanding. Along these lines, Franklin et al. (2014) developed a model that could explain why ectomycorrhizal symbiosis does not alleviate nitrogen limitation in boreal forests.

The efficiency by which forest trees and their EMF networks capture N to avoid leaching of N may depend on availability of other nutrients. Stevens et al. (1993) found enhanced N leaching from Sitka spruce stands which had developed K and P deficiency, while leaching was repressed upon fertilization with K and P. In the French Ardennes increased N leaching has been correlated with reduced availability of nutrients such as K, Mg and P (Jonard et al. 2012). Root (Gress et al. 2007) and EMM (Wallander and Thelin 2008) growth were enhanced in P-rich microsites in P-poor Norway spruce forests with high N input. This enhanced growth may explain the reduced N leaching after P fertilization. Slow release P fertilizer has been used in Finland (Aarnio et al. 2003) and this approach may reduce N leaching from forests with low P availability.

In another recent paper, Blanes et al. (2012) was able to separate the effect of autotrophs (roots + EMF) and saprotrophs on N retention by combining fertilization treatments with root-exclusion and isotope labeling. They found enhanced N retention in N-saturated forests after P fertilization, mainly due to a better N uptake by EM roots as shown by a ^{15}N labeling experiment. Trenching verified that N retention was also enhanced among saprotrophic organisms after P fertilization, probably by fungi since short-lived bacteria with much lower C:N ratios are less likely to be important for N retention (Högberg et al. 2011). Interestingly the P effect on the N retention by saprotrophs was only found in trenched plots, suggesting superior N retention capacity by the autotrophic (root + EMF) compared to the saprotrophic organisms.

N leaching from standing forest can be an effect of impairment of EMM growth as discussed above. But it can also be an effect of a changed EM community to species with lower capacity to take up N. Kjøller et al. (2012) found enhanced N leaching and a drastically changed EMF community when moving towards a forest edge that is more exposed to N deposition compared to more protected areas within the forest. The EMF community at the forest edge was dominated by smooth root tips with low capacity to form extensive mycelia networks and presumably low capacity to retain N. Lilleskov et al. (2002) suggested a shift in the EMF community of N-saturated forests to species more efficient in P uptake, which could lead to less efficient N uptake and more N leaching. Although smooth root tips probably are inefficient in P uptake, other N tolerant species such as *Paxillus involutus* are known to have extremely high P uptake rates (Colpaert et al. 1999), and it would be interesting to see if the N uptake rates from such species are reduced under conditions of high N input, which would allow more N to leach. Gorissen and Kuyper (2000) have demonstrated that nitrophilic (N tolerant) species such as *Laccaria bicolor* retain more N in the fungal biomass while nitrophobic (N sensitive) *Suillus bovinus* deliver more N to the host plant when studied in a pot experiment. If nitrophilic species can reduce N uptake by retaining it in their biomass rather than transferring it to the host plant, they may tolerate N better by spending less C on N assimilation. This would allow them to spend more C on EMM growth under excess N, as suggested in the hypothesis presented by Wallander (1995). In support for this hypothesis, Gruffman et al. (2012) found recently that, contrary to inorganic N, organic N fertilizer (based on amino acids) C did not impair ectomycorrhizal colonization of Norway spruce roots. The reason is probably a lower C cost for the fungus when amino acids are taken up instead of inorganic N, and this will result in more C available for fungal growth. Organic N fertilizers were as efficient as inorganic fertilizers and were recommended in nurseries to improve mycorrhiza formation (Gruffman et al. 2012).

3.5.2 The Importance of EMM for C Sequestration

Recent results suggest that EMM contribute significantly to SOM formation. Godbold et al. (2006) used ingrowth cores with soil that had a different C isotopic signature than the colonizing EMM, and found that EMM accounted for an accumulated input of 1000 g C m^{-2} over a period of ~2.5 years, which corresponded to 62 % of new soil C in a poplar plantation in Italy. Wallander et al. (2011) used a similar approach by amending ^{13}C-enriched maize compost material to mesh bags and found a lower (100 g C m^{-2}), but still significant, C input to a Norway spruce forest soil through EMM over a period of two years. Furthermore, by using ^{14}C dating of SOM at different soil depth and a modeling approach in boreal forests in northern Sweden, Clemmensen et al. (2013) concluded that the majority (70 %) of the C in the upper 20 cm of the soil in later successional forests originated from roots and associated EMM while this figure declined to 47 % in stands at earlier stages of successions. This was explained by impaired degradation of fungal residues in later successional forests. It is possible that different EM and ericoid mycorrhizal species contribute differently to SOM formation by producing compounds that are more or less recalcitrant (Clemmensen et al. 2015). Fernandez et al. (2013) demonstrated recently that root tips formed by *Cenococum geophilum* persisted 4–10 times longer than other EMF species in the soil which suggest that this species is resistant to decay and may contribute significantly to C sequestration. In support for this Dahlberg et al. (1997) found that sclerotia formed by this fungus could make up 400 kg ha^{-1} in a Swedish Norway spruce forest soil. Some recent results highlight the capacity of many EM species to degrade or modify SOM, which may lead to enhanced decomposition and reduced C sequestration (Chapela et al. 2001; Talbot et al. 2008; Courty et al. 2010). These effects are however debated. Treseder et al. (2006) could not demonstrate any C uptake by EM fungi from ^{14}C-labelled litter added to an oak forest but work at natural ^{14}C abundance of fungal proteins suggested uptake of C (as amino acids or oligopeptides) by several taxa of EMF (Hobbie et al. 2013). Talbot et al. (2008) suggested that EM fungi released C as a side effect when removing N-rich compounds and Lindahl et al. (2007) demonstrated an increasing C:N ratio of SOM in the lower part of the organic horizon where EMF dominate, indicating preferential uptake of N-rich compounds. Old SOM may also be released through priming when labile C is exuded by roots and associated EMF (Dijkstra and Cheng 2007). Peroxidase-encoding genes have been identified among a wide range of EMF suggesting that these fungi degrade SOM in a similar way as white rot fungi (Bödecker et al. 2009) and other EMF seem to use the Fenton reaction to modify SOM to obtain N-rich compounds in a similar way as brown rot fungi (Rineau et al. 2012; Lindahl and Tunlid 2015).

The contribution of EM fungi to SOM formation depends on production, turnover and recalcitrance of EMM in the soil. When studied in microcosm systems, the mycelium of several long-distance types (e.g. *Suillus* spp., *Paxillus involutus*) are known to spread rapidly to colonize nutrient-rich organic patches, but disappear after a couple of weeks when the nutrients are exhausted (Bending and Read 1995). Other types form perennial mycelial mats in the soil (Ingham et al. 1991; Kluber et al. 2010) that presumably affect SOM formation differently than more short-lived types. Many short-distance and contact types produce very little mycelium but contribute to accumulation of SOM since they decompose much more slowly than non-mycorrhizal roots (Langley and Hungate 2003). In contrast, Koide et al. (2011) found that non-mycorrhizal roots decomposed faster than EM roots in a similar experiment. The reason for this controversy is not known but could be related to differences in species composition. Mycorrhizal mycelia decomposed fast when incubated in forest soil, 40–80 % of the mass remained after 1 month (Fernandez and Koide 2012). Bahr et al. (2015) found that most of the mycelium (90 %) produced in mesh bags incubated in forest soil degraded within a year. Although the rate of decomposition is of relevance for SOM formation, the most important aspect for long-term C sequestration is the proportion of the litter material that remain in the soil for longer time periods. For pine needle litter, Berg et al. (2010) demonstrated that up to 17–53 % of the mass remained after 3–5 years of incubation in litter bags. At the time of harvest decomposition rate had approached zero. If such high amounts of remaining material also exist among EM fungi, and if different species vary in this respect, the composition of the EM community could have a fundamental role in SOM formation. Another aspect recently highlighted is that the molecular structure of SOM does not alone control the long-term decomposition of SOM (Schmidt et al. 2011). Instead, the degree of protection from decomposition in the soil was suggested to be a more important regulator (Schmidt et al. 2011). Molecules can be protected inside soil aggregates and on mineral surfaces (Sollins et al. 2009), and one challenge for future research is to sort out the role of mycorrhizal fungi in this respect. The interactions between EMF and minerals have been reviewed recently (Hoffland et al. 2004; Finlay et al. 2009), but the possible role of these processes in C sequestration has been largely neglected. On the other hand, the decomposition of forest humus, with low amounts of mineral surfaces, was extremely slow in late successional forests in boreal forest chronosequence (Clemmensen et al. 2013), suggesting the molecular structure of SOC to be the most important factor in determining decomposition rates in these forests.

3.6 Conclusions

Accurate data on production, biomass and turnover of ectomycorrhzial mycelium are essential for improving the ability of ecosystem models to predict nutrient leaching and C sequestration in forest ecosystems. The composition of the EM community appears to have a fundamental role on N retention and turnover of EMM and more research on the influence of individual species on these processes is urgently needed. A possible way forward could be to classify EMF into functional groups along the lines of Agerer (2001). Apart from morphological characters, other criteria's such as nitrophilic/nitrophobic, enzyme production, sensitivity to disturbance etc. could be useful when defining the groups. There is increasing evidence that the EMM of mycorrhizal fungi play a key role in C cycling in ecosystems. This was highlighted in a recent paper from a boreal forest chronosequence in Sweden which suggests that belowground litter of roots and rhizosphere fungi, with EMF being the most prominent, contributed up to 70 % of the C sequestered in SOM (Clemmensen et al. 2013). The importance of EMF for the C cycling in forests has been the topic of two other recent reviews (Cairney 2012; Ekblad et al. 2013). Having other focuses than the present review, we therefore recommend the reading of these for a more complete cover of the subject.

Acknowledgements We thank COST (European Cooperation in Science and Technology) for financial and coordinative support to the Cost Action FP0803: Belowground carbon turnover in European forests.

References

Aarnio T, Raty M, Martikainen PJ (2003) Long-term availability of nutrients in forest soil derived from fast and slow-release fertilizers. Plant Soil 252:227–239

Aber J, McDowell W, Nadelhoffer K, Magill A, Berntson G, Kamakea M, McNulty Currie W, Rustad L, Fernandez I (1998) Nitrogen saturation in temperate forest ecosystems—hypotheses revisited. Bioscience 48:921–934

Agerer R (2001) Exploration types of ectomycorrhizae—a proposal to classify ectomycorrhizal mycelial systems according to their patterns of differentiation and putative ecological importance. Mycorrhiza 11:107–114

Agerer R (2006) Fungal relationships and structural identity of their ectomycorrhizae. Mycol Prog 5:67–107

Akselsson C, Westling O, Alveteg M, Thelin G, Fransson A-M, Hellsten S (2006) The influence of N load and harvest intensity on the risk of P limitation in Swedish forest soils. Sci Total Environ 404:284–289

Arnebrant K (1994) Nitrogen amendments reduce the growth of ectramatrical mycelium. Mycorrhiza 5:7–15

Bahr A, Ellström M, Akselsson C, Ekblad A, Mikusinska A, Wallander H (2013) Growth of ectomycorrhizal fungal mycelium along a Norway spruce forest nitrogen deposition gradient and its effect on nitrogen leakage. Soil Biol Biochem 59:38–48

Bahr A, Ellström M, Bergh J, Wallander H (2015) Nitrogen leaching and ectomycorrhizal nitrogen retention capacity in a Norway spruce forest fertilized with nitrogen and phosphorus. Plant Soil 390:323–335

Bending GD, Read DJ (1995) The structure and function of the vegetative mycelium of ectomycorrhizal plants V. Foraging behaviour and translocation of nutrients from exploited litter. New Phytol 130:401–409

Berg B, Davey MP, De Marco A, Emmett B, Faituri M, Hobbie SE, Johansson MB, Liu C, McClaugherty Norell L, Rutigliano FA, Vesterdal L, De Santo AV (2010) Factors influencing limit values for pine needle litter decomposition: a synthesis for boreal and temperate pine forest systems. Biogeochem 100:57–73

Berner C, Johansson T, Wallander H (2012) Long term effect of apatite on ectomycorrhizal growth and community structure. Mycorrhiza 22:615–621

Berner C (2013) The effect of forest nutrient status on ectomycorrhzial growth and community structure in response to minerals. Dissertation, Lund University, Sweden

Blanes CA, Emmett BA, Vingela B, Carreira JA (2012) Alleviation of P limitation makes tree roots competitiove for N against microbes in a N-saturated forest: a test through P fertilization and N-15 labelling. Soil Biol Biochem 48:51–59
Blum JD, Klaue A, Nezat CA, Driscoll CT, Johnson CE, Siccama TG, Eagar C, Fahey TJ, Likens GE (2002) Mycorrhizal weathering of apatite as an important calcium source in base-poor forest ecosystems. Nature 417:729–731
Bödecker ITM, Nygren CMR, Taylor AFS, Olsson Å, Lindahl B (2009) Class II peroxidase-encoding genes are present in a phylogenetically wide range of ectomycorrhial fungi. ISME J 3:1387–1395
Boström B, Comstedt D, Ekblad A (2007) Isotope fractionation and ^{13}C enrichment in soil profiles during the decomposition of soil organic matter. Oecologia 153:89–98
Cairney JWG (2012) Extramatrical mycelia of ectomycorrhizal fungi as moderators of carbon dynamics in forest soil. Soil Biol Biochem 47:198–208
Cakmak I, Kirkby EA (2008) Role of magnesium in carbon partitioning and alleviating photooxidative damage. Physiol Plant 133:692–704
Chapela IH, Osher LJ, Horton TR, Henn MR (2001) Ectomycorrhizal fungi introduced with exotic pine plantations induce soil carbon depletion. Soil Biol Biocehm 33:1733–1740
Clemmensen K, Bahr A, Ovaskainen O, Dahlberg A, Ekblad A, Wallander H, Stenlid J, Finlay R, Wardle D, Lindahl B (2013) Roots and associated fungi drive long-term carbon sequestration in boreal forest. Science 339:1615–1618
Clemmensen K, Finlay R, Dahlberg A, Stenlid J, Wardle D, Lindahl B (2015) Carbon sequestration is related to mycorrhizal fungal community shifts during long-term succession in boreal forests. New Phytol 205:1525–1536
Colpaert JV, van Tichelen KK, van Assche JA JA (1999) Short-term phosphorus uptake rates in mycorrhizal and non-mycorrhizal roots of intact *Pinus sylvestris* seedling. New Phytol 143:589–597
Courty P-E, Buée M, Diedhiou AG, Frey-Klett P, Le Tacon F, Rineau F, Turpault M-P, Uroz S, Garbaye J (2010) The role of ectomycorrhizal communities in forest ecosystem processes: new perspectives and emerging concepts. Soil Biol Biochem 42:679–698
Coutts MP, Nicoll BC (1990) Growth and survival of shoots, roots, and mycorrhizal mycelium in clonal Sitka spruce during the first growing season after planting. Can J For Res 20:861–868
Dahlberg A (2001) Community ecology of ectomycorrhizal fungi: an advancing interdisciplinary field. New Phytol 150:555–562
Dahlberg A, Jonsson L, Nylund J-E (1997) Species diversity and distribution of biomass abobe and below ground among ectomycorrhizal fungi in an old-growth Norway spruce forest in south Sweden. Can J Bot 75:1323–1335
Dawes MA, Hagedorn F, Handa IT, Streit K, Ekblad A, Rixen C, Körner C, Hättenschwiler S (2013) An alpine treeline in a carbon dioxide-rich world: synthesis of a nine-year free-air carbon dioxide enrichment study. Oecologia 171:623–637
De la Varga H, Águeda B, Ágreda T, Martínez-Peña F, Parladé J, Pera J (2013) Seasonal dynamics of *Boletus edulis* and *Lactarius deliciosus* extraradical mycelium in pine forests of central Spain. Mycorrhiza 23:391–402
Del Vecchio TA, Gehring CA, Cobb NS, Whitham TG (1993) Negative effects of scale insect herbivory on the ectomycorrhizae of juvenile pinyon pine. Ecology 74:2297–2302
Dijkstra FA, Cheng W (2007) Interactions between soil and tree roots accelerate long-term soil carbon decomposition. Ecol Lett 10:1046–1053
Ekblad A, Wallander H, Carlsson R, Huss-Danell K (1995) Fungal biomass in roots and extramatrical mycelium in relation to macronutrients and plant biomass of ectomycorrhizal *Pinus sylvestris* and *Alnus incana*. New Phytol 131:443–451
Ekblad A, Wallander H, Godbold DL, Johnson D, Baldrian P, Björk RG, Cruz C, Epron D, Kieliszewska-Rokicka B, Kjöller R, Kraigher H, Matzner E, Neumann J, Plassard C (2013) The production and turnover of extramatrical mycelium of ectomycorrhizal fungi in forest soils: role in carbon cycling. Plant Soil 366:1–27

Emmett BA (2007) Nitrogen saturation of terrestrial ecosystems: some recent findings and their implications for our conceptual framework. Wat Air Soil Pollut Focus 7:99–109

Ericsson T (1995) Growth and shoot:root ratio of seedlings in relation to nutrient availability. Plant Soil 168–169:205–214

Fernandez CW, Koide RT (2012) The role of chitin in decomposition of ectomycorrhizal fungal litter. Ecology 93:24–28

Fernandez CW, McCormack ML, Hill JM, Pritchard SG, Koide RT (2013) On the persistence of *Cenococcum geophilum* ectomycorrhizas and its implications for forest carbon and nutrient cycles. Soil Biol Biochem 65:141–143

Fernandez CW, Kennedy PG (2015) Moving beyond the black-box: fungal traits, community structure, and carbon sequestartion in forest soil. New Phytol 205:1378–1380

Finlay R, Wallander H, Smits M, Holmstrom S, van Hees P, Lian B, Rosling A (2009) The role of fungi in biogenic weathering in boreal forest soils. Fungal Biol Rev 23:101–106

Franklin O, Högberg P, Ekblad A, Ågren G (2003) Pine forest floor carbon accumulation in response to N and PK additions: bomb ^{14}C modelling and respiration studies. Ecosystems 6:644–658

Franklin O, Näsholm T, Högberg P, Högberg MN (2014) Forest trapped in nitrogen limitation—an ecological market perspective on ectomycorrhizal symbisois. New Phytol 203:657–666

Fransson PMA, Anderson IC, Alexander IJ (2007) Ectomycorrhizal fungi in culture respond differently to increased carbon availability. FEMS Microbiol Ecol 61:246–257

Fransson P (2012) Elevated CO_2 impacts ectomycorrhiza-mediated forest soil carbon flow: fungal biomass production, respiration and exudation. Fungal Ecol 5:85–98

Futter MN, Ring E, Högbom L, Entenmann S, Bishop K (2010) Consequences of nitrate leaching following stem-only harvesting of Swedish forests are dependent on spatial scale. Environ Pollut 158:3552–3559

Gafur A, Schutzendubel A, Langenfeld-Heyser R, Fritz E, Polle A (2004) Compatible and incompetent *Paxillus involutus* isolates for ectomycorrhiza formation in vitro with poplar (*Populus x canescens*) differ in H_2O_2 production. Plant Biol 6:91–99

Godbold DL, Hoosbeek MR, Lukac M, Cotrufo MF, Janssens IA, Ceulemans R, Polle A, Velthorst EJ, Scarascia-Mugnozza G, De Angelis P, Miglietta F, Peressotti A (2006) Mycorrhizal turnover as a dominant process for carbon input into soil organic matter. Plant Soil 281:15–24

Gorissen A, Kuyper TW (2000) Fungal species-specific responses of ectomycorrhizal scots pine (*Pinus sylvestris*) to elevated CO_2. New Phytol 146:163–168

Gress SE, Nichols TD, Northcraft CC, Peterjohn WT (2007) Nutrient limitation in soils exhibiting differing nitrogen availabilities: what lies beyond nitrogen saturation. Ecology 88:110–130

Gruffman L, Ishida T, Nordin A, Näsholm T (2012) Cultivation of Norway spruce and scots pine on organic nitrogen improves seedling morphology and field performance. For Ecol Manage 276:118–124

Gundersen P, Callesen I, De Vries W (1998) Nitrate leaching in forest ecosystems is related to forest floor C/N ratios. Environ Pollut 102:403–407

Hagerberg D, Thelin G, Wallander H (2003) The production of ectomycorrhizal mycelium in forests: relation between forest nutrient status and local mineral sources. Plant Soil 252:279–290

Hendricks JJ, Mitchell RJ, Kuehn KA, Pecot SD, Sims SE (2006) Measuring external mycelia production of ectomycorrhizal fungi in the field: the soil matrix matters. New Phytol 171:179–186

Hobbie EA, Sanchez FS, Rygiewicz PT (2012) Control of isotopic patterns in saprophytic and ectomycorrhizal fungi. Soil Biol Biochem 48:60–68

Hobbie EA, Ouimette EG, Schuur D, Kierstead J, Trappe K, Bendiksen Ohenoja E (2013) Radiocarbon evidence for the mining of organic nitrogen from soil by mycorrhizal fungi. Biogeochem 114:381–389

Hoffland E, Kuyper TW, Wallander H, Plassard C, Gorbushina A, Haselwandter K, Holmström S, Landeweert R, Lundström U, Rosl ing A, Sen R, Smits M, van Hees P, van Breemen N (2004) The role of fungi in weathering. Front Ecol Environ 2:258–264

Högberg P, Johannisson C, Yarwood S, Callesen I, Näsholm T, Myrold DD, Högberg M (2011) Recovery of ectomycorrhiza after 'nitrogen saturation' of a conifer forest. New Phytol 189:515–525

Högberg P, Nordgren A, Buchmann N, Taylor AFS, Ekblad A, Högberg MN, Nyberg G, Ottosson-Löfvenius M, Read DJ (2001) Large-scale forest girdling shows that current photosynthesis drives soil respiration. Nature 411:789–792

Högberg P, Plamboeck AH, Taylor AFS, Fransson PMA (1999) Natural ^{13}C abundance reveals trophic status of fungi and host-origin of carbon in mycorrhizal fungi in mixed forests. PNAS 96:8534–8539

Hyvönen R, Ågren G, Linder S, Persson T, Cotrufo F, Ekblad A, Freeman M, Grelle A, Janssens I, Jarvis PG, Kellomäki S, Lindroth A, Loustau D, Lundmark T, Norby R, Oren R, Pilegaard K, Ryan M, Sigurdsson B, Strömgren M, van Oijen M, Wallin G (2007) The likely impact of elevated [CO_2], nitrogen deposition, increased temperature and management on carbon sequestration in temperate and boreal forest ecosystems: a literature review. New Phytol 173:463–480

Ingham ER, Griffiths RP, Cromack K, Entry JA (1991) Comparison of direct vs fumigation incubation microbial biomass estimates from ectomycorrhizal mat and non-mat soils. Soil Biol Biochem 23:465–471

Jonard M, Legout A, Nicolas M, Dambrine E, Nys C, Ulrich E, van der Perre R, Ponette Q (2012) Deterioration of Norway spruce vitality despite a sharp decline in acid deposition: a long-term integrated perspective. Glob Change Biol 18:711–725

Kalliokoski T, Pennanen T, Nygren P, Sievänen R, Helmisaari H-S (2010) Belowground interspecific competition in mixed boreal forests: fine root and ectomycorrhiza characteristics along stand developmental stage and soil fertility gradients. Plant Soil 330:73–89

Kårén O, Nylund J-E (1997) Effects of ammonium sulphate on the community structure and biomass of ectomycorrhizal fungi in a Norway spruce stand in southwestern Sweden. Can J Bot 75:1628–1642

Kimmins JP (2004) Forest ecology, a foundation for sustainable forest management and environmental ethics in forestry. Prentice Hall, New Jersey

Kjøller R, Nilsson L-O, Hansen K, Schmidt IK, Vesterdal L, Gundersen P (2012) Dramatic changes in ectomycorrhizal community composition, root tip abundance and mycelial production along a stand-scale nitrogen deposition gradient. New Phytol 194:278–286

Kluber LA, Tinnesand KM, Caldwell BA, Dunham SM, Yarwood RR, Bottomley PJ, Myrold DD (2010) Ectomycorrhizal mats alter forest soil biogeochemistry. Soil Biol Biogeochem 42:1607–1613

Koide RT, Fernandez CW, Peoples MS (2011) Can ectomycorrhizal colonization of *Pinus resinosa* roots affect their decomposition? New Phytol 191:508–514

Koide RT, Malcolm GM (2009) N concentration controls decomposition rates of different strains of ectomycorrhizal fungi. Fungal Ecol 2:197–202

Korkama T, Fritze H, Pakkanen A, Pennanen T (2007) Interactions between extraradical ectomycorrhizal mycelia, microbes associated with the mycelia and growth rate of Norway spruce (*Picea abies*) clones. New Phytol 173:798–807

Kuikka K, Harma E, Markkola A, Rautio P, Roitto M, Saikkonen K, Ahonen-Jonnart U, Finlay R, Tuomia J (2003) Severe defoliation of scots pine reduces reproductive investment by ectomycorrhizal symbionts. Ecology 84:2051–2061

Langley JA, Hungate BA (2003) Mycorrhizal controls on belowground litter quality. Ecology 84:2302–2312

Lilleskov EA, Fahey TJ, Horton TR, Lovett GM (2002) Belowground ectomycorrhizal fungal community change over a nitrogen deposition gradient in Alaska. Ecology 83:104–115

Lilleskov EA, Hobbie EA, Horton TR (2011) Conservation of ectomycorrhizal fungi: exploring the linkages between functional and taxonomic responses to anthropogenic N deposition. Fungal Ecol 4:174–183

Lindahl BD, Ihrmark K, Boberg J, Trumbore SE, Högberg P, Stenlid J, Finlay RD (2007) Spatial separation of litter decomposition and mycorrhizal nitrogen uptake in a boreal forest. New Phytol 173:611–620

Lindahl BD, Tunlid A (2015) Ectomycorrhizal fungi—potential organic matter decomposers, yet not saprotrophs. New Phytol 205:1443–1447

Mikusinska A, Tryggve P, Taylor AFS, Ekblad A (2013) Effects of ingrowth bag-size and presence of soil animals on ectomycorrhizal extraradical mycelia production and isotopic composition in a Norway spruce forest. Soil Biol Biochem 66:154–162

Mejstrik V (1989) Ectomycorrhizas and forest decline. Agric Ecosyst Environ 28:325–337

Näsholm T, Högberg P, Franklin O, Metcalfe D, Keel SG, Campbell C, Hurry V, Linder S, Högberg M (2013) Are ectomycorrhizal fungi alleviating or aggravating nitrogen limitation of tree growth in boreal forest? New Phytol 198:214–221

Nilsson LO, Wallander H (2003) The production of external mycelium by ectomycorrhizal fungi in a Norway spruce forest was reduced in response to nitrogen fertilization. New Phytol 158:409–416

Nilsson LO, Bååth E, Falkengren-Grerup U, Wallander H (2007) Growth of ectomycorrhizal mycelia and composition of soil microbial communities in oak forest soils along a nitrogen deposition gradient. Oecologia 153:375–384

Nilsson LO, Wallander H, Gundersen P (2012) Changes in microbial activities and biomasses over a forest floor gradient in C-to-N ratio. Plant Soil 355:75–86

Nohrstedt HÖ, Arnebrant K, Bååth E, Söderström B (1989) Changes in carbon content, respiration rate, ATP content, and microbial biomass in nitrogen-fertilized pine forest soil in Sweden. Can J For Res 19:323–328

Parrent JL, Vilgalys R (2007) Biomass and compositional responses of ectomycorrhizal fungal hyphae to elevated CO_2 and nitrogen fertilization. New Phytol 176:164–174

Perez-Moreno J, Read DJ (2001) Mobilization and transfer of nutrients from litter to tree seedlings via the vegetative mycelium of ectomycorrhizal plants. New Phytol 145:301–309

Peter M, Ayer F, Egli S (2001) Nitrogen addition in a Norway spruce stand altered macromycete sporocarp production and below-ground ectomycorrhizal species composition. New Phytol 149:311–325

Pritchard SG, Strand AE, McCormack ML, Davis MA, Oren R (2008) Mycorrhizal and rhizomorph dynamics in a loblolly pine forest during 5 years of free-air-CO_2-enrichment. Glob Change Biol 14:1–13

Read DJ (1992) The mycorrhizal mycelium. In: Allen MF (ed) Mycorrhizal functioning. Chapman and Hall, London, pp 102–133

Read DJ, Leake JR, Perez-Moreno J (2004) Mycorrhizal fungi as drivers of ecosystem processes in heathland and boreal forest biomes. Can J Bot 82:1243–1263

Rineau F, Roth D, Shah F, Smits M, Johansson T, Canbäck B, Bjarke Olsson P, Persson P, Nedergard Grell M, Lindquist E, Grioriev IV, Lange L, Tunlid A (2012) The ectomycorrhizal fungus *Paxillus involutus* converts organic matter in plant litter using a trimmed brown-rot mechanism involving fenton chemistry. Environ Microbiol 14:1477–1487

Rouhier H, Read DJ (1998) Plant and fungal responses to elevated atmospheric carbon dioxide in mycorrhizal seedlings of *Pinus sylvestris*. Environ Exp Bot 40:237–246

Schmalholz M, Hylander K (2009) Succession of bryophyte assemblages following clear-cut logging in boreal southern-central Sweden—does retrogressive succession occur? Can J For Res 39:1871–1880

Sims SE, Hendricks JJ, Mitchell RJ, Kuehn KA, Pecot SD (2007) Nitrogen decreases and precipitation increases ectomycorrhizal extramatrical mycelia production in a longleaf pine forest. Mycorrhiza 17:299–309

Schmidt MWI, Torn MS, Abiven S, Dittmar T, Guggenberger G, Janssens IA, Kleber M, Kögel-Knaber I, Lehmann J, Manning DAC, Nannipieri P, Rasse DP, Weiner S, Trumbore SE (2011) Persistence of soil organic matter as an ecosystem property. Nature 478:49–56

Schulze ED (1989) Air pollution and forest decline in a spruce (*Picea abies*). Science 244:776–783

Smith SE, Read DJ (2008) Mycorrhizal symbiosis, 3rd edn. Academic Press, London

Stevens PA, Harrison AF, Jones HE, Williams TG, Hughes S (1993) Nitrate leaching from a Sitka spruce plantation and the effect of fertilization with phosphorus and potassium. For Ecol Manag 58:233–247

Söderström B, Read DJ (1987) Respiratory activity of intact and excised ectomycorrhizal mycelial systems growing in unsterilized soil. Soil Biol Biochem 19:231–236

Sollins P, Kramer MG, Swanston C, Lajtha K, Filley T, Aufdenkampe AK, Wagai R, Bowden RD (2009) Sequential density fractionation across soils of contrasting mineralogy: evidence for both microbial- and mineral-controlled soil organic matter stabilization. Biogeochem 96:209–231

Talbot JM, Allison SD, Treseder KK (2008) Decomposers in disguise: mycorrhizal fungi as regulators of soil C dynamics in ecosystems under global change. Funct Ecol 22:955–963

Taylor AFS, Martin F, Read DJ (2000) Fungal diversity in ectomycorrhizal communities of Norway spruce (*Picea abies* (L.) Karst.) and beech (*Fagus sylvatica* L.) along north-south transects in Europe. Ecol Stud 142:343–365

Taylor AFS, Fransson PM, Högberg P, Högberg MN, Plamboeck AH (2003) Species level patterns in ^{13}C and ^{15}N abundance of ectomycorrhizal and saprotrophic fungal sporocarps. New Phytol 159:757–774

Thelin G, Sverdrup H, Holmqvist J, Rosengren U, Lindén M (2002) Assessment of nutrient sustainability in Norway spruce and mixed Norway spruce-oak stands at Jämjö. In: Sverdrup H, Stjernquist I (eds) Developing principles for sustainable forestry in Southern Sweden, Kluwer Academic Publishers, Dordrecht, pp 337–354

Treseder KK, Allen MF, Ruess RW, Pregitzer KS, Hendrick RL (2005) Lifespans of fungal rhizomorphs under nitrogen fertilization in a pinyon-juniper woodland. Plant Soil 270:249–255

Treseder KK, Torn MS, Masiello CA (2006) An ecosystem-scale radiocarbon tracer to test use of litter carbon by ectomycorrhizal fungi. Soil Biol Biochem 38:1077–1082

Tedersoo L, May TW, Smith ME (2010) Ectomycorrhizal lifestyle in fungi: global diversity, distribution, and evolution of phylogenetic lineages. Mycorrhiza 20:217–263

Wallander H (1995) A new hypothesis to explain allocation of dry-matter between mycorrhizal fungi and pine seedlings in relation to nutrient supply. Plant Soil 168:243–248

Wallander H, Ekblad A, Bergh J (2011) Growth and carbon sequestration by ectomycorrhizal fungi in intensively fertilized Norway spruce forests. For Ecol Manag 262:999–1007

Wallander H, Ekblad A, Godbold DL, Johnson D, Bahr A, Baldrian P, Björk RG, Kieliszewska-Rokicka B, Kjøller R, Kraigher H, Plassard C, Rudawska M (2013) Evaluation of methods to estimate production, biomass and turnover of ectomycorrhizal mycelium in forests soils—a review. Soil Biol Biochem 57:1034–1047

Wallander H, Göransson H, Rosengren U (2004) Production, standing biomass and natural abundance of ^{15}N and ^{13}C in ectomycorrhizal mycelia collected at different soil depths in two forest types. Oecologia 139:89–97

Wallander H, Johansson U, Sterkenburg E, Brandström M, Lindahl B (2010) Production of ectomycorrhizal mycelium peaks during canopy closure in Norway spruce forests. New Phytol 187:1124–1134

Wallander H, Nilsson LO, Hagerberg D, Bååth E (2001) Estimation of the biomass and production of external mycelium of ectomycorrhizal fungi in the field. New Phytol 151:753–760

Wallander H, Nilsson L-O, Hagerberg D, Rosengren U (2003) Direct estimates of C:N ratios of ectomycorrhizal mycelia collected from Norway spruce forest soils. Soil Biol Biochem 35:997–999

Wallander H, Nylund J-E (1992) Effects of excess nitrogen and phosphorus starvation on the extramatrical mycelium of ectomycorrhizas of *Pinus sylverstris* L. New Phytol 120:495–503

Wallander H, Thelin G (2008) The stimulating effect of apatite on ectomycorrhizal growth diminishes after PK fertilization. Soil Biol Biochem 40:2517–2522

Wallenda T, Kottke I (1998) Nitrogen deposition and ectomycorrhizas. New Phytol 139:169–187

Weigt R, Raidl S, Verma R, Agerer R (2012a) Erratum to: exploration type-specific standard values of extramatrical mycelium—a step towards quantifying ectomycorrhizal space occupation and biomass in natural soil. Mycol Prog 11:349–350

Weigt R, Raidl S, Verma R, Agerer R (2012b) Exploration type-specific standard values of extramatrical mycelium—a step towards quantifying ectomycorrhizal space occupation and biomass in natural soil. Mycol Prog 11:287–297

Weigt RB, Raidl S, Verma R, Rodenkirchen H, Göttlein A, Agerer R (2011) Effects of twice-ambient carbon dioxide and nitrogen amendment on biomass, nutrient contents and carbon costs of Norway spruce seedlings as influenced by mycorrhization with *Piloderma croceum* and *Tomentellopsis submollis*. Mycorrhiza 21:375–391

Chapter 4
Nutrient Dynamics in Arbuscular Mycorrhizal Networks

Iver Jakobsen and Edith C. Hammer

Abstract Transport and exchange of nutrients is a key feature to the function of arbuscular mycorrhiza (AM) and therefore also to the function of common mycorrhizal networks (CMNs). These networks establish between two or more plant individuals and one or more extraradical mycelia (EM). Complex networks with many nodes and linkages can be observed in sterile cultures and are probably common in nature. This chapter aims to describe how the nutrient dynamics of the CMNs influence plant competition. The discussion will concentrate on the rather variable access of the individual plants to the nutrient pool in the EM. Plant-to-plant transfer of nutrients via one or more connective EM does not appear to occur in significant quantities except in special cases. Competition between individual plants is in general asymmetric such that larger individuals will obtain a disproportionate share of a limited resource and suppress the growth of smaller individuals. Our major challenge is to unravel whether nutrient dynamics in CMNs will result in even stronger or in more relaxed competition. We investigated common outcomes in competition studies including: (a) adult plant-seedling combinations; (b) intraspecific competition of plant populations of similar age and (c) interspecific competition of plants of similar age. Results from root organ culture models indicate that AM fungi transfer phosphorus (P) to roots representing the strongest carbon (C) source strength. This is in accordance with the results of ^{32}P-aided studies of P translocation in CMNs with large plants and seedlings. An evaluation

I. Jakobsen (✉) · E.C. Hammer
Department of Chemical and Biochemical Engineering, Technical University of Denmark, Copenhagen, Denmark
e-mail: ivja@plen.ku.dk

I. Jakobsen
Department of Plant and Environmental Sciences, Section for Plant and Soil Science, University of Copenhagen, Thorvaldsensvej 40-3, 1871 Frederiksberg C, Denmark

E.C. Hammer
Department of Biology, Microbial Ecology, Lund University, Sölvegatan 37, 223 62 Lund, Sweden
e-mail: edith.hammer@biol.lu.se

T.R. Horton (ed.), *Mycorrhizal Networks*,
Ecological Studies 224, DOI 10.1007/978-94-017-7395-9_4

of these results together with CMN studies not involving tracers leads us to suggest that nutrients in the EM are distributed to the plants in accordance with their size or C source strength. In consequence, the CMN may rather amplify than relax the competition towards establishing seedlings. Our model implies, however, that the suppression of the seedling is only temporary and is likely to shift to a typical mycorrhizal growth response when the large plant become senescent or is grazed. In conspecific plant populations we predict that AM fungal networks confer an advantage to plant individuals that are for various reasons slightly larger than their neighbors. Interspecific plant competition is more difficult to predict as different combinations of species-specific traits may either amplify or relax nutrient competition in a CMN in addition to effects of size differences. Future research needs are discussed including the need to investigate roles of CMNs in interplant transfer of plant signals and allelochemicals.

Keywords Arbuscular mycorrhizal networks · Connective mycelium · Carbon–phosphorus exchange · Sharing of phosphorus in mycelium · Plant competition

4.1 Introduction

Nutrient exchange is a key function in associations between arbuscular mycorrhizal (AM) fungi and their host plants. The extraradical mycelium (EM) of the fungi takes up mineral nutrients from the soil and releases some of these nutrients at the fungus-plant interface while the plant in reverse provides carbon (C) to the fungus. Common mycorrhizal networks (CMN) are assumed to be the normal condition in nature and form when the EM engage with the roots of other plants or when individual mycelia fuse (Olsson et al. 2002; Giovannetti et al. 2004; Rosendahl 2008; Giovannetti et al. Chap. 2, this volume). This actual connection of different EMs can be directly observed in sterile cultures, but usually needs to be assumed in soil systems where hyphal pathways are obscured. However, isotope probing provided good evidence that those CMN hyphal connections actually exist also in natural soil systems (Merrild et al. 2013). CMNs can involve both con- and heterospecific plant individuals at a broad range of ontogenetic stages. The CMNs also contribute to more closed nutrient cycles as their nutrient pool will be protected against loss by leaching (Asghari et al. 2005; Van der Heijden 2010; Asghari and Cavagnaro 2012).

Information on the structure of CMNs is now emerging (Beiler et al. 2010; Montesinos-Navarro et al. 2012; Chagnon et al. 2012; Toju et al. 2014; Torrecillas et al. 2014) and it has been suggested that CMNs are fundamental agents in ecosystems by providing important pathways for various ecological interaction processes (Bever et al. 2010; Simard et al. 2012; Simard et al. Chap. 5, this volume). This chapter aims to provide a platform for understanding the nutrient dynamics in CMNs of AM plants. Focus will be on plant phosphorus (P) nutrition, but CMNs are probably important for other nutrients as well. Key processes in nutrient exchange and transport in AM-CMNs will be discussed including C-P exchange at the symbiotic interface when there is more than one symbiotic partner on each side. Plant-to-plant transfer processes will also be critically evaluated.

It is widely assumed that seedlings getting connected into existing networks will get immediate access to nutrients acquired by the fungus (Van der Heijden and Horton 2009) and that competition between plants in a CMN gets more relaxed (Wagg et al. 2011a, b). We challenge this assumption and discuss how the nutrient pool in the mycelium of the CMN is shared among the interlinked plant individuals. We attempt to identify factors controlling the C nutrition of the mycelium and its transfer of mineral nutrients to a specific plant member of the CMN. Finally, we will discuss the impact of networks on competition between plants of different and plants of similar life stage and we will investigate whether effects differ between inter- and intraspecific pairs. Our literature survey of CMN effects on plant-plant competition comprises 149 different cases from 62 papers and the results are summarized in Tables 4.1, 4.2, 4.3 and 4.4.

4.2 Nutrient Transport and Exchange in Extraradical Mycelium of Solitary Plants

The structure and function of EM of solitary plants has been reviewed in detail (e.g. Leake et al. 2004) and here we focus on aspects of particular importance to nutrient dynamics in CMNs. Nutrient exchange takes place in associations between roots and AM fungi that grow in the root cortex as intraradical mycelium (IM) and in the soil as extraradical mycelium (EM). The IM develops arbuscules inside cortex cells and the plant-fungus nutrient exchange takes place predominantly across the intimate interface between the arbuscules and the cell membrane (Harrison et al. 2002; Helber et al. 2011). The fungus takes up plant hexoses that are converted into lipids and used in proliferation of the IM and the EM. The C availability to roots influences EM biomass, anastomosis frequency and spore production (Olsson et al. 2014). The EM develops short-lived, branched absorbing structures (Bago et al. 1998) that become more frequent at high P concentrations in the medium (Olsson et al. 2014). Spatially, the EM typically extends at a rate of 1–4 mm day^{-1} and can reach maximum distances of at least 20 cm (Jakobsen et al. 1992b; Smith et al. 2000; Jansa et al. 2003; Thonar et al. 2011). Fungi of the genera *Gigaspora* or *Scutellospora* do not spread far into the soil and take up P in competition with P uptake at the root surface (Schnepf et al. 2008). These fungi commonly form only few anastomoses within the individual mycelium (Giovannetti et al. 2006; Voets et al. 2006). In contrast, *Acaulospora* and *Rhizophagus* (syn: *Glomus*) fungi are more likely to reach beyond P depletion zones around the roots as their mycelium front continually moves and anastomoses (Giovannetti et al. 2004; Voets et al. 2006; Schnepf et al. 2008). This agrees with time course studies of P uptake at

Table 4.1 Performance of seedlings linking into a CMN of a larger AM plant

Donor species and tracer usage	Seedling species and age at harvest	Growth of seedling in response to CMN as compared to			CMN effects on nutrient concentrations in shoots of seedlings	Author's suggested mechanism(s) behind CMN-effect on seedling	Reference
		Absence of mycorrhiza	Colonized solitary seedling	Non-host plants			
Predominantly negative effects of CMN on seedlings							
Sorghum vulgare	*S. vulgare; Brassica oleraceae* 112 days	0	–	0	Decreased for P, N, K, Ca	Competition for soil nutrients	Ocampo (1986)
Plantago lanceolata Festuca rubra	*P. lanceolata*;	+	ND	0	ND	Mutualism	Francis and Read (1995)
	7 ruderal species 21, 42 and 63 days	–				Competition for soil nutrients or antagonism	
Prunella vulgaris;	*P. vulgaris* 60 days	–	–	ND	ND	Resource pre-emption	Moora and Zobel (1996)
Fragaria vesca		+				The interspecific interaction depends on MGR of each species	
Hypericon perforatum	*H. perforatum* 65 days	–, no root contact	–	ND	ND	Competition for soil nutrients	Moora and Zobel (1998)
		0, with root contact					
Sibbaldia procumbens	*S. procumbens; Antennaria dioica; Campanula rotundifolia; Solidago virgaurea* 37–52 days	0	–, in 10 out of 12 plant-AMF combinations	ND	ND	Competitive interactions within the CMN overwhelm any potential benefits to seedling	Kytoviita et al. (2003)

(continued)

Table 4.1 (continued)

Donor species and tracer usage	Seedling species and age at harvest	Growth of seedling in response to CMN as compared to			CMN effects on nutrient concentrations in shoots of seedlings	Author's suggested mechanism(s) behind CMN-effect on seedling	Reference
		Absence of mycorrhiza	Colonized solitary seedling	Non-host plants			
Community dominated by *Melinis repens*	*Bidens pilosa* 80 days	0	ND	ND	ND	Clipping of donor plants increased invasion by *B. pilosa*	Stampe and Daehler (2003)
Gnaphalium norvegicum ± simulated grazing	*G. norvegicum* 28–35 days	0	–	ND	Decreased shoot N	Enhanced competition for mycorrhiza-mediated resources	Pietikainen and Kytoviita (2007)
Trifolium subterraneum; P. lanceolata $^{13}CO_2$ to donors	*T.subterraneum; P. lanceolata* 4, 7, 12 and 20 days	–	ND	ND	ND	Competition for light and EM depletion of soil nutrients	Nakano-Hylander and Olsson (2007)
Tripleurospermum inodorum	*T. inodorum; Sisymbrium loeselii* (non-host) 60 days	–	ND	0	Increased for shoot P in both species; decreased for P content	Pre-emption of soil N	Janouskova et al. (2011)
Arisaema triphyllum; Maianthemum racemosum	*A. triphyllum; M. racemosum* 42 days (in 3rd annual growth cycle)	0, *At* –, *Mr* with *At* 0, *Mr* with *Mr*	ND	ND	*At*: no change; *Mr*: increased for P with both donor species	Differences in patterns and life span of root growth	Burke (2012)
Litsea glutinosa CMNs were severed or not	*Eucalyptus tetrodonta; Ceiba pentandra* 141 days	ND	–, *E. tetrodonta* 0, *C. pentandra*	ND	No significant change	Exacerbation of iron deficiency	Janos et al. (2013)

(continued)

Table 4.1 (continued)

Donor species and tracer usage	Seedling species and age at harvest	Growth of seedling in response to CMN as compared to			CMN effects on nutrient concentrations in shoots of seedlings	Author's suggested mechanism(s) behind CMN-effect on seedling	Reference
		Absence of mycorrhiza	Colonized solitary seedling	Non-host plants			
Trifolium pratense; Lolium multiflorum	*Arabidopsis thaliana* 42d	–	n.a.	n.a.	ND	EM depletion of nutrients; costly defense responses	Veiga et al. (2013)
Cucumis sativus ± simulated grazing ^{32}P to root free soil patch	*Solanum lycopersicon*; wild type and *rmc* 20, 25, 32 days	0	–	0 (*rmc*)	Decreased shoot P P increased after clipping	Competition for P in CMN	Merrild et al. (2013)
C. sativus	*S. lycopersicon* 17, 24, 31 days	–	–	ND	Decreased P (N/P ratio ≥20)	Competition for P in CMN	Merrild et al. (2013)
Predominantly positive effects of CMN on seedlings							
Mesocosms dominated by *Festuca ovina*; ± simulated grazing; $^{14}CO_2$ to donors	18 herbaceous species 365 days	+, 2 grasses and 10 forbs 0, 4 grasses	ND	+, 2 non-host forbs	ND	C export from *F. ovina* to seedlings AND enhanced mineral nutrient capture by seedlings	Grime et al. (1987)
Grassland	*P. lanceolata; F. ovina; Medicago lupulina* 5–60 days	ND	ND	ND	Shoot N:P decreases after 15–20 days	Increased access to P upon mycorrhiza formation	Read and Birch (1988)

(continued)

Table 4.1 (continued)

Donor species and tracer usage	Seedling species and age at harvest	Growth of seedling in response to CMN as compared to			CMN effects on nutrient concentrations in shoots of seedlings	Author's suggested mechanism(s) behind CMN-effect on seedling	Reference
		Absence of mycorrhiza	Colonized solitary seedling	Non-host plants			
Festuca idahoensis	*Centaurea maculosa* 49 days	+	ND	ND	ND	Increased invasive strength of *Centaurea* seedlings	Marler et al. (1999)
Bouteloua gracilis; F. idahoensis; B. gracilis had 3X the DW of *F. idahoensis*	*C. maculosa* 49 days	+, both donors	0, *B. gracilis* +, *F. idahoensis*	ND	ND	C transfer from *F. idahoensis* to *C. maculosa*	Carey et al. (2004)
Bromus erectus-dominated grassland mesocosms	*Bromus erectus*	+	ND	ND	Increased shoot P	P uptake from CMN	Van der Heijden (2004)
	Brachypodium pinn.	0					
	P. vulgaris	+					
	Trifolium pratense 273 days	+					
T. pratense + simulated grazing	*Tripleurospermum inodorum*	+	ND	Proportion of *Atriplex* (non-host) decreases	ND	*Atriplex* is suppressed by competition from mycorrhiza seedlings	Püschel et al. (2007)
	Calamagrostis epigejos	+					
	Atriplex sagittata 84 days	−					

(continued)

Table 4.1 (continued)

Donor species and tracer usage	Seedling species and age at harvest	Growth of seedling in response to CMN as compared to			CMN effects on nutrient concentrations in shoots of seedlings	Author's suggested mechanism(s) behind CMN-effect on seedling	Reference
		Absence of mycorrhiza	Colonized solitary seedling	Non-host plants			
Neutral or contrasting effects of CMN on seedlings							
P. lanceolata ^{32}P to soil	*P. lanceolata* 37 days	0	0	ND	Reduced N	Competition for soil nutrients	Eissenstat and Newman (1990)
Calamagrostis epigejos + simulated grazing	*C. epigejos* 70 and 168 days	− in nutrient deficient soil; otherwise 0 or +	ND	ND	ND	Competition for soil nutrients; no effect of disturbance of EM links on seedling growth	Malcova et al. (2001)
Paspalum notatum	*S. lycopersicon* 57–63 days	+ at low P − at high P	−	ND	Unchanged	C costs are not balanced by increased P uptake at high P	Sylvia et al. (2001)
Medicago truncatula *In vitro*	Con- and heterospecific pairs of *M. truncatula* (Mt) and *Silene vulgaris* (Sv) 3, 6, 9, 12 days	−, Mt-Mt +, Mt-Sv 0, Sv	ND	ND	ND	Negative effects due to interaction with donor? Positive effects due to increased development of EM?	Derelle et al. (2012)

(continued)

Table 4.1 (continued)

Donor species and tracer usage	Seedling species and age at harvest	Growth of seedling in response to CMN as compared to			CMN effects on nutrient concentrations in shoots of seedlings	Author's suggested mechanism(s) behind CMN-effect on seedling	Reference
		Absence of mycorrhiza	Colonized solitary seedling	Non-host plants			
Leymus chinensis	*L. chinensis*; *Stipa krylovii* 49 days	+	ND	ND	Increased P and N in N fertilized plants	CMN effects depend on seedling species and soil N level	Zhen et al. (2014)
	Artemisia frigida	–	ND	ND	Decreased P at intermediate N fertilization		
CMN studies in root organ cultures							
Daucus carota ^{13}C-glucose and ^{33}P in root cultures	*D. carota*, at high or low sucrose supply; root cultures	ND	ND	ND	Most ^{13}C to low C roots	C starved roots are storage units for fungal carbon; C rich roots are stronger sinks for ^{33}P in EM than C starved roots	Lekberg et al. (2010)
					Most ^{33}P to high C roots		
D. carota ^{33}P in root cultures	*D. carota;* root cultures, at high or low sucrose supply	ND	ND	ND	Most ^{33}P to high C roots	C rich roots are stronger sinks for ^{33}P in EM than C starved roots	Kiers et al. (2011)

CMN effects on seedlings are reported as negative (−), neutral (0) or positive (+). Hyphal connections between plant individuals are assumed but remain unconfirmed in several cases

n.a. not applicable

Table 4.2 Competition between coexisting plants as influenced by AMF network; summary of data in Tables 4.1, 4.3 and 4.4

AMF effects on plant-to-plant competition	Different life stage (target = seedling)		Similar life stage (two or more targets)	
	Intraspecific (from Table 4.1)	Interspecific (from Table 4.1)	Intraspecific (from Table 4.3)	Interspecific (from Table 4.4)
Amplified, −	8	19	10	20
Relaxed, +	3	21	0	25
Neutral, 0	7	10	1	28

different distances from the roots (Jakobsen et al. 1992a; Smith et al. 2000). Uptake of P at the fungus–soil interface is mediated by high affinity Pi transporters (Harrison and Van Buuren 1995) and their expression is influenced by the P concentration in the soil solution and by the host P status (Maldonado-Mendoza et al. 2001), but is apparently independent of the host C status (Olsson et al. 2006). Transport of P in EM can occur over 10–20 cm distance in semi-sterile soil (Jakobsen et al. 1992b; Jansa et al. 2003) and a limited number of field studies shows that also native communities of AM fungi colonize and take up P from root free soil compartments (Schweiger and Jakobsen 1999; Johnson et al. 2001). The EM also takes up and transports other nutrients to the plant including N and Zn (Johansen et al. 1992; Bürkert and Robson 1994; Govindarajulu et al. 2005; Leigh et al. 2009).

Part of the P pool in CMNs is transported to and released into the periarbuscular apoplast space from where it is absorbed by the plant via AM specific P_i transporters (Javot et al. 2007b). The mycorrhizal pathway can account for up to 100 % of plant P uptake (Pearson and Jakobsen 1993b; Ravnskov and Jakobsen 1995; Smith et al. 2003), which means that uptake at the root epidermis is very low either due to down-regulation of direct plant Pi transporters (Javot et al. 2007b; Yang et al. 2012; Grønlund et al. 2013) or to reduced P_i concentration in the rhizosphere soil solution. In return, the biotrophic AM fungi use indispensable sugar transporters to take up hexoses from the symbiotic apoplast (Helber et al. 2011) and this fungal C uptake is in the range of 5–20 % of host photosynthates (Paul and Kucey 1981; Jakobsen and Rosendahl 1990).

Plants are usually not C limited as they up-regulate photosynthesis in response to the C sink of the colonizing AM fungi (Wright et al. 1998a, b; Miller et al. 2002; Mortimer et al. 2008; Kaschuk et al. 2009, 2010; Correa et al. 2012). Further, AM fungi often have a surplus of P stored in spores or vesicles (Olsson et al. 2008, 2011; Hammer et al. 2011). Such observations led to the suggestion that mycorrhiza function is about exchange of luxury resources (Kiers and Van der Heijden 2006) and emerging evidence suggest that the C–P exchange is functionally coupled and bi-directionally controlled. It appears that plants have mechanisms to locally control C allocation that depends on the Pi homeostasis of the cell (Fitter 2006). The nature and sequence of events involved are not fully understood, but certain patterns are emerging. Hexose

Table 4.3 Influence of AM (and assumed CMNs) on inequality of individuals in plant populations

Species and harvest age	Type of experiment; treatments	AM-affected inequality among individuals; variable measured	AM-affected changes in P concentration	Author's suggested mechanism(s) behind CMN-induced change in equality of individuals	Reference
Otholobium hirtum; *Aspalathus linearis* 120 days	Pots; 1, 4, 8 and 16 plants × ± AMF inoc	−, shoot DW	Increased shoot P	Pre-emption of limiting belowground resources by fitter individuals	Allsopp and Stock (1992)
Vulpia ciliata	Field; natural population × ± benomyl	−, reproduction	ND	Plants in isolated patches grow large when mycorrhizal	Carey et al. (1992)
Abutilon theophrasti 22, 36, 50, 64, 78, 92 days	Fumigated field plots; 82,300 and 28,600 seeds m^{-2} × ± AMF inoc	−, reproduction	Increased seed P	Increased shoot branching and seed weight	Shumway and Koide (1995)
Trifolium subterraneum 56 days	Pots; 1, 6, 14, and 24* plants × ± AMF inoc. × 2 light levels *≈2500 plants m^{-2}	−, shoot DW at 6 plants per pot	Increased shoot P at 1 and 6 plants per pot	Pre-emption of limiting belowground resources by fitter individuals	Facelli et al. (1999)
T. subterraneum; 56 days	Pots; 250 and 1000 seeds m^{-2} × ± AMF inoc	−, shoot DW at 250 seeds m^{-2}	ND	Pre-emption of limited resources by fitter individuals	Facelli and Facelli (2002)
Plantago lanceolata; 140 days	Field; 4 and 64 plants m^{-2} × ± iproidone Pots; 20 and 61 plants m^{-2} × ± iproidone	0, shoot DW	ND	Intense competition for nutrients	Ayres et al. (2006)

(continued)

Table 4.3 (continued)

Species and harvest age	Type of experiment; treatments	AM-affected inequality among individuals; variable measured	AM-affected changes in P concentration	Author's suggested mechanism(s) behind CMN-induced change in equality of individuals	Reference
Medicago sativa 30, 60, 90, 120 days	Pots; 6000 and 17,500 seeds m^{-2} × ± benomyl	−, mortality (self thinning)	ND	Increased shoot competition at high mycorrhiza	Zhang et al. (2011a)
M. sativa; 126 days	Field; 10, 100, 1000 and 10,000 seeds m^{-2} × ± benomyl	−, mortality at lowest density	ND	Enhanced intraspecific competition	Zhang et al. (2011b)
M. sativa 160 days	Pots; 40 and 10,000 plants m^{-2} × 4 AMF + NM	−, mortality; varies with AMF species	ND	Self-thinning due to higher shoot competition	Zhang et al. (2011c)
Andropogon gerardii 94 days	Pots with CMN and plants in intact or rotated cone-tainers	−, shoot DW	Increased shoot P	Pre-emption of limiting belowground resources by fitter individuals	Weremijewicz and Janos (2013)

Increased or unchanged inequality are denoted as − or 0

Table 4.4 Influence of AM (and assumed CMNs) on competition between different species of same age

Interacting species	A. Target growth relative to absence of CMN (AM vs. NM in competition)	B. Target growth in CMN relative to its growth when solitary and AM	Author's suggested mechanism(s) behind CMN-effect on competition	Reference
Lolium perenne	–	–	Root length is reduced in Lp when AM	Fitter (1977)
Holcus lanatus	+	+		
L. perenne	0	0	Competition for P	Hall (1978)
Trifolium repens	+	–		
L. perenne	–	ND	Competition for P	Buwalda (1980)
T. repens	+			
Koeleria pyranidata	–	–	Competition for P	Hetrick et al. (1989)
Andropogon gerardii	+	0		
Setaria lutescens	0	0	Soil P depletion by Sl	Koide and Li (1991)
Abutilon theophrasti	0	–		
Medicago sativa	≥0	ND	Better nutrient balance in Ms. Modified allocation of EM nutrients	Hamel et al. (1992)
Bromus inermis	≤0			
Phleum pratense	≤0			
Elymus canadensis	–	(–)	Competition for P	Hartnett et al. (1993)
A. gerardii	+	–		
E. canadensis	0	0	Competition for P	Hetrick et al. (1994)
K. pyranidata	0	0		
A. gerardii	+	–		
H. lanatus	+ at few Dg	0	CMN benefit of each target decreases with increasing interspecies competition	West (1996), Watkinson and Freckleton (1997)
Dactylis glomerata	+ at few Hl	–		
Centauria maculosa	0	+	Inter-plant C transfer; competition for other resources	Marler et al. (1999)
Festuca idahoensis	–	–		
Centaurea melitensis	+	+	Inter-plant C transfer; changes in microbial communities	Callaway et al. (2001)
Nacella pulcra	–	–		

(continued)

Table 4.4 (continued)

Interacting species	A. Target growth relative to absence of CMN (AM vs. NM in competition)	B. Target growth in CMN relative to its growth when solitary and AM	Author's suggested mechanism(s) behind CMN-effect on competition	Reference
L. perenne	0	ND	Competition for P	Joner and Leyval (2001)
T. repens	+			
C. maculosa with	+	ND	Competition for P in hyphal network	Zabinski et al. (2002)
F. idohaensis	0			
Koleria cristata	0			
Pseodoroegneria spicata	0			
Achilla millefolium	+			
Gaillardia aristata	0			
C. melitensis	+, Np; −, Ab	+, Np; −, Ab	Inter-plant C transfer; changes in microbial communities	Callaway et al. (2003)
N. pulcra	−, Cm only	−, Cm only		
Avena barbata	−, Cm only	+, Cm only		
Citrus sinensis	0	ND	Better competition for P	Yao et al. (2005)
Stylosanthes gracilis	+			
L. perenne	0	ND	Soil nutrient conditions	Endlweber and Scheu (2007)
T. repens	0			
Lotus corniculatus	+	−	Not specified; presumably nutrients	Scheublin et al. (2007)
F. ovina	−	−		
P. lanceolata	0	−		
Capsicum annuum	−/+*	−	Intraspecific competition is amplified by AM in Ca and Zm (larger AM-induced growth depressions), but relaxed by AM in Cp	Schroeder-Moreno and Janos (2008)
Zea mays	−/+**	−/0		
Cucurbita pepo	−	−/0		

(continued)

Table 4.4 (continued)

Interacting species	A. Target growth relative to absence of CMN (AM vs. NM in competition)	B. Target growth in CMN relative to its growth when solitary and AM	Author's suggested mechanism(s) behind CMN-effect on competition	Reference
Agropyron smithii	–	ND	Competition for nutrients at low P. AM effect diminishes at high P	Collins and Foster (2009)
A. gerardii	+; 0 at high P			
Bouteloua curtipendula	+; 0 at high P			
Elymus canadensis	0			
Hordeum jubatum	–; + at high P			
K. pyramidata	–			
Schizachyrium scoparium	+; 0 at high P			
Sorgastrum nutans	+; 0 at high P			
Sporobolus heterolepis	+; 0 at high P			
Helianthus annuus	0	–	Nutrients were less available to weeds	Rinaudo et al. (2010)
Six weed species	–	– (strongly)		
Z. mays	0	–	Competition for nutrients	Veiga et al. (2011)
Nine weed species	–	– (strongly)		
Elymus nutans	+	–	Unrelated to nutrients?	Jin et al. (2011)
Ligularia virgaurea	–	+		
T. pratense	+	–	Competition for nutrients	Wagg et al. (2011a)
L. multiflorum	≤0	0		
Trifolium pratense	+	ND	Competition for nutrients	Wagg et al. (2011b)
Lolium multiflorum	≤0			
Linum usitatissimum	+	+	Lu obtains the most ER nutrients, but at a low C cost	Walder et al. (2012)
Sorghum bicolor	0	0		

(continued)

Table 4.4 (continued)

Interacting species	A. Target growth relative to absence of CMN (AM vs. NM in competition)	B. Target growth in CMN relative to its growth when solitary and AM	Author's suggested mechanism(s) behind CMN-effect on competition	Reference
Bothriochloa bladhii	ND	+	Bb induces reduced % AM in Ag and Sc. No non-AM treatment	Wilson et al. (2012)
A. gerardii	ND	−		
S. scoparium	ND	−		
Z. mays	0 (−for Ca)	−	Not discussed	Daisog et al. (2012)
Solanum nigrum	0 (−for Ca)	−		
Chenopodium album	0 (−for Sn)	0/+		
Taraxacum officinale	−	≥0	Dominants, To and Ac, were suppressed the most by AM	Mariotte et al. (2013)
Agrostis capillaris	−	−		
Prunella vulgaris	0	+		
Achillea millefolium	0(−)	≤0		
Triticum aestivum	0	−	Temporal separation of N and P uptake, complementary resource use	Qiao et al. (2015)
Vicia faba	+	−		

CMN effects on growth of target species was assessed by comparing to the corresponding interaction in absence of CMN (A) and to its growth when solitary and colonized (B)
−, +, and 0 denotes amplified, relaxed or unchanged competition to target species
*+ at low planting density of target species
**+ at low planting density of target species when grown with Ca

levels increase in colonized roots due to increased activity of sucrose synthase and acid invertase (Ravnskov et al. 2003; Schaarschmidt et al. 2007) and such response is parallel to the increased C allocation to and growth of roots encountering a P patch in soil (Drew 1975). Hence, a locally increased P concentration in colonized root cells was suggested to be the signal inducing C allocation to the symbiotic apoplast (Fitter 2006; Helgason and Fitter 2009). In accordance with this hypothesis, plants grown in split root systems allocate a greater share of C to the colonized than to the uncolonized root half (Koch and Johnson 1984).

Likewise, plants can trigger P transfer from the fungus. A higher C supply to colonized roots resulted in higher P concentrations and conversion of poly-Pi to Pi in the IM of the fungus (Bücking and Shachar-Hill 2005). The C supply also resulted in increased efflux of P from the IM (Solaiman and Saito 2001). Furthermore, an increased C supply to colonized roots stimulated the transport of N from EM to roots (Fellbaum et al. 2012). Currently it is not possible to unravel which partner takes the first step to establish the mutualistic C–P exchange (see Smith and Smith 2012). However, the fungus may be able to use plant cell wall sugars (Helber et al. 2011) and P reserves in AM fungal spores could serve as signals during early colonization (Hammer et al. 2011).

The highest possible gain per unit investment is most beneficial for each partner. Ample P supply usually reduces mycorrhizal colonization (Olsson et al. 1997; Breuillin et al. 2010; Balzergue et al. 2013) and this indicates that plants trade only when they are in need of nutrients. The fungal partner may also exert control over its P release such that hosts that cannot provide C to the fungus may gain insignificant amounts of P from the symbiosis (Hammer et al. 2011). However, some fungi, especially of the genera *Scutellospora* and *Gigaspora*, receive much plant C for a small P delivery in return (Pearson and Jakobsen 1993a; Lendenmann et al. 2011). It is unknown whether this presumably unbalanced symbiosis would still obtain a large share of plant C if it co-existed with a more P efficient fungus. However, other mycorrhizal functions than P delivery could contribute to a continued plant investment into a mycorrhizal partnership: the delivery of other nutrients than P (Bürkert and Robson 1994; Govindarajulu et al. 2005), P delivery at a different time course (Cavagnaro et al. 2005) or protection against pathogens (Newsham et al. 1995). These factors need to be kept in mind when interpreting C–P exchange studies since their effects may enhance or offset species C–P exchange patterns.

4.3 Formation of CMNs and C–P Exchange in a Multiple Partner Setting

Generalizations from nutrient exchange studies involving just a single plant and a single fungus should be avoided: "Populations have properties that individuals do not. Thus, the effects of mycorrhizal fungi on populations of plants are not simply the sum of their effects on the individuals within the population." This statement by Koide and Dickie (2002) extends to plant communities and in both cases a CMN forms when two or more plants become connected by the EM of one or more AM fungi. Connections derive from foraging of the EM for new carbon sources i.e. new roots or by the fusion of individual EMs via anastomosis, a feature which is treated in detail in another chapter in this volume (Giovannetti et al. Chap. 2, this volume). Evidence for nutrient translocation between two anastomosing mycelia, both in AM and ectomycorrhizal fungi (Mikkelsen et al. 2008; Wu et al. 2012), implies that the formation of a large CMN from the fusion of many small EMs of an individual genotype (Rosendahl 2008) enables a long-distance belowground translocation of nutrients. Grazing by collembola or other soil biota can severely disrupt (Johnson et al. 2005) nutrient flow in the mycelium and repair by anastomosis could be important to mitigate such disruptions.

Only a small number of ecophysiological studies have been performed on simple CMNs consisting of two or three individuals of either fungus or plant in combination with one individual of the other (Nakano-Hylander and Olsson 2007; Bever et al. 2009; Lekberg et al. 2010, Kiers et al. 2011; Walder et al. 2012; Merrild et al. 2013; Fellbaum et al. 2014). EM connections between different root patches can be directly observed in the clear gel of monoxenic root cultures. Such cultures allow precise control of the C strength of host roots via choice of sugar concentration in the growth medium and are therefore suitable to study C investment of root patches in a multi-partner setting. Experimental approaches involve compartmentation of the experimental plates into inoculum- or root compartments for roots of different C strength and EM compartments where P strength can be varied. Root cultures also allow for a precise control of P availability in the absence of interacting soil minerals. So far, P is the only nutrient which has been studied in root cultures with CMN. Interestingly, it is also possible to use these Petri plate model systems to establish CMNs between different whole plants when their sterile roots grow into a pre-established EM (Derelle et al. 2012).

Reciprocal control of the C-P trade is important for a CMN with many players on both sides. Thereby it may be avoided that non-contributing individuals will profit on those that contribute to the built-up of the CMN. As a result, non-contributing individuals will have less of a negative impact on the CMN. Recent studies on the C–P exchange between more than two partners indicate that both plant and fungus can differentiate between partners of different quality: given a choice, AM fungi allocate P towards host roots with the most generous C supply (Lekberg et al. 2010; Kiers et al. 2011; Fellbaum et al. 2014) and also allocate more C to the compartment with the highest P source. In four-compartment plates, ^{13}C from donor roots was distributed over the whole network including two receiver root patches and a compartment for EM only (Lekberg et al. 2010). However, most ^{13}C was allocated to the compartment containing C-starved roots where it stayed in the EM and the IM of the fungus, i.e. there was no transfer to the root itself.

Pot-grown plants being co-inhabited by several AM fungal species allocate most C to the fungus that in a previous experiment had been found to deliver the most P (Bever et al. 2009; Kiers et al. 2011). The hypothesis by Fitter could explain this phenomenon as the fungus delivering the most P would be perceived as the richest P patch and therefore receive the most C (Helgason and Fitter 2009). This C for P hypothesis is supported by other studies (Olsson et al. 2006; Javot et al. 2007a; Yang et al. 2012) but conclusive evidence is still lacking. Even though such reciprocal control of nutrient exchange takes place, the least P efficient of two or three co-occurring fungi still received a substantial amount of C (Bever et al. 2009; Kiers et al. 2011). This suggests that sanctions may be relaxed—and promiscuity selected for—to increase the likelihood of receiving benefits under shifting environments.

AM fungi colonize poor hosts within the mycorrhizal network to a similar extent as good hosts (Lekberg et al. 2010). From a mycocentric point of view, such unlimited colonization of seedlings or C-starved roots may be a fungal strategy to ensure a steady C supply because the C supply from other hosts will fluctuate over time. The IRM of C-starved roots had higher vesicle densities and enhanced concentrations of storage lipids (Lekberg et al. 2010) and such host roots providing little C may represent a predator-free C depository for the fungus at little or no cost. The C-starved roots received a high proportion of ^{13}C from a donor root, but the isotope stayed in the fungal lipid structures of the IM.

In conclusion, C transfer at the symbiotic interface occurs in the direction of the fungal nutrient source while mineral nutrient transfer occurs in the direction of the plant C source. In the CMN context, two important questions emerge: (1) can significant nutrient transfer occur in the opposite direction i.e. C from AM fungus to plant and minerals from plant to AM fungus; (2) what determines the C supply from and the mineral nutrient gain for the individual plant in a shared CMN? These questions will be treated in the following sections.

4.4 Interplant Nutrient Transfer

An abundant literature on plant-to-plant transfer of nutrients in CMNs of AM systems primarily reflects a view that 'the function of CMNs is to provide pathways for movement or transfer of nutrients from one plant to another' (He et al. 2009). However, critical examination of the published data does not support such general statement, at least not in arbuscular mycorrhizal systems. Although the nutrient translocation processes in AM fungal hyphae are bidirectional (Uetake et al. 2002; Nielsen et al. 2002) it is obvious that net translocation of carbon and of mineral nutrients occurs in principally opposite directions. Carbon is mainly transferred to the growing tips of the EM in order to sustain the branched absorbing structures and to maximize the recruiting of new C sources i.e. roots. Mineral nutrients are translocated in the opposite direction i.e. towards the host plant (Johansen et al. 1993).

Early studies reported CMN-mediated interplant transfer of carbon fed to one plant as $^{14}CO_2$ (Francis and Read 1984; Finlay and Read 1986a; Grime et al. 1987). Most of the ^{14}C in the receiver plant was found in the roots which include the fungal IM and any levels in shoots were small; hence the data provided no evidence for any CMN-mediated net transfer of carbon between plants (Newman 1988). A similar conclusion was reached in a later review (Robinson and Fitter 1999) which included the report on ^{13}C transfer between two ectomycorrhizal species (Simard et al. 1997; but see also Simard et al., Chap. 5 this volume). In AM plant pairs, C from $^{13}CO_2$-labeled donor plants was not directed towards mycelium in receiver roots to any higher degree than towards other directions (Nakano-Hylander and Olsson 2007) and conclusive negative evidence for interplant C transfer in AM systems was provided using AM fungal root cultures and in vitro cultures of intact mycorrhizal systems (Pfeffer et al. 2004; Voets et al. 2008; Lekberg et al. 2010). The latter study demonstrated that transferred ^{13}C remained in the intraradical mycelium.

The multiple evidence against AM-mediated C transfer between autotrophic plants is moderated by one report of substantial C transfer from a bulbous plant to a tree seedling in the bud burst phase (Lerat et al. 2002). Otherwise, myco-heterotrophic plants represent an important case where a connective mycelium mediates C transfer from an autotrophic to a heterotrophic plant (Leake and Cameron 2010). Such C drain on the fungus has been clearly demonstrated in ectomycorrhizal networks (McKendrick et al. 2000; Gebauer and Meyer 2003) and similar evidence is emerging for myco-heterotrophs engaging with AM networks (Merckx et al. 2010; Courty et al. 2011). This has been based solely on analysis of patterns of natural abundance of ^{13}C and there is a need for studies involving pulse labeling with ^{13}C or ^{14}C. Myco-heterotrophs may be considered as 'cheaters' providing no advantage to the other partners in the network, but it is also possible that the carbon transfer to the heterotrophic plant is too small to impose any measurable costs to the mycorrhizal fungi (Smith and Read 2008; Merckx et al. 2009).

The N transfer from legumes to grasses via connecting EM has also been widely studied and also in this case there is little evidence for direct transfer (Newman 1988; Frey and Schüepp 1993; Johansen and Jensen 1996; Li et al. 2009). Instead, interplant transfer can be indirect by EM absorption and translocation of minerals released from exudating or dying roots (Newman and Ritz 1986). This is supported by abundant studies where root colonization by AM fungi increases N and P transfer from a dying to a living plant (Heap and Newman 1980; Newman and Eason 1989; Eason et al. 1991; Johansen and Jensen 1996) and this is probably associated with the decomposing root system functioning as a nutrient patch for the EM that was previously fed with C from the dying root system. Early field work suggested that AM fungal connections were responsible for observed patterns of distribution of ^{32}P from a source plant to its neighboring plants (Chiariello et al. 1982). However, as in the case of nitrogen, several controlled studies do not provide any evidence for direct P transfer from plant to fungus (e.g. Johansen and Jensen 1996).

Instead of being considered as a system for interplant nutrient transfer the CMN rather represents a system for potential resource uptake from a shared pool. The EM may on one hand obtain C from several plant individuals while on the other it serves as an efficient mesh for absorbing mineral nutrients to cover its own needs and to distribute the surplus among the plant components of the CMN. The challenge is to identify the factors which regulate the exchange taking place across the symbiotic interfaces at the plant individual level.

4.5 Sharing of EM Nutrients Between CMN Plants of Different Age

Although CMNs will undoubtedly influence nutrient uptake by the interlinked plants we still need a set of rules to determine whether a plant will benefit or not (see e.g. Van der Heijden and Horton 2009). The previous section emphasized that autotrophic AM plants do not obtain ecologically significant amounts of C, if any,

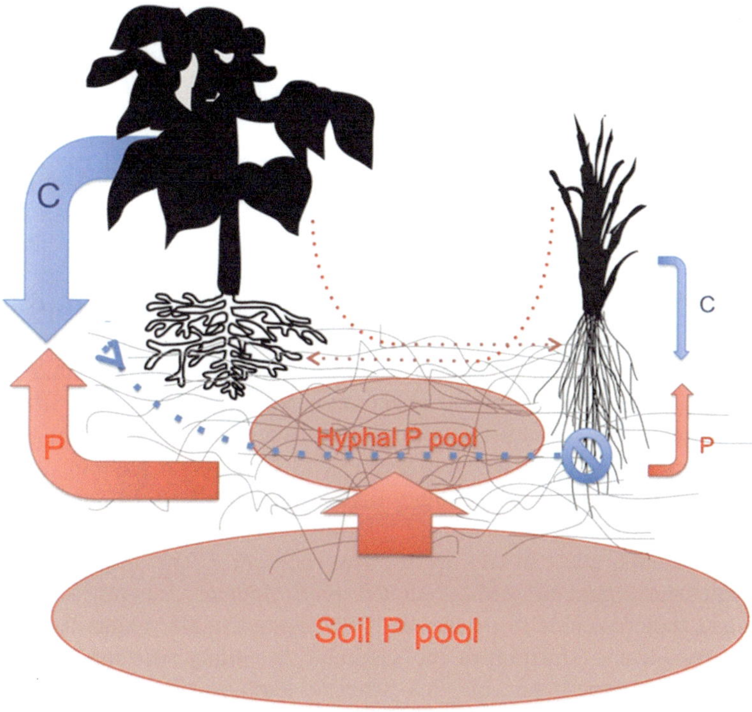

Fig. 4.1 Conceptual model of the nutrient dynamics of two plants being connected into a CMN: AM fungi take up P from the soil P pool and store it in the mycelium. Phosphorus is delivered to the plant that allocates the highest amount of C to the mycelium. This partitioning can depend on plant size and species specific carbon allocation traits and regulation presumably takes place at the symbiotic interface in the root. Inter-plant transfer of nutrients as P is minor (*dashed lines*). Carbon may be distributed into roots of receiver plants but commonly stays in the intraradical mycelium (indicated by the *stop symbol*)

from the fungus. Therefore, C issues of plants in the CMN mainly concern their relative supply of C to the shared EM. Likewise, there is no evidence for any significant transfer of mineral nutrients from plant to fungus and CMN functioning can accordingly be coined in this way: 'Instead of resources moving between plants, plants may be accessing resources from a hyphal network that functions as an additional pool of resources.' (Zabinski et al. 2002). Indeed, this additional pool would rather become an alternative pool when the direct uptake at the root epidermis is decreased by mycorrhizas (Smith et al. 2004). The key issue for mineral nutrients in CMNs is about sharing: which plants will have access to the nutrients in the common EM and what are the regulating mechanisms (Fig. 4.1).

Tracer isotopes are key tools for studying interplant sharing of mineral nutrients in a common EM. However, in controlled studies with CMNs, tracers of a nutrient (e.g. ^{13}C, ^{14}C, ^{15}N, ^{32}P, ^{33}P) have predominantly been applied to a donor plant (leaf or split-root application) in order to investigate their transfer to a receiver plants (see references in previous section). Such a procedure is inadequate for studying how nutrients in a mycelium are shared among the interlinked plants. Instead, the label should be applied to the EM in a soil patch as exemplified in a study using autoradiography to demonstrate transfer of ^{32}P from a labeled patch to the individual plants in a rhizobox (Finlay and Read 1986b).

Principles of nutrient dynamics in CMNs are most easily studied in pair-wise combinations of adult plants and seedlings which have contrasting source and sink strengths for C and mineral nutrients, respectively. According to the initial hypothesis of C or nutrient transfer, the older/bigger plant is often termed "donor" and seedlings are termed "receivers". Seedlings of non-hosts (plant species that commonly do NOT form AM symbiosis) can be included as controls to enable a separation between nutrient uptake via connection into the CMN and nutrient uptake via mass flow and diffusion in the soil. Non-host seedlings will likewise experience deficiencies from pre-empting of nutrients by the surrounding hyphae. Two other control treatments aid in assessing CMN effects on seedlings: seedlings grown without mycorrhizas or a solitary mycorrhizal seedling. In the first control, effects of EM links to other plants are intermingled with general effects of mycorrhizal colonization and can add up or offset each other. The second control will undergo the same general AM-associated physiological changes as the CMN seedling and therefore aids to determine if EM links to other plants have specific effects on growth and nutrition of the seedling. A solitary seedling is the proper control to a CMN-experiment, but this is not an evolutionary option in nature, as plants do not control their engagement in CMNs. The only way to avoid a CMN is to evolve being NM. Unfortunately, studies examining CMNs often differ in their use of control treatments and only few use all three types (Table 4.1).

4.5.1 Suppressed or Enhanced Seedling Growth

The rapid colonization of seedlings linking into CMNs (Birch 1986) led to the early suggestion that seedlings would gain access to nutrients in the EM and thereby get more competitive (Newman 1988; Read 1991; Gange et al. 1993). However, the situation turns out to be complex; a literature survey showed that effects of a CMN on seedlings could be positive, neutral or negative (Van der Heijden and Horton 2009) and reasons for this variation is little understood. Twenty-three studies investigating the impact of AM fungi on 65 cases of competition between seedlings and adult 'donor' plants are grouped in Table 4.1 according to their outcome: predominantly negative (13 studies), predominantly positive (6 studies) and neutral/contrasting effects on the seedlings (4 studies). Each study comprises one or several pairs of plants and some were performed in multispecies mesocosms with a dominating grass (e.g. Grime et al. 1987; Van der Heijden 2004). Two root culture experiments using isotopes to investigate effects of C source strength on P transfer in CMNs are also included in Table 4.1. Hyphal connections between plant individuals were rarely confirmed but are here assumed to have existed; in any case,

EM links were most likely present in cases where seedlings were planted in soil with a pre-established EM.

Table 4.1 includes examples of both intra- and interspecific adult-seedling combinations. In intraspecific combinations, seedlings most often grow less in the presence than in the absence of assumed EM links whereas the outcome is more balanced when seedling and adult are different species (Table 4.2). This agrees with Moora and Zobel (2010) and suggests that additional, species-specific factors may influence the outcome of the experiments. These factors are discussed in Sect. 4.5.2.

The summary in Table 4.2 is based on comparison of the CMN outcome to that of seedling-adult interactions in the absence of mycorrhiza. Suggested mechanisms behind the observed interactions are listed in Table 4.1 but they often remain unvalidated. Intraspecific seedling-donor interactions were neutral in seven cases and in two cases CMN seedlings were actually suppressed when compared to a solitary AM-colonized seedlings (Kytoviita et al. 2003; Pietikainen and Kytoviita 2007). This shows that the CMN seedlings were nutrient-limited in the same way as seedlings competing with large plants in the absence of mycorrhiza and it means that CMN seedlings were unable to access sufficient nutrients from the EM to unfold their mycorrhiza response potential. The magnitude of suppressive CMN effects on seedling growth is often similar in non-host and in host seedlings (Ocampo 1986; Francis and Read 1995; Janouskova et al. 2011) and nutrient concentrations in seedlings are similarly reduced (Ocampo 1986). Seedlings of the non-host species *Arabidopsis thaliana* were suppressed by pre-established *Trifolium pratense* plants; this suppression was amplified by the presence of an EM (Veiga et al. 2013; Table 4.1) that might have increased preemption of nutrients. However, activation of costly defense responses could also have been involved as suggested by the authors.

Preemption of nutrients or antagonism from the fungus was also suggested to account for growth suppression in seven ruderal seedlings, including four non-host species, when exposed to an EM from an adult *Plantago lanceolata* (Francis and Read 1995). In contrast, conspecific *P. lanceolata* seedlings responded positively to the CMN situation and represent one of two cases of AM-induced relaxed competition in intraspecific seedling adult combinations. Otherwise, positive effects of mycorrhizas on seedlings were reported predominantly in interspecific seedling-adult cases (Table 4.2) and enhanced shoot nutrient status was reported in two of these (Read and Birch 1988; Van der Heijden 2004). The 20 cases of relaxed competition were experimentally different from those showing amplified competition: 16 were carried out in natural or reconstructed grasslands (Read and Birch 1988; Grime et al. 1987; Van der Heijden 2004) and seedlings were grown for 9–12 months in the two latter studies. Such extended study period will better reflect the relevant time frame for evolutionary selection on plants as annuals will reach reproductive state within this time.

Various degrees of simulated grazing was applied in three of the studies (Grime et al. 1987; Van der Heijden 2004; Püschel et al. 2007) and it has been shown that such treatment relaxes the suppression of the seedling (see Sect. 4.5.2). The importance of the donor size is also suggested by a study where the invader

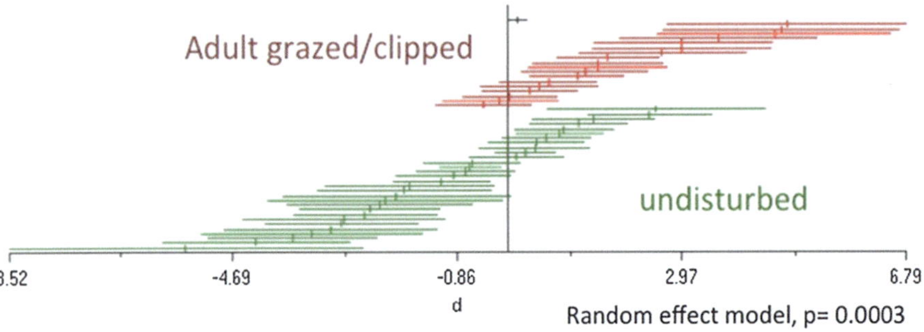

Fig. 4.2 Effect sizes of the presence of a CMN compared to NM situation on seedling DW growth next to an adult plant. The adult plant was present during the whole experiment or was eliminated by grazing or clipping at some time point. Positive effect sizes imply increased seedling DW compared to a NM situation, while negative effect sizes imply decreased seedling DW in a CMN situation. In undisturbed systems, seedlings tend to suffer from the connection into a CMN, while they profit after the adult plant disappears. Data was extracted from studies listed in Table 4.1. The effect size was calculated as Cohen's d with metawin: $d = \frac{\bar{\mu}_t - \bar{\mu}_c}{s_p}$, where $\bar{\mu}_t$ is the mean of the test treatment, $\bar{\mu}_c$ the mean of the control treatment and s_p is the pooled variance within the experiment $\left(s_p = \sqrt{\frac{(n_1-1)s_1^2 + (n_2-1)s_2^2}{n_1+n_2-2}}\right)$

Centauria maculosa was unaffected by *Bouteloua gracilis* but enhanced by the smaller *Festuca idahoensis* donor (Carey et al. 2004). One study showed variable seedling responses including positive ones that were probably caused by variation in soil nutrient status as disturbance of EM links had no effect on seedling growth (Malcova et al. 2001). Sorting the studied intraspecific and interspecific adult-seedling pairs in Table 4.1 according to the state of the adult at termination of the study (intact versus grazed or clipped), it becomes obvious that the presence of a CMN relaxes competition on seedlings only after the connected adult plant has been eliminated (Fig. 4.2). If adjacent to an intact adult plant, seedlings tend to suffer if connected to the adult plant via a CMN.

4.5.2 Mechanisms Regulating CMN Effects on Plants

Overall, Tables 4.1 and 4.2 confirm our hypothesis for intraspecific seedling-adult pairs i.e. that allocation of P (and other mineral nutrients) to CMN plants depends on their relative size or C source strength: Larger plants get the main share of the nutrients from the CMN as illustrated in Fig. 4.3a. Plant access to the CMN nutrient pool is presumably governed more by their C source strength than by their nutrient sink strength as suppressed seedlings of intact CMNs can show clear nutrient deficiencies (Fig. 4.4). Such nutrient deficiency would generate a strong nutrient concentration gradient from the EM to the apoplast of cortex cells in colonized

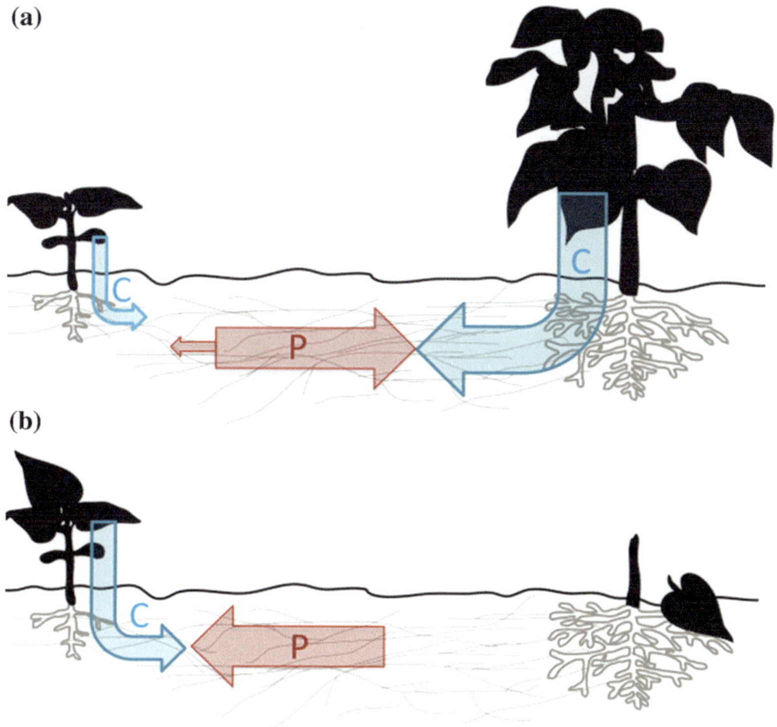

Fig. 4.3 A conceptual model of the time course in nutrient dynamics within a CMN between adult plants and seedlings: In the presence of a CMN, P is distributed to the plants according to their C source strength, leading to P deficiency of the seedling (**a**). During time, adult plants may become senescent or get grazed and after removal of the superior C source the seedling gets access to the P pool within the mycelium (**b**)

seedlings that would accordingly constitute a nutrient sink regardless of their size. Nevertheless, the seedlings do not seem to participate in the nutrient sharing of the CMN. However, this asymmetric C strength setting is not a fixed state as surrounding larger plants may be grazed or become senescent and this will remove the major C source of a CMN. Thus, simulated grazing resulted in a switch from depressed to enhanced seedling growth in two pot experiments and this corresponded to changes in their shoot nutrient status (Pietikainen and Kytoviita 2007; Merrild et al. 2013). Obviously, nutrient fluxes in the CMN were redirected towards the already connected seedling (Fig. 4.3b). The early linkage of seedlings into the CMN serves as a long-term insurance by enabling them to obtain rapid benefits from later temporary changes in C strength patterns of the CMN. This model helps to explain most results in Table 4.1, both the dominance of depressions in short term studies and growth enhancements in long-term studies.

The preferential translocation of P towards the strongest C source is directly demonstrated by radioactive tracers (Eissenstat and Newman 1990; Merrild et al.

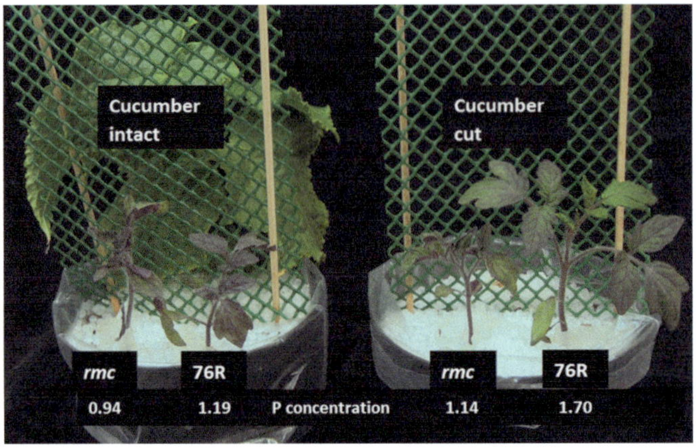

Fig. 4.4 Growth of tomato seedlings (**76R** wild type and *rmc* reduced mycorrhiza colonization mutant) in CMNs with intact cucumber plant and cucumber plant where shoot was cut 11 days earlier. Tomato seedlings grew in 25 μm mesh bags and green vertical mesh prevented competition for light between species. Notice the low shoot P concentrations and visual P deficiency symptoms in all but the greening 76R seedling in the pot with a cut cucumber shoot (Adapted from Merrild et al. 2013)

2013; Lekberg et al. 2010; Kiers et al. 2011). Merrild et al. (2013) included a ^{32}P labeled soil patch which could be accessed by the EM but not by roots. Simulated grazing induced an increased seedling uptake of ^{32}P that after 12 days was fourfold the uptake in the ungrazed treatment. This result is best explained by the changes in the relative C source strength of the plants.

Importantly, some studies demonstrate that AM enhances plant diversity mainly due to a growth reduction of grasses and a mycorrhizal benefit of forbs (e.g. Grime et al. 1987; Gange et al. 1993; Van der Heijden et al. 1998). These experimental communities were dominated by grasses which most likely were also the major C source for the assumed hyphal networks. Still, AM increased the growth of most subordinate plants, mostly at the expense of the grasses, and would therefore have obtained nutrients from the CMNs. This apparent disagreement with the nutrient-for-C hypothesis could have at least two possible explanations: first, these experiments were run over rather long growth periods; second, the larger grasses probably had a rather small sink strength for P in the CMN as ample P was already provided by their effective root systems. Under such P saturation of the major C source for the CMN conditions it seems likely that P in the hyphal network may become more easily accessible to the smaller plants.

The reported CMN-induced suppression of seedlings may contribute to explain results obtained in some early grassland studies of root competition (Cook and Ratcliff 1984; Snaydon and Howe 1986). Root interactions between grass seedlings and their neighbors strongly suppressed seedling growth relative to their growth in tubes inserted in the turf. The suppressed growth could be mitigated by high fertilizer doses. Besides, simulated grazing of the neighbors significantly increased seedling growth, but only when there was no physical barrier between roots of seedlings and roots of neighbors.

4.6 Intra- and Interspecific Competition Between Plants of Similar Age

Competition between plants of similar age is also markedly influenced by CMNs and will be discussed separately for intraspecific and interspecific interactions. Competition for nutrients between individual plants in a population is asymmetric such that larger individuals will obtain a disproportionate share of the resources (for their relative size) at the expense of growth of smaller individuals (Weiner 1990). It seems likely that mycorrhizas have a role in such asymmetric competition but information is limited. If nutrients or C were translocated between plants via CMNs (as suggested by e.g. Grime et al. 1987) this would relax competition and reduce differences in size. Experimental evidence rather proves the opposite.

In Table 4.3 we have listed the main results of studies of CMN effects on the inequality of individuals of conspecific populations consisting of a cohort of seedlings with an initial natural variation. Mycorrhizal colonization caused an increase in the coefficient of variation for size classes, reproduction or mortality in 9 of the 10 studies listed in Table 4.3. Koide and Dickie (2002) suggest that slight differences in size or fitness will result in small increases in the plant's C strength (Fig. 4.5). This will enable them to better take advantage of mycorrhizal colonization as the P flow will be directed to the larger C source and cause a positive feedback to increase their growth. The controlled studies in root organ cultures confirm this hypothesis (Lekberg et al. 2010; Kiers et al. 2011). Thus, the presence of a CMN increases asymmetric competition and accelerates competitive takeover by the larger plant individuals compared to a NM situation. In the absence of root system overlap, CMN also induced increased size inequality within populations of *Andropogon gerardii* and this was explained by positive feedbacks between mycorrhiza formation, mineral nutrient uptake and host growth (Weremijewicz and Janos 2013). One study in Table 4.3 shows no effect of CMNs on inequality of *P. lanceolata* grown in pot and field experiments (Ayres et al. 2006); the authors suggest that this could have been caused by a selection for equality of seedlings at the start of the experiment.

The results discussed above may be dependent on plant density. Inequality changes were observed only at lower plant densities in three studies (Facelli et al. 1999; Facelli and Facelli 2002; Zhang et al. 2011b). Low density plant populations benefit more from mycorrhizal presence when uptake of nutrients e.g. P is limited by diffusion processes. The reason is that nutrient depletion zones around the roots will constitute a relatively small soil volume leaving a large undepleted volume to be exploited by the EM. At high density of plant individuals, belowground competition for other resources than nutrients increases and starts overruling benefits of

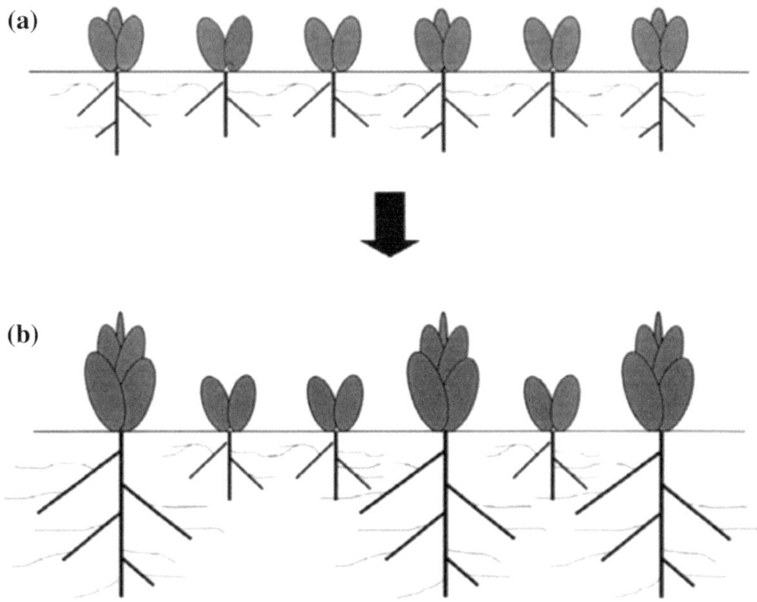

Fig. 4.5 A conceptual model of conspecific competition in a plant population of the same age cohort: initial minor variation in plant size (**a**) is increased by the presence of a CMN as asymmetric nutrient delivery in favour of the stronger C source plants increases their competitive advantage (**b**) (Adapted from Koide and Dickie 2002)

mycorrhizal colonization (see Facelli et al. 1999; Facelli and Facelli 2002; Table 4.3).

Studies using plants of different ages in a CMN as presented in Sect. 4.5 have amplified differences in initial size as part of their experimental design and interpretation can be unified into a model of increased competition caused by differing C source strength. The CMN competes for nutrients in the proximity of the less competitive plant and will deliver them to the dominant plant. This will strongly increase imbalance between individuals. Such CMN-linked inequality in populations could have significant ecological consequences if the original small size differences had a genetic basis. The presence of a CMN would shift the genetic representation of the next generation towards the more dominant plants as suggested by Facelli et al. (1999).

In summary, CMNs clearly amplify plant-plant competition in plant populations and this is in good accordance with the majority of CMN effects observed in intraspecific seedling-adult combinations (Table 4.2). The situation is much more variable for CMN effects on interspecific pairs of similarly aged plants (Table 4.4) and this is also in accordance with the pattern observed for interspecific seedling-adult combinations (Table 4.2). The main reason for this contrasting CMN effect on intra- and interspecific pairs is that different species may differ widely in their inherent P acquisition efficiencies. As a consequence grasses and small seeded legumes will respond differently to colonization by AM fungi. Therefore, mycorrhiza will in general relax the competition between such species as exemplified by mixtures of *Lolium perenne* and *Lotus corniculatus*, or *Triticum aestivum* and *Vicia faba*, where the mycorrhiza-responsive legume is strongly suppressed by the grass when non-mycorrhizal (Scheublin et al. 2007; Qiao et al. 2015, respectively). Such impact of mycorrhiza on the relative performance of pairs of plant individuals differing markedly in mycorrhiza dependency will occur even without the formation of a CMN. Interestingly, mycorrhiza may also influence competition between species that respond only weakly to mycorrhiza when grown alone. Thus, competition between *Holcus lanatus* and *L. perenne* was balanced in the absence AM fungi whereas *L. perenne* was markedly suppressed when root were colonized by AM fungi (Fitter 1977). The suppression of *L. perenne* was associated with a mycorrhiza-induced suppression of root growth in particular. The possible role of EM interlinking these plants remains unknown but cannot be excluded.

In most studies in Table 4.4 the observed CMN effects were ascribed mainly to competition for nutrients, P in particular. Some authors suggested interplant C transfer to play a role (Marler et al. 1999; Callaway et al. 2001, 2003) but more recent evidence renders that possibility as less likely (see Sect. 4.4). The presence of EM links between similarly aged individuals of different plant species often remain unconfirmed and some of the results in Table 4.4 could have been caused by mycorrhiza-induced changes in morphology and uptake properties of individual root systems. However, if EM links exist and if we assume that the carbon source strength is a major determinant for the relative ability of plant individuals to obtain nutrients from the EM, then we can also assume this mechanism to play a role in CMN effects on competition between different species of the same age.

The relative C supply of individuals of different plant species to the built-up of an interlinking EM has rarely been measured or compared to the relative nutrient uptake from the EM pool. One exception is the C–P and C–N exchange studies in model intercropping system with sorghum and flax that was established in rhizoboxes with root-free compartments (Walder et al. 2012, 2015). EM in one compartment was used to quantify the relative C allocation from each plant using differences in their natural abundance of ^{13}C while a compartment with younger EM contained ^{33}P and ^{15}N for assessment of their relative uptake into each plant. The flax plants being most mycorrhiza responsive received the highest proportion of N and P from the EM while the highest proportion of C was derived from the less responsive sorghum plants. Flax thus appeared to benefit the most from the CMN despite its relatively lower C investment in the EM. Future work to understand the background for this presumably unbalanced exchange of minerals for carbon must ensure that the plant species involved are both in their active growth phase at time of measurement.

Overall, more work is needed and future experiments should take more advantage of the powerful tools of tracer isotopes and should systematically compare directions of transfer in plant combinations representing a range of age and size differences. Such studies are important to increase our understanding of belowground nutrient dynamics in intercropping systems which will most often represent CMNs. Nutrient allocation patterns in Walder et al. (2012) differed between CMNs with two different AM fungi and other CMN studies with a range of AM fungi, either alone or in combination, confirm that species identity of the fungus matters (Van der Heijden 2004; Scheublin et al. 2007; Wagg et al. 2011a, b). This illustrates that the function of CMNs will almost certainly depend on the identity of both plants and fungi and the function of CMNs will furthermore be influenced by the experimental and/or environmental conditions (see Hoeksema Chap. 9, this volume).

4.7 Role of CMN in Non-nutritional Transport Processes

A structured belowground network connecting plant individuals is known for mycorrhizal fungi only and may therefore be unique. Saprotrophic fungi would not be driven to grow from root to root but have a more random hyphal network in the soil probably from nutrient patch to nutrient patch. There is emerging evidence that CMNs also may facilitate the distribution of other substances than mineral nutrients in the soil. Song et al. (2010) found that tomato seedlings produced defense products when their neighbor plant was infected by a pathogen and connected within a CMN, while this did not happen without the presence of a CMN. Transport of such stress-induced signals between neighboring plants also appeared to occur in response to herbivore attack (Babikova et al. 2013; Song et al. 2014) and could significantly increase the performance of individuals within a population. Pollutants may also be spread by mycorrhizal fungi as demonstrated by AMF-mediated transfer of radiocesium between *Medicago* plants (Gyuricza et al. 2010). However, only traces could be found in shoots and most of the transferred ^{134}Cs remained in the roots of the receiver plant.

Furthermore, there is now evidence that a CMN can increase the distribution of herbicides and allelopathic substances, i.e. toxins that are produced by a plant to increase its own performance by reducing competition with neighboring plants (Barto et al. 2011). Such production of allelochemicals is common among a wide range of plant species (Lambers et al. 2008). Concentrations of different toxins in root-free soil compartments were increased by a non-disturbed CMN and this resulted in reduced growth of bioassay plants (Barto et al. 2011; Achatz et al. 2014). It is not yet clear whether mycorrhizal hyphae would transport the compounds within their hyphae, e.g. due to passive uptake and release, or if the hyphal network conducts passive water flow in the soil. If these findings prove to be general, a CMN would again amplify the performance of an already competitive plant individual. This also implies that CMN-induced growth depressions of seedlings might be caused by toxins rather than by competition for nutrients. Allelopathic compounds can be very different chemicals and little is known on their effect on AMF themselves. Some may be toxic to AMF as glucosinolate compounds released from garlic mustard have strong biocidal effects on AM fungi (Stinson et al. 2006; Cantor et al. 2011).

4.8 Conclusions, Perspectives and Research Needs

Nutrient dynamics in networks of plants interlinked by AM fungi have a strong impact on both groups of participants: A fungal mycelium can potentially obtain C from several host plants, which on the other hand have potential access to mineral nutrients in the mycelium. While this is essential for the fungus due to its biotrophic lifestyle, the impact on the plants is more complex. Plant individuals share the access to the CMN nutrient pool and its P in particular that is sparsely available in the soil solution. The relative ability of plants to take up P from the CMN pool will influence their relative growth performance. We have focused on nutrient dynamics in CMNs with adults and seedlings where the competitive relationships are most severe (e.g. Weiner 1990; Cook and Ratcliff 1984) and we conclude that more than half of the relevant studies demonstrate that the adult-to-seedling competition is amplified and not relaxed as previously suggested (see e.g. Read 1991). It is also clear that a CMN-induced amplification of the competition is more frequent in intraspecific than in interspecific seedling-adult combinations. Importantly, the same conclusions can be drawn for CMN-effects on competition between plants of similar age: The competition between individuals is amplified in plant populations and either amplified or relaxed in interspecific combinations. Furthermore, the competitive outcome will be modulated by inherent characteristics of each plant and by the environmental conditions.

We propose the following model for nutrient dynamics in CMNs of AM fungi and their host plants. Here we consider simple networks with two or more plant individuals connected by a single AM fungus, but the model could be extended and tested also for more complex networks involving connective mycelia of different AM fungal genotypes.

1. The most important feature of CMNs is their pool of nutrients in the connective mycelia. The ability of the interconnected plants to access this pool will influence plant competition. Plant-to-plant transfer of nutrients via connective AM fungal mycelia does not occur in ecologically relevant amounts.
2. The connective EM is fed with C primarily from large, older plants that usually have an unrealised capacity for photosynthesis.
3. The P pool in the ERM is directed towards these plants representing major C source strengths.
4. Young plant individuals linking into the CMN are smaller and have a marginal C source strength, deliver relatively little C to the EM and receive only little P from the EM pool.
5. Such intraspecific plant size-dependent competition for the EM nutrient pool also dominates in plant populations with a minor initial variation in size of individuals. The presence of a CMN can promote asymmetric below-ground competition to the benefit of the initially larger individuals.
6. CMN-induced suppression of plant individuals is temporary and will shift to a typical mycorrhizal growth response when the larger plants become senescent or are grazed.

7. CMN effects on competition between different plant species of different or similar age is much more variable due to species dependent differences in e.g. responsiveness to mycorrhiza.

The various steps in the model can be tested in mesocosms with CMNs of increasing complexity. Tracer isotopes will be important tools to monitor the movement of specific nutrients in the network (see Walder et al. 2012; Merrild et al. 2013), ideally by dual use of tracers for carbon and mineral nutrients. In general, there is an urgent need to include analysis of nutrient concentrations in plants and in soil compartments in CMN studies. More than one AM fungus should be included in the CMNs in order to investigate how (a) the fungal identity and (b) a choice of fungal partners will influence plant competition. Compartmented rhizoboxes and dual labelling with ^{32}P and ^{33}P will aid the latter studies. There is also a need to design the experiments such that the possible interaction of effects from plant allelochemicals can be elucidated. Several control treatments are required for the proper interpretation of results. Growth in the CMN should be compared to the absence of AM fungi and to mycorrhizal responsiveness of solitary plants. Furthermore, either non-host plants or membranes inserted in the soil can be used to distinguish EM-mediated nutrient transfer from nutrient transport by mass flow and diffusion.

Future research must pay more attention to the nutrient dynamics of CMNs under field conditions. Translocation of nutrients from tracer isotope-labelled soil patches (Schweiger and Jakobsen 1999; Johnson et al. 2001) to individual plants should be studied in CMNs that have been mapped by e.g. high throughput sequencing (Öpik et al. 2009; Montesinos-Navarro et al. 2012). Analysis of changes in nutrient dynamics upon the experimental removal of a highly connected individual (a hub) would most likely help us to better understand the role of CMNs in plant communities and populations.

Acknowledgments This work was supported by the Danish Council for Independent Research (Technology and Production Sciences), grant 0602-01412B (IJ) and by The Swedish Research Council Formas (EH).

References

Achatz M, Morris EK, Müller F, Hilker M, Rillig MC (2014) Soil hypha-mediated movement of allelochemicals: arbuscular mycorrhizae extend the bioactive zone of juglone. Funct Ecol 28:1020–1029

Allsopp N, Stock WD (1992) Density dependent interactions between VA mycorrhizal fungi and even-aged seedlings of 2 perennial fabaceae species. Oecologia 91:281–287

Asghari HR, Cavagnaro TR (2012) Arbuscular mycorrhizas reduce nitrogen loss via leaching. PLoS ONE 7:151–155

Asghari HR, Chittleborough DJ, Smith FA, Smith SE (2005) Influence of arbuscular mycorrhizal (AM) symbiosis on phosphorus leaching through soil cores. Plant Soil 275:181–193

Ayres RL, Gange AC, Aplin DM (2006) Interactions between arbuscular mycorrhizal fungi and intraspecific competition affect size, and size inequality, of *Plantago lanceolata* L. J Ecol 94:285–294

Babikova Z, Gilbert L, Bruce TJA, Birkett M, Caulfield JC, Woodcock C, Pickett JA, Johnson D (2013) Underground signals carried through common mycelial networks warn neighbouring plants of aphid attack. Ecol Lett 16:835–843

Bago B, Azconaguilar C, Goulet A, Piche Y (1998) Branched absorbing structures (BAS): a feature of the extraradical mycelium of symbiotic arbuscular mycorrhizal fungi. New Phytol 139:375–388

Balzergue C, Chabaud M, Barker DG, Bécard G, Rochange SF (2013) High phosphate reduces host ability to develop arbuscular mycorrhizal symbiosis without affecting root calcium spiking responses to the fungus. Front Plant Sci 4(426):1–15

Barto EK, Hilker M, Mueller F, Mohney BK, Weidenhamer JD, Rillig MC (2011) The fungal fast lane: common mycorrhizal networks extend bioactive zones of allelochemicals in soils. PLoS ONE 6(11):1–7

Beiler KJ, Durall DM, Simard SW, Maxwell SA, Kretzer AM (2010) Architecture of the wood-wide web: *Rhizopogon* spp. Genets link multiple Douglas-fir cohorts. New Phytol 185:543–553

Bever JD, Richardson SC, Lawrence BM, Holmes J, Watson M (2009) Preferential allocation to beneficial symbiont with spatial structure maintains mycorrhizal mutualism. Ecol Lett 12:13–21

Bever JD, Dickie IA, Facelli E, Facelli JM, Klironomos J, Moora M, Rillig MC, Stock WD, Tibbett M, Zobel M (2010) Rooting theories of plant community ecology in microbial interactions. Trends Ecol Evol 25:468–478

Birch CPD (1986) Development of VA mycorrhizal infection in seedlings in semi-natural grassland turf. In: Gianinazzi-Pearson V, Gianinazzi S (eds) Proceedings of the 1st European symposium on mycorrhizas. INRA, Paris, pp 233–239

Breuillin F, Schramm J, Hajirezaei M, Ahkami A, Favre P, Druege U, Hause B, Bucher M, Kretzschmar T, Bossolini E, Kuhlemeier C, Martinoia E, Franken P, Scholz U, Reinhardt D (2010) Phosphate systemically inhibits development of arbuscular mycorrhiza in *Petunia hybrida* and represses genes involved in mycorrhizal functioning. Plant J 64:1002–1017

Bücking H, Shachar-Hill Y (2005) Phosphate uptake, transport and transfer by the arbuscular mycorrhizal fungus *Glomus intraradices* is stimulated by increased carbohydrate availability. New Phytol 165:899–912

Burke DJ (2012) Shared mycorrhizal networks of forest herbs: does the presence of conspecific and heterospecific adult plants affect seedling growth and nutrient acquisition? Botany 90:1048–1057

Bürkert B, Robson A (1994) Zn-65 uptake in subterranean clover (*Trifolium subterraneum* L) by 3 vesicular-arbuscular mycorrhizal fungi in a root-free sandy soil. Soil Biol Biochem 26:1117–1124

Buwalda JG (1980) Growth of a clover-ryegrass association with vesicular arbuscular mycorrhizas. N Z J Agric Res 23:379–383

Callaway RM, Newingham B, Zabinski CA, Mahall BE (2001) Compensatory growth and competitive ability of an invasive weed are enhanced by soil fungi and native neighbours. Ecol Lett 4:429–433

Callaway RM, Mahall BE, Wicks C, Pankey J, Zabinski C (2003) Soil fungi and the effects of an invasive forb on grasses: Neighbor identity matters. Ecology 84:129–135

Cantor A, Hale A, Aaron J, Traw MB, Kalisz S (2011) Low allelochemical concentrations detected in garlic mustard-invaded forest soils inhibit fungal growth and AMF spore germination. Biol Invasions 13:3015–3025

Carey PD, Fitter AH, Watkinson AR (1992) A field study using the fungicide benomyl to investigate the effect of mycorrhizal fungi on plant fitness. Oecologia 90:550–555

Carey EV, Marler MJ, Callaway RM (2004) Mycorrhizae transfer carbon from a native grass to an invasive weed: evidence from stable isotopes and physiology. Plant Ecol 172:133–141

Cavagnaro TR, Smith FA, Smith SE, Jakobsen I (2005) Functional diversity in arbuscular mycorrhizas: exploitation of soil patches with different phosphate enrichment differs among fungal species. Plant Cell Environ 28:642–650

Chagnon PL, Bradley RL, Klironomos JN (2012) Using ecological network theory to evaluate the causes and consequences of arbuscular mycorrhizal community structure. New Phytol 194:307–312

Chiariello N, Hickman JC, Mooney HA (1982) Endomycorrhizal role for interspecific transfer of Phosphorus in a community of annual plants. Science 217:941–943

Collins CD, Foster BL (2009) Community-level consequences of mycorrhizae depend on phosphorus availability. Ecology 90:2567–2576

Cook SJ, Ratcliff D (1984) A study of the effects of root and shoot competition on the growth of green panic (*Panicum maximum* var Trichoglume) seedlings in an existing grassland using root exclusion tubes. J Appl Ecol 21:971–982

Correa A, Gurevitch J, Martins-Loucao MA, Cruz C (2012) C allocation to the fungus is not a cost to the plant in ectomycorrhizae. Oikos 121:449–463

Courty PE, Walder F, Boller T, Ineichen K, Wiemken A, Rousteau A, Selosse MA (2011) Carbon and nitrogen metabolism in mycorrhizal networks and mycoheterotrophic plants of tropical forests: a stable isotope analysis. Plant Physiol 156:952–961

Daisog H, Sbrana C, Cristani C, Moonen A-C, Giovannetti M, Bàrberi P (2012) Arbuscular mycorrhizal fungi shift competitive relationships among crop and weed species. Plant Soil 353:395–408

Derelle D, Declerck S, Genet P, Dajoz I, Van Aarle IM (2012) Association of highly and weakly mycorrhizal seedlings can promote the extra- and intraradical development of a common mycorrhizal network. FEMS Microbiol Ecol 79:251–259

Drew MC (1975) Comparison of effects of a localized supply of phosphate, nitrate, ammonium and potassium on growth of seminal root system, and shoot, in barley. New Phytol 75:479–490

Eason WR, Newman EI, Chuba PN (1991) Specificity of interplant cycling of phosphorus—The role of mycorrhizas. Plant Soil 137:267–274

Eissenstat DM, Newman EI (1990) Seedling establishment near large plants—Effects of vesicular-arbuscular mycorrhizas on the intensity of plant competition. Funct Ecol 4:95–99

Endlweber K, Scheu S (2007) Interactions between mycorrhizal fungi and Collembola: effects on root structure of competing plant species. Biol Fertil Soils 43:741–749

Facelli E, Facelli JM (2002) Soil phosphorus heterogeneity and mycorrhizal symbiosis regulate plant intra-specific competition and size distribution. Oecologia 133:54–61

Facelli E, Facelli JM, Smith SE, McLaughlin MJ (1999) Interactive effects of arbuscular mycorrhizal symbiosis, intraspecific competition and resource availability on *Trifolium subterraneum* cv. Mt. Barker. New Phytol 141:535–547

Fellbaum CR, Gachomo EW, Beesetty Y, Choudhari S, Strahan GD, Pfeffer PE, Kiers ET, Bucking H (2012) Carbon availability triggers fungal nitrogen uptake and transport in arbuscular mycorrhizal symbiosis. P Natl Acad Sci USA 109:2666–2671

Fellbaum CR, Mensah JA, Cloos AJ, Strahan GE, Pfeffer PE, Kiers ET, Bücking H (2014) Fungal nutrient allocation in common mycorrhizal networks is regulated by the carbon source strength of individual host plants. New Phytol 203:646–656

Finlay RD, Read DJ (1986a) The structure and function of the vegetative mycelium of ectomycorrhizal plants 1. Translocation of ^{14}C labeled carbon between plants interconnected by a common mycelium. New Phytol 103:143–156

Finlay RD, Read DJ (1986b) The structure and function of the vegetative mycelium of ectomycorrhizal plants 2. The uptake and distribution of phosphorus by mycelial strands interconnecting host plants. New Phytol 103:157–165

Fitter AH (1977) Influence of mycorrhizal infection on competition for phosphorus and potassium by 2 grasses. New Phytol 79:119–125

Fitter AH (2006) What is the link between carbon and phosphorus fluxes in arbuscular mycorrhizas? A null hypothesis for symbiotic function. New Phytol 172:3–6

Francis R, Read DJ (1984) Direct transfer of carbon between plants connected by mycorrhizal mycelium. Nature 307:53–56
Francis R, Read DJ (1995) Mutualism and antagonism in the mycorrhizal symbiosis, with special reference to impacts on plant community structure. Can J Bot 73:1301–1309
Frey B, Schüepp H (1993) A role of vesicular-arbuscular (VA) mycorrhizal fungi in facilitating interplant nitrogen transfer. Soil Biol Biochem 25:651–658
Gange AC, Brown VK, Sinclair GS (1993) Vesicular arbuscular mycorrhizal fungi—A determinant of plant community structure in early succession. Funct Ecol 7:616–622
Gebauer G, Meyer M (2003) ^{15}N and ^{13}C natural abundance of autotrophic and myco-heterotrophic orchids provides insight into nitrogen and carbon gain from fungal associations. New Phytol 160:209–223
Giovannetti M, Sbrana C, Avio L, Strani P (2004) Patterns of below-ground plant interconnections established by means of arbuscular mycorrhizal networks. New Phytol 164:175–181
Giovannetti M, Avio L, Fortuna P, Pellegrino E, Sbrana C, Strani P (2006) At the root of the wood wide web: self recognition and non-self incompatibility in mycorrhizal networks. Plant Signal Behavior 1:1–5
Govindarajulu M, Pfeffer PE, Jin HR, Abubaker J, Douds DD, Allen JW, Bucking H, Lammers PJ, Shachar-Hill Y (2005) Nitrogen transfer in the arbuscular mycorrhizal symbiosis. Nature 435:819–823
Grime JP, Mackey JML, Sillier SH, Read DJ (1987) Floristic diversity in a model system using experimental microcosms. Nature 328:420–422
Grønlund M, Albrechtsen M, Johansen IE, Hammer E, Nielsen TH, Jakobsen I (2013) The interplay between P uptake pathways in mycorrhizal peas: a combined physiological and gene-silencing approach. Physiol Plant 149:234–248
Gyuricza V, Thiry Y, Wannijn J, Declerck S, de Boulois HD (2010) Radiocesium transfer between *Medicago truncatula* plants via a common mycorrhizal network. Environ Microbiol 12:2180–2189
Hall IR (1978) Effects of endomycorrhizas on the competitive ability of white clover. N Z J Agric Res 21:509–515
Hamel C, Furlan V, Smith DL (1992) Mycorrhizal effects on interspecific plant competition and nitrogen transfer in legume-grass mixtures. Crop Sci 32:991–996
Hammer EC, Pallon J, Wallander H, Olsson PA (2011) Tit for tat? A mycorrhizal fungus accumulates phosphorus under low plant carbon availability. FEMS Microbiol Ecol 76:236–244
Harrison MJ, Van Buuren ML (1995) A phosphate transporter from the mycorrhizal fungus *Glomus versiforme*. Nature 378:626–629
Harrison MJ, Dewbre GR, Liu J (2002) A phosphate transporter from *Medicago truncatula* involved in the acquisition of phosphate released by arbuscular mycorrhizal fungi. Plant Cell 14:1–17
Hartnett DC, Hetrick BAD, Wilson GWT, Gibson DJ (1993) Mycorrhizal influence on intraspecific and interspecific neighbor interactions among co-occurring prairie grasses. J Ecol 81:787–795
He XH, Xu MG, Qiu GY, Zhou JB (2009) Use of N-15 stable isotope to quantify nitrogen transfer between mycorrhizal plants. J Plant Ecol 2:107–118
Heap AJ, Newman EI (1980) The influence of vesicular-arbuscular mycorrhizas on phosphorus transfer between plants. New Phytol 85:173–179
Helber N, Wippel K, Sauer N, Schaarschmidt S, Hause B, Requena N (2011) A versatile monosaccharide transporter that operates in the arbuscular mycorrhizal fungus glomus sp is crucial for the symbiotic relationship with plants. Plant Cell 23:3812–3823
Helgason T, Fitter AH (2009) Natural selection and the evolutionary ecology of the arbuscular mycorrhizal fungi (Phylum Glomeromycota). J Exp Bot 60:2465–2480
Hetrick BAD, Wilson GWT, Hartnett DC (1989) Relationship between mycorrhizal dependence and competitive ability of two tallgrass prairie grasses. Can J Bot 67:2608–2615

Hetrick BAD, Hartnett DC, Wilson GWT, Gibson DJ (1994) Effects of mycorrhizae, phosphorus availability, and plant-density on yield relationships among competing tallgrass prairie grasses. Can J Bot 72:168–176

Jakobsen I, Rosendahl L (1990) Carbon flow into soil and external hyphae from roots of mycorrhizal cucumber plants. New Phytol 115:77–83

Jakobsen I, Abbott LK, Robson AD (1992a) External hyphae of vesicular-arbuscular mycorrhizal fungi associated with *Trifolium subterraneum* L. 2. Hyphal transport of 32P over defined distances. New Phytol 120:509–516

Jakobsen I, Abbott LK, Robson AD (1992b) External hyphae of vesicular-arbuscular mycorrhizal fungi associated with *Trifolium subterraneum*. 1: Spread of hyphae and phosphorus inflow into roots. New Phytol 120:371–380

Janos DP, Scott J, Aristizábal C, Bowman DMJS (2013) Arbuscular-Mycorrhizal networks inhibit *Eucalyptus tetrodonta* seedlings in rain forest soil microcosms. PLoS ONE 8(1–11):e57716

Janouskova M, Rydlova J, Puschel D, Szakova J, Vosatka M (2011) Extraradical mycelium of arbuscular mycorrhizal fungi radiating from large plants depresses the growth of nearby seedlings in a nutrient deficient substrate. Mycorrhiza 21:641–650

Jansa J, Mozafar A, Frossard E (2003) Long-distance transport of P and Zn through the hyphae of an arbuscular mycorrhizal fungus in symbiosis with maize. Agronomie 23:481–488

Javot H, Penmetsa RV, Terzaghi N, Cook DR, Harrison MJ (2007a) A *Medicago truncatula* phosphate transporter indispensable for the arbuscular mycorrhizal symbiosis. P Natl Acad Sci USA 104:1720–1725

Javot H, Pumplin N, Harrison MJ (2007b) Phosphate in the arbuscular mycorrhizal symbiosis: transport properties and regulatory roles. Plant Cell Environ 30:310–322

Jin L, Zhang G, Wang X, Dou C, Chen M, Lin S, Li Y (2011) Arbuscular mycorrhiza regulate inter-specific competition between a poisonous plant, *Ligularia virgaurea*, and a co-existing grazing grass, *Elymus nutans*, in Tibetan Plateau Alpine meadow ecosystem. Symbiosis 55:29–38

Johansen A, Jensen ES (1996) Transfer of n and p from intact or decomposing roots of pea to barley interconnected by an arbuscular mycorrhizal fungus. Soil Biol Biochem 28:73–81

Johansen A, Jakobsen I, Jensen ES (1992) Hyphal transport of ^{15}N labelled nitrogen by a vesicular—arbuscular mycorrhizal fungus and its effect on depletion of inorganic soil-N. New Phytol 122:281–288

Johansen A, Jakobsen I, Jensen ES (1993) External hyphae of vesicular-arbuscular mycorrhizal fungi associated with *Trifolium subterraneum* L.3. Hyphal transport of ^{32}P and ^{15}N. New Phytol 124:61–68

Johnson D, Leake JR, Read DJ (2001) Novel in-growth core system enables functional studies of grassland mycorrhizal mycelial networks. New Phytol 152:555–562

Johnson D, Krsek M, Wellington EMH, Stott AW, Cole L, Bardgett RD, Read DJ, Leake JR (2005) Soil invertebrates disrupt carbon flow through fungal networks. Science 309:1047

Joner EJ, Leyval C (2001) Influence of arbuscular mycorrhiza on clover and ryegrass grown together in a soil spiked with polycyclic aromatic hydrocarbons. Mycorrhiza 10:155–159

Kaschuk G, Kuyper TW, Leffelaar PA, Hungria M, Giller KE (2009) Are the rates of photosynthesis stimulated by the carbon sink strength of rhizobial and arbuscular mycorrhizal symbioses? Soil Biol Biochem 41:1233–1244

Kaschuk G, Leffelaar PA, Giller KE, Alberton O, Hungria M, Kuyper TW (2010) Responses of legumes to rhizobia and arbuscular mycorrhizal fungi: a meta-analysis of potential photosynthate limitation of symbioses. Soil Biol Biochem 42:125–127

Kiers ET, Van der Heijden MGA (2006) Mutualistic stability in the arbuscular mycorrhizal symbiosis: exploring hypotheses of evolutionary cooperation. Ecology 87:1627–1636

Kiers ET, Duhamel M, Beesetty Y, Mensah JA, Franken O, Verbruggen E, Fellbaum CR, Kowalchuk GA, Hart MM, Bago A, Palmer TM, West SA, Vandenkoornhuyse P, Jansa J, Bucking H (2011) Reciprocal rewards stabilize cooperation in the mycorrhizal symbiosis. Science 333:880–882

Koch KE, Johnson CR (1984) Photosynthate partitioning in split-root citrus seedlings with mycorrhizal and nonmycorrhizal root systems. Plant Physiol 75:26–30

Koide RT, Dickie IA (2002) Effects of mycorrhizal fungi on plant populations. Plant Soil 244:307–317

Koide RT, Li M (1991) Mycorrhizal fungi and the nutrient ecology of three oldfield annual plant species. Oecologia 85:403–412

Kytoviita MM, Vestberg M, Tuom J (2003) A test of mutual aid in common mycorrhizal networks: established vegetation negates benefit in seedlings. Ecology 84:898–906

Lambers H, Chapin FS III, Pons TL (2008) Plant physiological ecology. Springer, Berlin 610 pp

Leake JR, Cameron DD (2010) Physiological ecology of mycoheterotrophy. New Phytol 185:601–605

Leake JR, Johnson D, Donnelly DP, Muckle GE, Boddy L, Read DJ (2004) Networks of power and influence: the role of mycorrhizal mycelium in controlling plant communities and agroecosystem functioning. Can J Bot 82:1016–1045

Leigh J, Hodge A, Fitter AH (2009) Arbuscular mycorrhizal fungi can transfer substantial amounts of nitrogen to their host plant from organic material. New Phytol 181:199–207

Lekberg Y, Hammer EC, Olsson PA (2010) Plants as resource islands and storage units—Adopting the mycocentric view of arbuscular mycorrhizal networks. FEMS Microbiol Ecol 74:336–345

Lendenmann M, Thonar C, Barnard RL, Salmon Y, Werner RA, Frossard E, Jansa J (2011) Symbiont identity matters: carbon and phosphorus fluxes between *Medicago truncatula* and different arbuscular mycorrhizal fungi. Mycorrhiza 21:689–702

Lerat S, Gauci R, Catford JG, Vierheilig H, Piche Y, Lapointe L (2002) C-14 transfer between the spring ephemeral Erythronium americanum and sugar maple saplings via arbuscular mycorrhizal fungi in natural stands. Oecologia 132:181–187

Li YF, Ran W, Zhang RP, Sun SB, Xu GH (2009) Facilitated legume nodulation, phosphate uptake and nitrogen transfer by arbuscular inoculation in an upland rice and mung bean intercropping system. Plant Soil 315:285–296

Malcova R, Albrechtova J, Vosatka M (2001) The role of the extraradical mycelium network of arbuscular mycorrhizal fungi on the establishment and growth of *Calamagrostis epigejos* in industrial waste substrates. Appl Soil Ecol 18:129–142

Maldonado-Mendoza IE, Dewbre GR, Harrison MJ (2001) A phosphate transporter gene from the extra-radical mycelium of an arbuscular mycorrhizal fungus *Glomus intraradices* is regulated in response to phosphate in the environment. Mol Plant-Microb Interact 14:1140–1148

Mariotte P, Meugnier C, Johnson D, Thébault A, Spiegelberger T, Buttler A (2013) Arbuscular mycorrhizal fungi reduce the differences in competitiveness between dominant and subordinate plant species. Mycorrhiza 23:267–277

Marler MJ, Zabinski CA, Callaway RM (1999) Mycorrhizae indirectly enhance competitive effects of an invasive forb on a native bunchgrass. Ecology 80:1180–1186

McKendrick SL, Leake JR, Read DJ (2000) Symbiotic germination and development of myco-heterotrophic plants in nature: transfer of carbon from ectomycorrhizal Salix repens and Betula pendula to the orchid Corallorhiza trifida through shared hyphal connections. New Phytol 145:539–548

Merckx V, Bidartondo MI, Hynson NA (2009) Myco-heterotrophy: when fungi host plants. Ann Bot 104:1255–1261

Merckx V, Stöckel M, Fleischmann A, Bruns TD, Gebauer G (2010) ^{15}N and ^{13}C natural abundance of two mycoheterotrophic and a putative partially mycoheterotrophic species associalted with arbuscular mycorrhizal fungi. New Phytol 188:590–596

Merrild M, Ambus P, Rosendahl S, Jakobsen I (2013) Common arbuscular mycorrhizal networks amplify competition for phosphorus between seedlings and established plants. New Phytol 200:229–240

Mikkelsen BL, Rosendahl S, Jakobsen I (2008) Underground resource allocation between individual networks of mycorrhizal fungi. New Phytol 180:890–898

Miller RM, Miller SP, Jastrow JD, Rivetta CB (2002) Mycorrhizal mediated feedbacks influence net carbon gain and nutrient uptake in *Andropogon gerardii*. New Phytol 155:149–162
Montesinos-Navarro A, Segarra-Moragues JG, Valiente-Banuet A, Verdu M (2012) The network structure of plant-arbuscular mycorrhizal fungi. New Phytol 194:536–547
Moora M, Zobel M (1996) Effect of arbuscular mycorrhiza on inter- and intraspecific competition of two grassland species. Oecologia 108:79–84
Moora M, Zobel M (1998) Can arbuscular mycorrhiza change the effect of root competition between conspecific plants of different ages? Can J Bot 76:613–619
Moora M, Zobel M (2010) Arbuscular mycorrhizae and plant-plant interactions. Impact of invisible world on visible patterns. In: Pugnaire F (ed) Positive plant interactions and community dynamics. CRC Press, Boca Raton, pp79–98
Mortimer PE, Perez-Fernandez MA, Valentine AJ (2008) The role of arbuscular mycorrhizal colonization in the carbon and nutrient economy of the tripartite symbiosis with nodulated *Phaseolus vulgaris*. Soil Biol Biochem 40:1019–1027
Nakano-Hylander A, Olsson PA (2007) Carbon allocation in mycelia of arbuscular mycorrhizal fungi during colonisation of plant seedlings. Soil Biol Biochem 39:1450–1458
Newman EI (1988) Mycorrhizal links between plants—Their functioning and ecological significance. Adv Ecol Res 18:243–270
Newman EI, Eason WR (1989) Cycling of nutrients from dying roots to living plants, including the role of mycorrhizas. Plant Soil 115:211–215
Newman EI, Ritz K (1986) Evidence on the pathways of phosphorus transfer between vesicular-arbuscular mycorrhizal plants. New Phytol 104:77–87
Newsham KK, Fitter AH, Watkinson AR (1995) Arbuscular mycorrhiza protect an annual grass from root pathogenic fungi in the field. J Ecol 83:991–1000
Nielsen JS, Joner EJ, Declerck S, Olsson S, Jakobsen I (2002) Phospho-imaging as a tool for visualisation and nonnvasive measurement of P transport dynamics in arbuscular mycorrhizas. New Phytol 154:809–820
Ocampo JA (1986) Vesicular arbuscular mycorrhizal infection of host and nonhost plants—Effect on the growth-responses of the plants and competition between them. Soil Biol Biochem 18:607–610
Olsson PA, Baath E, Jakobsen I (1997) Phosphorus effects on the mycelium and storage structures of an arbuscular mycorrhizal fungus as studied in the soil and roots by analysis of fatty acid signatures. Appl Environ Microbiol 63:3531–3538
Olsson PA, Jakobsen I, Wallander H (2002) Foraging and resource allocation strategies of mycorrhizal fungi in a patchy environment. In: Van der Heijden MGA, Sanders IR (eds) Mycorrhizal ecology. Springer, Berlin, pp 93–116
Olsson PA, Hansson MC, Burleigh SH (2006) Effect of P availability on temporal dynamics of carbon allocation and *Glomus intraradices* high-affinity P transporter gene induction in arbuscular mycorrhiza. Appl Environ Microbiol 72:4115–4120
Olsson PA, Hammer EC, Wallander H, Pallon J (2008) Phosphorus availability influences elemental uptake in the mycorrhizal fungus *Glomus intraradices*, as revealed by particle-induced X-ray emission analysis. Appl Environ Microbiol 74:4144–4148
Olsson PA, Hammer EC, Pallon J, Van Aarle IM, Wallander H (2011) Elemental composition in vesicles of an arbuscular mycorrhizal fungus, as revealed by PIXE analysis. Fung Biol 115:643–648
Olsson O, Olsson PA, Hammer EC (2014) Phosphorus and carbon availability regulate structural composition and complexity of AM fungal mycelium. Mycorrhiza. doi:10.1007/s00572-014-0557-8
Öpik M, Metsis M, Daniell TJ, Zobel M, Moora M (2009) Large-scale parallel 454 sequencing reveals host ecological group specificity of arbuscular mycorrhizal fungi in a boreonemoral forest. New Phytol 184:424–437
Paul EA, Kucey RMN (1981) Carbon flow in plant microbial associations. Science 213:473–474
Pearson JN, Jakobsen I (1993a) Symbiotic exchange of carbon and phosphorus between cucumber and three arbuscular mycorrhizal fungi. New Phytol 124:481–488

Pearson JN, Jakobsen I (1993b) The relative contribution of hyphae and roots to phosphorus uptake by arbuscular mycorrhizal plants, measured by dual labelling with ^{32}P and ^{33}P. New Phytol 124:489–494

Pfeffer PE, Douds DD, Bucking H, Schwartz DP, Shachar-Hill Y (2004) The fungus does not transfer carbon to or between roots in an arbuscular mycorrhizal symbiosis. New Phytol 163:617–627

Pietikainen A, Kytoviita MM (2007) Defoliation changes mycorrhizal benefit and competitive interactions between seedlings and adult plants. J Ecol 95:639–647

Püschel D, Rydlova J, Vosatka M (2007) Mycorrhiza influences plant community structure in succession on spoil banks. Basic Appl Ecol 8:510–520

Qiao X, Bei S, Li C, Dong Y, Li H, Christie P, Zhang F, Zhang J (2015) Enhancement of faba bean competitive ability by arbuscular mycorrhizal fungi is highly correlated with dynamic nutrient acquisition by competing wheat. Sci Rep 5:8122

Ravnskov S, Jakobsen I (1995) Functional compatibility in arbuscular mycorrhizas measured as hyphal P transport to the plant. New Phytol 129:611–618

Ravnskov S, Wu Y, Graham JH (2003) Arbuscular mycorrhizal fungi differentially affect expression of genes coding for sucrose synthases in maize roots. New Phytol 157:539–545

Read DJ (1991) Mycorrhizas in ecosystems. Experientia 47:376–391

Read DJ, Birch CPD (1988) The effects and implications of disturbance of mycorrhizal mycelial systems. P Roy Soc Edinb B 94:13–24

Rinaudo V, Bàrberi P, Giovannetti M, Van der Heijden MGA (2010) Mycorrhizal fungi suppress aggressive agricultural weeds. Plant Soil 333:7–20

Robinson D, Fitter A (1999) The magnitude and control of carbon transfer between plants linked by a common mycorrhizal network. J Exp Bot 50:9–13

Rosendahl S (2008) Communities, populations and individuals of arbuscular mycorrhizal fungi. New Phytol 178:253–266

Schaarschmidt S, Gonzalez M-C, Roitsch T, Strack D, Sonnewald U, Hause B (2007) Regulation of arbuscular mycorrhization by carbon. The symbiotic interaction cannot be improved by increased carbon availability accomplished by root-specifically enhanced invertase activity. Plant Physiol 143:1827–1840

Scheublin TR, Van Logtestijn RSP, Van der Heijden MGA (2007) Presence and identity of arbuscular mycorrhizal fungi influence competitive interactions between plant species. J Ecol 95:631–638

Schnepf A, Roose T, Schweiger P (2008) Impact of growth and uptake patterns of arbuscular mycorrhizal fungi on plant phosphorus uptake—A modelling study. Plant Soil 312:85–99

Schroeder-Moreno MS, Janos DP (2008) Intra- and inter-specific density affects plant growth responses to arbuscular mycorrhizas. Botany 86:1180–1193

Schweiger PF, Jakobsen I (1999) Direct measurement of arbuscular mycorrhizal phosphorus uptake into field-grown winter wheat. Agron J 91:998–1002

Shumway DL, Koide RT (1995) Size and reproductive inequality in mycorrhizal and nonmycorrhizal populations of *Abutilon theophrasti*. J Ecol 83:613–620

Simard SW, Perry DA, Jones MD, Myrold DD, Durall DM, Molina R (1997) Net transfer of carbon between ectomycorrhizal tree species in the field. Nature 388:579–582

Simard SW, Beiler KJ, Bingham MA, Deslippe JR, Philip LJ, Teste FP (2012) Mycorrhizal networks: mechanisms, ecology and modelling. Fung Biol Rev 26:39–60

Smith SE, Read DJ (2008) Mycorrrhizal symbiosis, 3rd edn. Academic Press, London

Smith SE, Smith FA (2012) Fresh perspectives on the roles of arbuscular mycorrhizal fungi in plant nutrition and growth. Mycologia 104:1–13

Smith FA, Jakobsen I, Smith SE (2000) Spatial differences in acquisition of soil phosphate between two arbuscular mycorrhizal fungi in symbiosis with *Medicago truncatula*. New Phytol 147:357–366

Smith SE, Smith FA, Jakobsen I (2003) Mycorrhizal fungi can dominate phosphate supply to plants irrespective of growth responses. Plant Physiol 133:16–20

Smith SE, Smith FA, Jakobsen I (2004) Functional diversity in arbuscular mycorrhizal (AM) symbioses: the contribution of the mycorrhizal P uptake pathway is not correlated with mycorrhizal responses in growth or total P uptake. New Phytol 162:511–524

Snaydon RW, Howe CD (1986) Root and shoot competition between established ryegrass and invading grass seedlings. J Appl Ecol 23:667–674

Solaiman MZ, Saito A (2001) Phosphate efflux from intraradical hyphae of *Gigaspora margarita* in vitro and its implication for phosphorus translocation. New Phytol 151:525–533

Song Y, Zeng RS, Xu JAF, Li J, Shen XA, Yihdego WG (2010) Interplant communication of tomato plants through underground common mycorrhizal networks. PLoS ONE 5:e13324

Song YY, Ye M, Li C, He X, Zhu-Salzman K, Wang RL, Su YJ, Luo SM, Zeng RS (2014) Hijacking common mycorrhizal networks for herbivore-induced defence signal transfer between tomato plants. Scientific Reports 4:3915

Stampe ED, Daehler CC (2003) Mycorrhizal species identity affects plant community structure and invasion: a microcosm study. Oikos 100:362–372

Stinson KA, Campbell SA, Powell JR, Wolfe BE, Callaway RM, Thelen GC, Hallett SG, Prati D, Klironomos JN (2006) Invasive plant suppresses the growth of native tree seedlings by disrupting belowground mutualisms. PLoS Biol 4:727–731

Sylvia DM, Alagely AK, Chellemi DO, Demchenko LW (2001) Arbuscular mycorrhizal fungi influence tomato competition with bahaiagrass. Biol Fertil Soils 34:448–452

Thonar C, Schnepf A, Frossard E, Roose T, Jansa J (2011) Traits related to differences in function among three arbuscular mycorrhizal fungi. Plant Soil 339:231–245

Toju H, Guimaraes PR, Olesen JM, Thompson JN (2014) Assembly of complex plant-fungus networks. Nat Comm 5:5273

Torrecillas E, Alguacil MD, Roldan A, Diaz G, Montesinos-Navarro A, Torres MP (2014) Modularity reveals the tendency of arbuscular mycorrhizal fungi to interact differently with generalist and specialist plant species in gypsum soils. Appl Environ Microbiol 80:5457–5466

Uetake Y, Kojima T, Ezawa T, Saito M (2002) Extencive tubular vacuolesystem in an arbuscular mycorrhizal fungus, *Gigaspora margarita*. New Phytol 154:761–768

Van der Heijden MGA, Klironomos JN, Ursic M, Moutoglis P, Streitwolf-Engel R, Boller T, Wiemken A, Sanders IR. (1998) Mycorrhizal fungal diversity determines plant biodiversity, ecosystem variability and productivity. Nature 396:69–72

Van der Heijden MGA (2004) Arbuscular mycorrhizal fungi as support systems for seedling establishment in grassland. Ecol Letters 7:293–303

Van der Heijden MGA (2010) Mycorrhizal fungi reduce nutrient loss from model grassland ecosystems. Ecology 91:1163–1171

Van der Heijden MGA, Horton TR (2009) Socialism in soil? The importance of mycorrhizal fungal networks for facilitation in natural ecosystems. J Ecol 97:1139–1150

Veiga RSL, Jansa J, Frossard E, Van der Heijden MGA (2011) Can arbuscular mycorrhizal fungi reduce the growth of agricultural weeds? PLoS ONE 6(12):e27825

Veiga RSL, Faccio A, Genre A, Pieterse CM, Bonfante P, Van der Heijden MGA (2013) Arbuscular mycorrhizal fungi reduce growth and infect roots of the non-host plant *Arabidopsis thaliana*. Plant Cell Environ 36:1926–1937

Voets L, de la Providencia IE, Declerck S (2006) Glomeraceae and Gigasporaceae differ in their ability to form hyphal networks. New Phytol 172:185–188

Voets L, Goubau I, Olsson PA, Merckx R, Declerck S (2008) Absence of carbon transfer between *Medicago truncatula* plants linked by a mycorrhizal network, demonstrated in an experimental microcosm. FEMS Microbiol Ecol 65:350–360

Wagg C, Jansa J, Stadler M, Schmid B, Van der Heijden MGA (2011a) Mycorrhizal fungal identity and diversity relaxes plant-plant competition. Ecology 92:1303–1313

Wagg C, Jansa J, Schmid B, Van der Heijden MGA (2011b) Belowground biodiversity effects of plant symbionts support aboveground productivity. Ecol Lett 14:1001–1009

Walder F, Niemann H, Natarajan M, Lehmann MF, Boller T, Wiemken A (2012) Mycorrhizal networks: goods of plants shared under unequal terms of trade. Plant Physiol 159:789–797

Walder F, Brule D, Koegel S, Wiemken A, Boller T, Courty PE (2015) Plant phosphorus acquisition in a common mycorrhizal network: regulation of phosphate transporter genes of the Pht1 family in sorghum and flax. New Phytol 205:1632–1645

Watkinson AR, Freckleton RP (1997) Quantifying the impact of arbuscular mycorrhiza on plant competition. J Ecol 85:541–545

Weiner J (1990) Asymmetric competition in plant-populations. Trends Ecol Evol 5:360–364

Weremijewicz J, Janos DP (2013) Common mycorrhizal networks amplify size inequality in *Andropogon gerardii* monocultures. New Phytol 198:203–213

West HM (1996) Influence of arbuscular mycorrhizal infection on competition between *Holcus lanatus* and *Dactylis glomerata*. J Ecol 84:429–438

Wilson GWT, Hickman KR, Williamson MM (2012) Invasive warm-season grasses reduce mycorrhizal root colonization and biomass production of native prairie grasses. Mycorrhiza 22:327–336

Wright DP, Read DJ, Scholes JD (1998a) Mycorrhizal sink strength influences whole plant carbon balance of *Trifolium repens* L. Plant Cell Environ 21:881–891

Wright DP, Scholes JD, Read DJ (1998b) Effects of VA mycorrhizal colonization on photosynthesis and biomass production of *Trifolium repens* L. Plant Cell Environ 21:209–216

Wu B, Maruyama H, Teramoto M, Hogetsu T (2012) Structural and functional interactions between extraradical mycelia of ectomycorrhizal *Pisolithus* isolates. New Phytol 194:1070–1078

Yang S-Y, Grønlund M, Jakobsen I, Grotemeyer MS, Rentsch D, Miyao A, Hirochika H, Kumar CS, Sundaresan V, Salamin N, Catausan S, Mattes N, Heuer S, Gheyselinck J, Paszkowski U (2012) Non-redundant control of rice arbuscular mycorrhizal symbiosis by two members of the phosphate transporter 1 (PHT1) gene family. Plant Cell 24:4236–4251

Yao Q, Zhu HH, Chen JZ, Christie P (2005) Influence of an arbuscular mycorrhizal fungus on competition for phosphorus between sweet orange and a leguminous herb. J Plant Nutr 28:2179–2192

Zabinski CA, Quinn L, Callaway RM (2002) Phosphorus uptake, not carbon transfer, explains arbuscular mycorrhizal enhancement of *Centaurea maculosa* in the presence of native grassland species. Funct Ecol 16:758–765

Zhang Q, Tang JJ, Chen X (2011a) Plant mortality varies with arbuscular mycorrhizal fungal species identities in a self-thinning population. Biol Letters 7:472–474

Zhang QA, Xu LM, Tang JJ, Bai MG, Chen X (2011b) Arbuscular mycorrhizal mediation of biomass-density relationship of *Medicago sativa* L. under two water conditions in a field experiment. Mycorrhiza 21:269–277

Zhang QA, Zhang L, Weiner J, Tang JJ, Chen X (2011c) Arbuscular mycorrhizal fungi alter plant allometry and biomass-density relationships. Ann Bot 107:407–413

Zhen LN, Yang GW, Yang HJ, Chen YL, Liu N, Zhang YJ (2014) Arbuscular mycorrhizal fungi affect seedling recruitment: a potential mechanism by which N deposition favors the dominance of grasses over forbs. Plant Soil 375:127–136

Chapter 5
Resource Transfer Between Plants Through Ectomycorrhizal Fungal Networks

Suzanne Simard, Amanda Asay, Kevin Beiler, Marcus Bingham, Julie Deslippe, Xinhua He, Leanne Philip, Yuanyuan Song and François Teste

Abstract Carbon (C), nutrients and water (H_2O) have been known for five decades to flow between plants through ectomycorrhizal (EM) networks. This flux has the potential to affect plant and fungal performance and resource distribution within

Co-authors listed in alphabetical order.

S. Simard (✉)
Department of Forest and Conservation Sciences, Faculty of Forestry,
The University of British Columbia, 3rd Floor, Forest Sciences Center,
#3601-2424, Main Mall, British Columbia, CA V6T-1Z4, Canada
e-mail: suzanne.simard@ubc.ca

A. Asay
Department of Forest and Conservation Sciences, Faculty of Forestry,
The University of British Columbia, 2nd Floor, Forest Sciences Center,
#2621-2424, Main Mall, British Columbia, CA V6T-1Z4, Canada
e-mail: amanda.k.asay@gmail.com

K. Beiler
John-Schehr-Str. 20A, 10407 Berlin, Germany
e-mail: kevin.beiler@gmail.com

M. Bingham
2321 SE Ladd, Portland, OR 97214, USA
e-mail: binghm93@gmail.com

J. Deslippe
School of Biological Sciences, Victoria University of Wellington,
726 New Kirk Building, Wellington 6012, New Zealand
e-mail: Julie.Deslippe@vuw.ac.nz

X. He
Centre of Excellence for Soil Biology, College of Resources and Environment,
Southwest University, 2 Tiansheng Road, Beibei, Chongqing 400715, China
e-mail: hexinhua@swu.edu.cn; xinhua.he@uwa.edu.au

X. He
School of Plant Biology, University of Western Australia, 35 Stirling Highway,
Crawley, Perth, WA 6009, Australia

T.R. Horton (ed.), *Mycorrhizal Networks*,
Ecological Studies 224, DOI 10.1007/978-94-017-7395-9_5

communities. We asked two questions: (1) What are the pathways and mechanisms for C, nutrient and H_2O fluxes between plants through EM networks? (2) What are the magnitude, fate and importance of C, nutrient and H_2O transfer among EM plants? Mycorrhizal networks provide a distinct pathway for resource fluxes among plants and mycorrhizal fungi, partitioning them away from other competing soil microbes and plant roots in the soil matrix, and potentially providing a competitive advantage (or disadvantage) for some individuals involved in the network. Carbon and nutrients flow symplastically and apoplastically through mycorrhizal symbionts, hyphae and rhizomorphs along source-sink gradients across the networking mycelia and plant community. EM networks can also facilitate the hydraulic redistribution of soil or plant water following water potential gradients. Carbon fluxes through EM networks have been shown to supply 0–10 % of autotrophic, up to 85 % of partial myco-heterotrophic (MH), and 100 % of fully MH plant C. This C supply has been loosely associated with the increased survival and growth of autotrophic plants, but has been shown to be essential for the survival of MH plants. Network-mediated nitrogen (N) fluxes between N_2-fixing and non-N_2-fixing plants have supplied up to 40 % of receiver N, and this has been associated with increased plant productivity. Fluxes between non-N_2-fixing plants have supplied <5 % of receiver N. Hydraulic redistribution involving EM fungi has supplied up to 50 % of plant water; this has been shown as essential for plant survival in some cases. However, uncertainty remains as to how much of this water transfers through EM networks. Phosphorus transfer through EM networks has not been adequately demonstrated. Overall, this review chapter shows that resource fluxes though EM networks are sufficiently large in some cases to facilitate plant establishment and growth. Resource fluxes through EM networks may thus serve as a method for interactions and cross-scale feedbacks in the development of plant-microbial communities. The outcome of resource transfer through EM networks for the stability of terrestrial ecosystems depends upon the environmental context.

L. Philip
Forest Sciences, University of British Columbia, Vancouver, Canada

Y. Song
State Key Laboratory for Conservation and Utilization of Subtropical Agro-bioresources, Key Laboratory of Tropical Agro-environment, Ministry of Agriculture, South China Agricultural University, Guangzhou 510642, China

Y. Song
Institute of Genetics and Developmental Biology of the Chinese Academy of Sciences, Beijing 100101, China

F. Teste
School of Plant Biology, The University of Western Australia, 35 Stirling Highway, Crawley, Perth, WA 6009, Australia
e-mail: francois.teste@uwa.edu.au

F. Teste
Grupo de Estudios Ambientales, IMASL-CONICET, Universidad Nacional de San Luis, Av. Ejercito de los Andes 950, D5700HHW San Luis, Argentina

Keywords Carbon transfer · Nitrogen transfer · Hydraulic redistribution · Resource sharing · Facilitation · Ectomycorrhiza

5.1 Introduction

The discovery five decades ago that C, nutrients and water can move between plants through interlinking networks of mycorrhizal fungi has been both exciting and challenging for scientists (Björkman 1960; Chiarello et al. 1982; Brownlee et al. 1983; Francis and Read 1984). On one hand, resource sharing through mycorrhizal networks raises the intriguing possibility that mycorrhizal network facilitation of neighbours is important in the organization of ecosystems. While the role of competition has long been recognized, facilitation is increasingly understood as an important force driving ecological and evolutionary processes (Connell and Slatyer 1977; Bertness and Callaway 1994; Butterfield 2009; van der Heijden and Horton 2009). Mycorrhizal associations are understood as important in the evolution and co-evolution of individual plant and fungal species (Smith and Read 2008), but only recently has discussion and experimentation begun on the potential role mycorrhizal networks may play in selection (Perry 1998; Wilkinson 1998; Whitham et al. 2006; Schweitzer et al. 2008; van der Putten 2009; File et al. 2011).

On the other hand, direct sharing of resources among plants via mycorrhizal networks challenges our understanding of the role of competition versus facilitation in ecology and evolution, where plants have traditionally been viewed as discrete individuals competing for scarce resources (Pringle 2009). In this context, why would either a plant or a fungus give up C or nutrients to its neighbour? Is the mycorrhizal symbiosis and facilitation by mycorrhizal networks simply a moderation of competitive interactions among plants and fungi involved in the network? Or are these facilitative interactions part of a more complex, fluid set of ever-changing multi-trophic interactions? In any of these cases, mycorrhizal networks can be considered agents of self-organization because they provide avenues for cross-scale competitive and facilitative interactions and feedbacks from which the structure and function of complex adaptive ecosystems emerge (Bascompte et al. 2003; Anand et al. 2010; Parrott 2010; Simard et al. 2013). Understanding the role of C, nutrient and water transfer via mycorrhizal networks in self-organizing and evolutionary processes in complex adaptive systems may help with implementing conservation practices that promote biodiversity and maintain ecosystem processes.

In this chapter, we review evidence for the mechanisms, magnitude and significance of C, nutrient and H_2O transfers among EM plants through EM networks. This chapter builds on previous reviews on this subject (Newman 1988; Newman et al. 1992; Simard et al. 2002, 2012; He et al. 2003; Leake et al. 2004; Simard and Durall 2004; Selosse et al. 2006; van der Heijden and Horton 2009). The amount of research that has been conducted on resource transfers in EM networks is considerably less than that in arbuscular mycorrhizal (AM) systems, probably because of the greater longevity and size of many EM plant species (and hence greater difficulty in experimentation), and the lower agricultural importance of EM plants (Molina et al. 1992). Similar reasons underlie the even greater paucity of research on nutrient fluxes through ericoid mycorrhizal (ERM) networks (Leake and Cameron 2010). Most research in EM networks has focused on C, nitrogen (N) and, more recently, H_2O transfers between EM tree species. Scarce are direct studies on phosphorus (P) transfers, but include two in AM networks (Eason and Newman 1990a, b; Eissenstat and Newman 1990) and two in EM tree species (Finlay and Read 1986; Perry et al. 1989a, b). Recently, transfer of plant hormones, defence signals (Song et al. 2010; Babikova et al. 2013), allelopathic chemicals (Barto et al. 2011), nutrient analogues (arsenic, cesium and rubidium) (Meding and Zasoski 2008; Gyuricza et al. 2010) and genetic material (Giovannetti et al. 2004; Giovannetti et al. 2005) have been discovered in AM networks, but their importance in EM systems has yet to be investigated. The transfer of any one of these compounds through a mycorrhizal network would increase its zone of activity, with potentially important consequences for plant interactions in natural ecosystems.

We ask two questions: (1) What are the pathways and mechanisms for C, nutrient and H_2O fluxes between plants through mycorrhizal networks? (2) What is the magnitude, fate and importance of C, N, P and H_2O transfer among EM autotrophic, myco-heterotrophic and partial myco-heterotrophic plants? We also discuss the role of resource transfers through EM networks in the context of complex adaptive systems theory. Our review primarily addresses forest trees because they have been the subject of the most EM network research.

5.2 Pathways and Mechanisms

5.2.1 Pathways

Mycorrhizal networks form when the hyphae of a single mycorrhizal fungal individual, or genet, link together two or more plants of the same or different species. In natural plant communities, mycorrhizal networks typically involve several different mycorrhizal fungal species, and may also involve more than one plant species (Beiler et al. 2010; Bahram et al. 2011). Mycorrhizal networks are considered ubiquitous in natural ecosystems and have been documented in boreal and temperate forests and woodlands (Roy et al. 2008; Beiler et al. 2010), tropical forests and woodlands (Onguene and Kuyper 2002; Mangan et al. 2010), mediterranean and sclerophyllous woodlands and chaparral (Richard et al. 2005; Tedersoo et al. 2008), woodland savannah (Dickie et al. 2004), grasslands (Gai et al. 2009), and Arctic tundra (Deslippe and Simard 2011).

With molecular, microscopic imaging and isotopic innovations, evidence for the existence of ectomycorrhizal networks has become practically unequivocal. For example, ectomycorrhizal hyphae turnaround is currently estimated at 46 days, rhizomorphs at 11 months, and EM root tips at 1 year but up to 6 years (Bledsoe et al. 2014). Based on these EMF characteristics, mycorrhizal networks are very dynamic but not so dynamic that they could not play important roles in interplant resource transfer. One important research need is to initiate long-term field studies in which the dynamic nature of mycorrhizal networks is monitored and modeled (Bledsoe et al. 2014).

Nutrient movement among plants has been recognized for nearly a century, but it is only recently that scientists have implicated processes beyond decomposition and mineralization (Read et al. 1985; He et al. 2003). Today we recognize that nutrients can be transferred directly among plants through mycorrhizal networks, as distinct from movement through root grafts and rhizomes (Fraser et al. 2006; Philip et al. 2010). The relative importance of these pathways will determine the strength, direction and outcome of interactions among plants and soil organisms.

A mycorrhizal network pathway between plants is unique because it compartmentalizes transferred compounds away from potential disruptions, such as uptake by competing plant roots or mycorrhizal networks, competition and degradation by soil organisms, chemical decomposition, or adsorption to soil particles (Newman 1988; Schimel and Bennett 2004; Philip et al. 2010). Consequently a mycorrhizal network may allow for greater, faster and more sustained transfer of compounds between interconnected plants. Plants and fungi that are linked in a network should benefit from a greater acquisition of these compounds, which may result in a competitive advantage over non-networked individuals. Interplant resource transfer that occurs directly through rhizomes or root grafts provides similar benefits to mycorrhizal network transfer, but is limited to intraspecific transfer only (Graham and Bormann 1966; Fraser et al. 2006). Resource transfers among plants can also occur via root or mycorrhizal fungal uptake of water or root exudates in soils, which are influenced by soil water (which is affected by soil structure, soil porosity, mineral composition and energy potential gradients), organic matter (which adsorbs compounds and reduces their availability), and soil organisms (which cause immobilisation, mineralisation and transport of soil nutrients) (Perry et al. 1989a, b; Rillig and Mummey 2006; Philip et al. 2010).

5.2.2 Mechanisms

The mechanisms by which resources transfer from plant-to-plant through EM networks are still not well understood and likely vary with the plant functional groups (e.g., evergreen or deciduous trees) and fungal species involved (e.g., short to long distance explorers; Agerer 2001), the compounds being transported (e.g., amino acids, proteins, carbohydrates, inorganic molecules, water, or anything dissolved in water) and the nutritional mode of the plant (e.g., autotrophic or mycoheterotrophic

(MH)). Regardless of these variations, the literature consistently reports that EM networks transport C and nutrients among symbionts and soil patches along source-sink or energy potential gradients (see following section). In soils, EM networks grow where the fungus forages for resources. These resources may then be transported to its actively growing parts (e.g., other parts of the mycelium) and to growing parts of networked plant symbionts (e.g. expanding leaves).

In natural ecosystems, resource transfers through EM networks are highly complex because they can involve many plant and fungal individuals and species (Horton and Bruns 2001; Taylor 2002; Dickie et al. 2004; Nara 2006; Twieg et al. 2007). These networks can involve plants that differ in size, age, physiology and nutrient status (e.g. old (source) trees and growing (sink) seedlings, or N_2-fixers (source) to non-N_2-fixers (sink)), and may transport resources among plants along existing source-sink gradients. The networking fungi and plants interact to govern the magnitude, direction, fate and consequences of resource transfers. The magnitude, direction, and timing of resource transfers through mycorrhizal networks have important consequences for plant communities, and may determine plant establishment or growth, intra- and interspecific competition or facilitation, and stand dynamics and succession (Fraser et al. 2006; Teste et al. 2009; Deslippe and Simard 2011; see also previous reviews).

5.2.3 Carbon and Nutrients

Carbon and nutrients are transported through EM fungal networks in cytosol. In many Basidiomycetes, solutes, cytoplasm and small organelles can move from fungal cell to fungal cell through perforations in the dolipore septa, but the dolipore structure prevents intercellular movement of nuclei (Heaton et al. 2012). In Ascomycetes, pores in the septa allow cytosol and organelle movement across the intact mycelium, but this can be blocked by Woronin bodies or organelles following damage (Jedd 2011; Heaton et al. 2012). Thus, the EM fungi appear to have potential for sophisticated regulation of this transport through occlusion of the septal pores (Jedd 2011). Nutrients move in fungal cytosol by diffusion and active transport during cell expansion at growing mycelia fronts, where apical tip growth, branching, anastomosis and colonization of new plants occur (Heaton et al. 2012). Diffusion and active transport mechanisms appear sufficient to explain transport where resources are rich (e.g., at growing mycelial fronts or new colonization frontiers) and distances are short (Heaton et al. 2012). These mechanisms likely predominate in contact or short-distance explorer ectomycorrhizal fungal species (e.g., *Laccaria, Thelephora*, *Wilcoxina*) (Agerer 2006). They are likely sufficient to explain interplant transfer mechanisms where new seedlings establish in close proximity to refuge plants and become rapidly connected by ruderal contact or short-distance explorers (Teste et al. 2010; see Nara, Chap. 6, this volume). These mechanisms are also likely active even where hyphae explore soils in the absence of seedlings.

Some EM fungi also form more complex chords, strands or rhizomorphs (hereafter referred to collectively as rhizomorphs). These 'long distance explorer' fungi can not only exploit nutrients over short distances, but are capable of long-distance exploration of disparate resource patches or formation of connections between distant ectomycorrhizal plants (Cairney 2005; Agerer 2006; Beiler et al. 2010; Lilleskov et al. 2011). These rhizomorphs are capable of rapid, efficient high-volume resource transfer (Agerer 2006). This type of transfer is possible because the central inner medulla of the rhizomorph contains several hollow vessel hyphae 10–15 μm in diameter with few septal pores (Duddridge et al. 1980; Brownlee et al. 1983; Cairney 2005). The outer medulla contain loosely packed hyphae up to 2 μm in diameter, which could serve in bidirectional transport of C and nutrients to growing fungal fronts or interconnected plant sinks (observed by Tlalka et al. 2008; Heaton et al. 2012). Bidirectional transfer also occurs within an anti-parallel circulation system within the inner medulla of the rhizomorphs (Fricker et al. 2007). The hydrophobic outer surface of the rhizomorph inhibits leakage of solutes and thus facilitates mass flow to areas of local exudation, evaporation, transpiration or growth (Agerer 2006; Hobbie and Agerer 2010). The network is dynamic and adaptive; it simultaneously grows and regresses to exploit new soil resource patches and colonize new plant roots, thus developing patterns that efficiently exploit and transport new and old nutrient and C sources (Boddy and Jones 2007). Fungi appear to move resources over long-distances predominantly by advective mass flow driven by differences in energy potential along source-sink gradients generated by inter-plant nutrient or physiological differences or by fungal growth (but also other mechanisms proposed, see Heaton et al. 2012).

The identity, size and hydrophobicity of compounds transferred through mycorrhizal networks is important for understanding mechanisms of interplant resource transfer and how the nutrients are used by receiver plants. A wide range of macromolecules and ions have been shown to transport along source-sink gradients through rhizomorphs and other EM hyphae, including fungal carbohydrates (e.g., trehalose, mannitol, arabitol and erythritol, see below), protein monomers (e.g., glutamine and glycine), lipids, N ions (NH_4^+ or NO_3^-) and phosphates (Martin et al. 1986; Smith and Smith 1990; Bago et al. 2002; Nehls et al. 2007). The transport of amino acids would be important in driving N and C transport along nutrient gradients through networks. In mycorrhizal networks that connect autotrophic plants, leaf photosynthetic activity likely generates N and C sinks for amino acids, which move via the xylem into the aboveground tissues of the receiver plants (Martin et al. 1986; Teste et al. 2010; Deslippe and Simard 2011; Deslippe et al. 2015). The transport of amino acids, ions and other metabolites across the plasma membranes of Hartig net hyphae and host plant cells involves complex exchange mechanisms (Smith and Smith 1990). For example, 20 proteins involved in nutrient transport across the fungal-host membrane have been characterized in ectomycorrhizal fungi over the last decade (for a review, see Lucic et al. 2008). Martin et al. (2008) have found an especially large number (approximately 500) of membrane-bound transporter proteins in their characterization of the *Laccaria bicolor* genome. They report that this large number is likely important for adaptation of the mycelium to rapidly changing conditions (Marschner and Bell 1994). Our understanding of the pathways and mechanisms of C and nutrient assimilation and transport in mycorrhizal fungi, and the key genes involved, is rapidly increasing (Deveau et al. 2008; Witzany 2012).

In autotrophic plants, carbohydrates are delivered across the symbiotic interface from the host plant to the fungus. Sucrose is delivered by plant root rhizodermal and cortical cells into the apoplast at the plant-fungal interface in the symbiosis (Nehls et al. 2007). Upon release to the symbiotic apoplast, sucrose is cleaved by the plant extracellular invertase (EM and ERM fungi lack invertase) and the monosaccharides glucose and fructose are taken up by the fungus (Nehls et al. 2007). This process requires several essential transport mechanisms across fungal and plant membranes. Once transported across the membrane into the fungus, the monosaccharides are quickly converted via glycolysis into the short-term storage fungal carbohydrates, trehalose, mannitol, arabitol and erythritol, or the long-term storage compound glycogen (Nehls et al. 2007). The short-term compounds are more abundant in fungal hyphae and are likely transported long distances to different parts of the mycelium, thus generating a strong C source and sink gradient through the mycorrhizal network. It is possible that the rapidity of the conversion of monosaccharides to fungal carbohydrates results in a strong sink between the plant apoplast and fungal cytosol that enables mycorrhizal networks to compete more strongly than non-mycorrhizal roots for plant carbohydrate.

The question still remains, however: why would a fungus give up C, its most limiting resource, in the opposite direction to an autotrophic plant? The movement of C from fungi into plants through EM networks is likely to accompany N during the transport of amino acids. Glutamine and glycine, the primary amino acids through which N is transferred from EM fungi to their hosts, contain five C atoms for every two N atoms, reflecting the high-energy cost of N assimilation (Martin et al. 1986; Taylor et al. 2004). When glutamine or glycine are delivered in high quantities from the mycelium to the plant (Yang et al. 2010), the plant would receive a significant C subsidy in addition to N, while the fungus still receives its most limiting resource, C, from the plant (when photosynthetic). Most of this C, however, is likely consumed by root respiration (Taylor et al. 2004).

Myco-heterotrophic (MH) and partial MH plants are dependent on networking fungi to deliver C and to a lesser degree N from nearby autotrophic plants (Zimmer et al. 2007; Selosse and Roy 2009). Nitrogen may move along with carbon, particularly via xylem in the transpiration stream (in partial MH plants) and/or as organic nitrogen compounds (Abuzinadah and Read 1989; Teste et al. 2010). These plants are characterized by leaflessness, reduced leaf size or number of leaves, variegated leaves and lack of (or low levels) of chlorophyll, all diagnostic of a dependency on fungi for plant C (Merckx et al. 2009; Selosse and Roy 2009). Since they have no or limited photosynthetic capacity, a C sink develops throughout the plant. How the C is delivered from the fungus to the MH or partial MH plant is still poorly understood, however, as there have been relatively few studies on metabolite transport pathways in MH plants (Leake and Cameron 2010). Existing studies have revealed the involvement of fungal pelotons and penetration pegs that release nutrients on digestion by the MH plant, and a range of potential candidate metabolites have been identified that might be involved in fungal-to-plant C and nutrient fluxes.

5.2.4 Water

Water (and any of the above compounds that can be carried by water) can also move through multiple belowground pathways (Richards and Caldwell 1987; Brooks et al. 2006; Neumann and Cardon 2012). This occurs when water is transported passively along a water potential gradient from moist to drier regions, such as from deeper ground water to depleted surface soil layers during the day, or from moist surface soils to deeper drier soils at night, or horizontally along water potential gradients caused by irrigation or natural spatial heterogeneity in soils, plants and hyphae. During the day, plant transpiration drives water potential gradients from soil to mycorrhizal hyphae, hyphae to roots, and finally roots to leaves. In arid environments, when soil water potential frequently dips below the wilting point, mycorrhizal hyphae provide the crucial bridge between plant roots and soil micropores (Egerton-Warburton et al. 2003; Bornyasz et al. 2005; Querejeta et al. 2007; Warren et al. 2007; Allen 2007; Allen 2009; Augé et al. 2014). However, when water potential gradients are limited by stomatal closure, such as at night or during drought, water can move in the opposite direction via hydraulic redistribution. Here, water moves from plant roots into the hyphae of the mycorrhizal network, and will ultimately be redistributed to soils, or subsequently be taken up by interconnected plants. The latter occurs when the water potential of the plant vascular tissue is higher than that of surrounding soils, mycorrhizal fungi or connected sink plants (Allen 2009). Water movement through plants, fungi and soils along water potential gradients can happen even after seasonal plant senescence (Leffler et al. 2005).

Mycorrhizal networks have been shown to facilitate hydraulic redistribution, whereby water moves via mass flow along water potential gradients on the hyphal surfaces or inside the mycorrhizal network via cytoplasmic streaming (Plamboeck et al. 2007; Warren et al. 2008; Bingham and Simard 2011; Xu et al. 2015). Direct transport of water inside the hyphal strands or rhizomorphs has been observed in some but not all ectomycorrhizal species tested (Duddridge et al. 1980; Brownlee et al. 1983; Plamboeck et al. 2007; Xu et al. 2015). Rhizomorphs are thought to preferentially transfer water mainly through the apoplast at relatively similar rates (when corrected for different diameters) to root xylem because they have lost their septa (Duddridge et al. 1980; Querejeta et al. 2003). Rhizomorphs or strands are also considered more important than finer hyphae in hydraulic redistribution because resistance to flow is related to the fourth power of the hyphal pathway radius, such that smaller pathways increasingly limit the distance of potential water transport (Duddridge et al. 1980; Warren et al. 2008). Regardless of whether water moves on the inside or outside of fungal structures, however, the flow rates should still be orders of magnitude higher than through the soil matrix. Thus, water, and any resource that can be carried by water, has the potential to move through multiple pathways, with greatest transfer efficiency through mycorrhizal networks and root grafts. The transport of water through mycorrhizal networks is likely important in the water relations of recipient plants under water stress, in maintaining fungal hyphal viability, and in facilitating acquisition of nutrients (Querejeta et al. 2007; Allen 2009; see below).

5.2.5 Regulating Cheating

Source-sink or energy potential gradients among plants, fungi and soils are all important in regulating C, nutrient and water transfers in EM networks. The linking of multiple plants and fungi in a single network provides the opportunity for interspecific interactions (mutualisms and competition), leading to stability in communities (Perry 1995; Simard et al. 2013). The importance of mycorrhizal networks in stability of plant and fungal communities is increasingly evident (Dickie et al. 2004; Booth 2004; Nara 2006; Teste et al. 2009; Simard 2009; van der Heijden and Horton 2009; Booth and Hoeksema 2010). The involvement of multiple plant and fungal species generates the possibility for cheating within the network; that is for exploitation without paying the true cost of joining. This question was addressed in a recent study by Kiers et al. (2011), who found that an arbuscular mycorrhizal fungal species delivering more P to their host in turn received more C from that particular host, and vice versa. This provided evidence for bidirectional control of the mutualism, where the best fungal and plant partners were rewarded for their transfers, thus enforcing the stability of the cooperation. Similar bidirectional control is evident in ectomycorrhizal systems (Finlay 1989; Büking and Heyser 2001). For example, *Pinus sylvestris* provided *Suillus bovinus* more ^{14}C-labelled assimilate than *Suillus grevillei* or *Boletinus cavipes* (which notably are *Larix*-specific fungi and thus perhaps only compatible in the laboratory), allowing *Suillus bovinus* to colonize more extensively, supposedly at the benefit of increased nutrient supply (Finlay 1989). These studies suggest that non-contributing plants and fungi could thus be deprived of nutrients or C by their partner symbionts. As a result, selection pressure should maintain network parasites or cheaters as a relatively small proportion of the total community.

How can the network be protected against hostile takeover by cheaters? The results of Kiers et al. (2011) support the argument of Perry (1995) that mycorrhizal networks are resistant to take-over by cheaters. That is, (1) plants can curtail colonization of fungi that do not deliver nutrients by withholding C, and (2) fungi can mediate the flow of C along source-sink gradients through nutrient supply, meeting their own needs while delivering C and nutrients only to plants that reciprocate in kind. Studies show that this delivery can occur at no cost to C- or nutrient-rich plants (i.e., those in luxury consumption) involved in the network (Perry et al. 1989a, b; Simard et al. 1997a, b; Perry 1998). This balancing act between plant and fungal symbionts involved in a network can serve an evolutionary advantage: it stabilizes the food supply for the fungus over multiple plants, or vice versa, the nutrient supply for the plant over multiple fungi. This trading system becomes very complex, however, when considering that the interacting community can involve hundreds if not thousands of plant and fungal species, as well as the millions of bacteria and other organisms they interact with. Sorting out market trading of C and nutrients in such complex systems could be achieved through modeling stoichiometry (Johnson 2009), but undoubtedly the interactions are even more complex when considering the role of hormonal and defence signalling in plant health and performance. How this complexity of interactions affects selection has yet to be explored (Whitham et al. 2006; File et al. 2011), but there is growing evidence that parents can nurse their own kin through active secretion of soluble-signaling chemicals recognized by roots (Biedrzycki et al. 2010; Asay 2013; Pickles et al. submitted).

5.3 Magnitude, Fate and Importance of C, Nutrient or H_2O Transfers

The functioning of EM networks is highly responsive to local plant and fungal communities and environmental conditions. As a consequence, the magnitude, fate and importance of C, N and H_2O transfer depends on source-sink gradients governed by factors such as physiological, nutrient or water status of donor and receiver plants, degree or dependency of these plants on mycorrhization, the fungal species involved in the network, or nutrient or water status of patchy soil environments. This has been confirmed in many studies through experimental manipulation of plant photosynthetic capacity (e.g., shading, defoliation or seasonal variation), nutrient status (e.g., fertilization or use of nodulating and non-nodulating plant pairs), or water status (irrigation or induced drought). To a lesser degree, the mycorrhizal fungal network has also been experimentally manipulated through inoculations with different fungal species (Arnebrant et al. 1993; Ekblad and Huss-Danell 1995; Ek et al. 1996; Wu et al. 2001; He et al. 2004, 2005; Egerton-Warburton et al. 2007; Querejeta et al. 2003, 2012) and the use of mesh that allows certain fungal exploration types to join the network (Teste et al. 2009; Bingham and Simard 2013).

The significance of resource transfer through networks to individual plants or plant communities has revolved around three main issues. First, it is important to determine whether there is a net gain in resources by one plant over another (Newman 1988; Newman et al. 1992); however, most studies have examined only one-way transfers and only a handful have used dual isotope labeling to quantify bi-directional or net transfer (Johansen and Jensen 1996; Simard et al. 1997a, b; He et al. 2004, 2005; Teste et al. 2010; Philip et al. 2006). Second, whether resources are transferred to receiver fungal, root or shoot tissues has been considered of primary importance. Fitter et al. (1999) suggests that transfer must occur to plant tissues and that retention of transferred nutrients in fungal tissue alone is insufficient to signify significance of resource transfer to networked plants. Others argue, however, that any C supplied to the network by a plant will subsidise the small inputs of receiver plants that are C-limited through shading or other factors (Perry 1995; Simard and Durall 2004; van der Heijden and Horton 2009). This may allow plants under stress to allocate more resources for survival, growth and reproduction, while benefiting from increased nutrient uptake from the large mycorrhizal network that is being supported by other plants (Teste et al. 2009; Bingham and Simard 2011). Third, it is important to determine whether resource transfers through networks affect plant or fungal establishment, physiology, growth or species interactions in the field. These field experiments will provide the basis for elucidating the importance of networks to succession and plant or fungal community stability. Some studies have found that instantaneous measures of resource transfers were related to these processes (Leake et al. 2004; Teste et al. 2009, 2010; Cameron et al. 2010; Preiss et al. 2010; Bingham and Simard 2011), but there still have been no studies that have firmly established cause-and-effect relationships using long-term labelling experimentation. In the following subsections we discuss what is known about the magnitude, fate and importance of C, N, P and H_2O transfers between plants.

5.3.1 Carbon

Mycorrhizal networks are unequivocally critical to the survival and growth of MH plants. Full MH plants gain 100 % of their C from fungi that establish mycorrhizal networks with nearby autotrophic plants (Fig. 5.1). This has been confirmed empirically with the analysis of stable isotopes and calculation of fungal-derived C gains with isotope mixing models (Leake et al. 2004; Preiss et al. 2010). Severing the network between autotrophic and full MH plants has resulted in loss of MH biomass (McKendrick et al. 2000), and ultimately may cause death. Partial MH plants gain up to 85 % of their C from mycorrhizal networks (Selosse and Roy 2009; Bougoure et al. 2010). Partial MH plants acquiring life-sustaining C from mycorrhizal networks have very low photosynthetic rates compared to autotrophic plants (Cameron et al. 2009). Net C transfer between autotrophic plants has typically been small (<10 %) compared to the C fixed via photosynthesis (Simard et al. 1997a, b; Philip et al. 2010; Teste et al. 2010; Deslippe and Simard 2011), and compared to the mycelial-derived C in MH and partial MH plants (Table 5.1). In almost all studies, the C that is transferred through EM networks has been transferred to both shoots and roots of receiver autotrophic and MH plants (Table 5.1), contrasting with the majority of AM studies (Watkins et al. 1996; Fitter et al. 1999; Selosse et al. 2006). Gross C transfer through mycorrhizal networks are likely underestimated due to respiratory C loss (Girlanda et al. 2011).

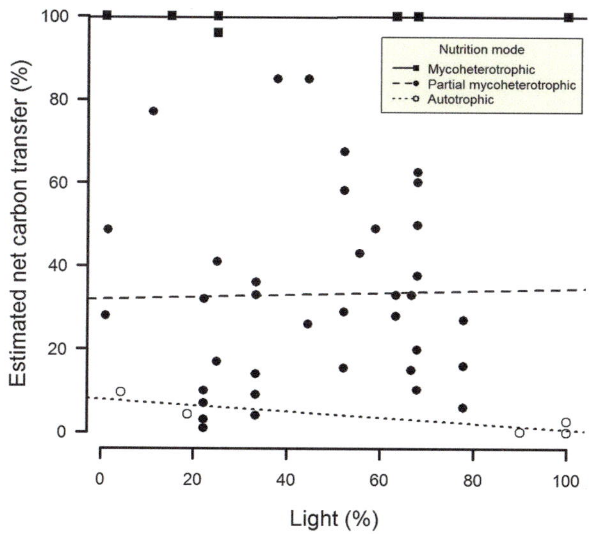

Fig. 5.1 Estimated net C gain via MNs in myco-heterotrophic (MH), partial myco-heterotrophic (partial MH), and autotrophic (AU) plants with light intensity. Data consolidated from Bidartondo (2005), Tedersoo et al. (2007), Teste et al. (2010), Simard et al. (1997a, b), Lerat et al. (2002) Motomura et al. (2010), Hynson and Bruns (2010) and others. Included here are field studies with plants associating with EM fungi that estimated net C transfer with dual (^{14}C–^{13}C) labelling or calculated net C gain via MNs with stable isotope analyses and stable isotope mixing models (Preiss et al. 2010). A notable exception is Lerat et al. (2002), who calculated net C gain between AM plants in the field with ^{14}C. None of the simple linear regressions were statistically significant; MH: $R^2 = 0.01$, $P = 0.66$; partial MH: $R^2 = 0.001$, $P = 0.89$; AU: $R^2 = 0.75$, $P = 0.06$. Net C transfer in partial MH plants as a group (pyroloids, green orchids) is important, reaching in some cases 85 % of all C acquired (Selosse and Roy 2009); however, this relationship is highly variable. Within the same partial MH genus (e.g., *Cephalanthera* green orchids) a strong relationship does exist between light intensity and the magnitude of C gained from fungi via MNs (Preiss et al. 2010). Reproduced, with permission, from Simard et al. (2012)

Studies have been conducted examining C transfer between autotrophic plants through EM networks. Because these have been reviewed by several others (Simard et al. 2002, 2012; Simard and Durall 2004; Selosse et al. 2006; van der Heijden and Horton 2009), we have chosen to highlight three case studies by our group in arid steppe interior Douglas-fir (*Pseudotsuga menziesii* var. *glauca*) forests, moist temperate mixed forests and the Arctic tundra of western North America.

Arid Steppe Interior Douglas-fir Forests

In the arid steppe region of interior British Columbia, much of the forested area is dominated by interior Douglas-fir growing in relatively pure multi-aged stands. The moderately shade tolerant interior Douglas-fir regenerates under its own canopy in small gap disturbances created by wind throw, insect infestations and root diseases (Simard 2009). Beiler et al. (2010) and Beiler et al. (2015) used

Table 5.1 Summary of published studies examining C transfer through ectomycorrhizal networks

Donor plant	Receiver plant	Networking fungus	Element transferred	C % transferred	Receiver tissue	Source-sink	Reference
Betula papyrifera	*Pseudotsuga menziesii* *Thuja plicata* (control)	Field soil in field	Pulsed ^{14}C and ^{13}C	3 % full light 10 % shade	Roots and shoots	Shading of *P. menziesii* receiver	Simard et al. (1997a)
Betula papyrifera	*Pseudotsuga menziesii*	Field soil in lab	Pulsed ^{14}C and ^{13}C	5 % in full light	Roots and shoots	Natural differences in N and C	Simard et al. (1997b)
Pinus densiflora	*Pinus densiflora*	*Pisolithus tinctorius*	Pulsed ^{14}C	Not determined	Fungal	Covered receivers with foil	Wu et al. (2001)
Pseudotsuga menziesii	*Pseudotsuga menziesii*	Field soil	Pulsed ^{13}C	<1 %	Roots and shoots	Smaller size of receiver	Teste et al. (2009)
Pseudotsuga menziesii	*Pseudotsuga menziesii*	Field soil	Pulsed ^{13}C and ^{14}C	<1 %	Roots and shoots	Smaller size of receiver	Teste et al. (2010)
Betula nana	*Betula nana*	Field soil	Pulsed ^{13}C	4.1 %	Roots and shoots	No gradient tested	Deslippe et al. (2011)
Pseudotsuga menziesii	*Pseudotsuga menziesii*	Field soil	Pulsed ^{13}C	<1 %, but not affected by MN	Roots and shoots	Smaller size of receiver	Bingham and Simard (2011)

microsatellite DNA markers to determine that most trees in these uneven-aged forests were interconnected by a complex network of the EM fungi, *Rhizopogon vesiculosus* and *R. vinicolor*, where most of the young trees were linked to large, old hub trees. These forests are ideal model systems for examining the ecological significance of mycorrhizal networks because of the simple tree species composition and self-regenerating stand development pattern. In more complex communities involving a greater diversity of tree and fungal species, which are more typical of forests world-wide, the networks and the interactions they mediate may be more complex than inferred by our model system.

In these forests, Teste et al. (2009) and Schoonmaker et al. (2007) investigated the effects of mycorrhizal networks on C, N and H_2O transfer from older trees to seedlings established by seed or planting of nursery-stock seedlings, where the presence and composition of EM networks were controlled using mesh bags of different pore sizes (i.e., 0.5-, 35-, 250-μm pores, or without mesh). Older saplings were pulse labeled with C ($^{13}CO_2$), N ($^{15}NH_4$–$^{15}NO_3$) or ^2H_2O to quantify resource transfer. Increasing access to the EM network (with increasing mesh size) improved seedling survival (Fig. 5.2). Surviving seedlings in the 250-μm mesh were colonized by a more

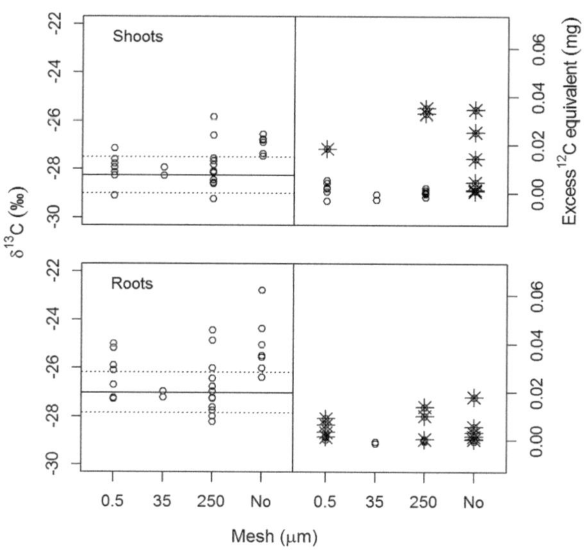

Fig. 5.2 Shoot and root $\delta^{13}C$ and excess ^{12}C-equivalent of Douglas-fir seedlings grown (from seed) in mesh treatments at 0.5 m away from labeled donor Douglas-fir trees in the field. The $\delta^{13}C$ background mean (*solid line*) and 99 % confidence intervals (*dotted line*) appear as *horizontal lines* on the *left portion* of the figure. Values above the 99 % confidence interval are considered enriched and are marked as *asterisks* on the *right portion* of the figure. Contingency table analysis of the frequency of ^{13}C-enriched seedlings (shoots + roots) only showed that the no mesh treatments differed from the 35 and 250 μm mesh treatments (i.e., the 0.5 μm was not different than the no mesh, $P = 0.19$). Reproduced, with permission, from Teste et al. (2009)

complex fungal community than those in mesh of smaller pore sizes, and this community was comprised of multiple long-distance exploration types (Agerer 2001, 2006; Teste et al. 2009). The seedlings with access to the EM network also received small amounts of C, N and H_2O transferred from the older trees (Schoonmaker et al. 2007; Teste et al. 2009). In general, the growth rate of receiver trees was related to the amount of C transferred to seedlings (Teste et al. 2010), suggesting that faster-growing receiver seedlings had enhanced sink strength. Linking into the network of older trees was important to establishment of seedlings from seed, but had no effect on planted seedlings, presumably because planted seedlings were comparatively large and replete in nutrients. In a separate field experiment, Teste et al. (2010) used dual ^{13}C–^{14}C labeling and similar mesh treatments to test whether increasing soil disturbance affected the magnitude of net C transfer from saplings to new seedlings. They found that net transfer increased with disturbance and receiver seedling growth rate, and access to mycorrhizal networks, but only where seedlings were associated with a diverse mycorrhizal community. Hence, increasing complexity of the mycorrhizal community (that included a highly efficient EMF species) combined with receiver seedling growth rates was important in generating sink gradients and driving C transfer.

In the same interior Douglas-fir forests, Bingham and Simard (2012a, b) applied a test of the stress-gradient hypothesis (Maestre et al. 2009), postulating that EM networks were most important where EM tree seedlings were establishing under high climatic drought stress. In an experiment using mesh bags with varying pore size (as above), Douglas-fir seedlings were planted at several distances from conspecific mature trees across a climatic moisture gradient. They found that growth of networked seedlings increased most when they were in close proximity to trees in dry climates, after adjusting for total N (Bingham and Simard 2012a). In a related study, they used isotope labelling to demonstrate that hydraulic redistribution of water occurs from seedlings able to access a reservoir of water to seedlings unable to access this same reservoir (Bingham and Simard 2011). These findings suggest that trees, which generate deeper taproots under dry conditions, access water unavailable to establishing seedlings, and that this water is then hydraulically redistributed to the EM mycelia and seedling symbionts.

The interior Douglas-fir forests are experiencing increasing drought stress as the climate changes in this region, resulting in increased insect attack and disease incidence (Woods et al. 2010). Previous studies have shown that plants damaged by herbivores or clipping increase export of C, P or defence signals to neighbouring plants through mycorrhizal networks (Eason and Newman 1990a, b; Waters and Borowicz 1994; Song et al. 2010). We have found that defoliation of interior Douglas-fir results in C and signal exports through ectomycorrhizal networks to neighboring healthy ponderosa pine seedlings, resulting in increased defense enzymes on the foliage of the ponderosa pine (Song et al. 2015). Thus, neighboring seedlings appear to benefit from increased resource supply and defense chemistry, directly transferred from dying neighbours. This process may be important in facilitating tree species shifts as climate changes, where mortality of large trees opens forest gaps for the establishment of new seedlings of more drought-tolerant species. In all of the above experiments, C has been transferred into both shoots and roots of receiver seedlings. Together, these studies suggest that EM networks may be important to the resilience of interior Douglas-fir forests to stresses caused by drought, soil disturbance and herbivory.

Moist Temperate Mixed Interior Douglas-fir Forests

The mixed Douglas-fir–paper birch (*Betula papyrifera*) stands in the moister, warmer climate of British Columbia are more productive and regenerate more readily after disturbance than the arid steppe interior Douglas-fir forests, and here mycorrhizal networks also play a role in forest establishment. In the rich tree species mixtures that include interior Douglas-fir, paper birch and western redcedar (*Thuja plicata*), establishment success of regenerating Douglas-fir has been greater where seedlings are linked into the EM network of older interior Douglas-fir and paper birch trees (Simard et al. 1997a, b). The mycorrhizal root systems of paper birch, which often survive fire, pathogen infections or clearcutting, are particularly important for subsequent establishment of EM networks by providing a diverse and rapid source of fungal inoculum for regenerating seedlings (Twieg et al. 2007).

In clearcuts, Douglas-fir seedlings have benefited not only from mycorrhizal fungal colonization, but also from C transferred from paper birch through networks, particularly where Douglas-fir is shaded (Simard et al. 1997b). Using dual ($^{13}CO_2$–$^{14}CO_2$) pulse-labelling, Simard et al. (1997b) examined transfer between EM interior Douglas-fir and paper birch seedlings in the field. There was bidirectional C transfer, but Douglas-fir received a net C gain from paper birch. Net gain by Douglas-fir averaged 6 % of C isotope uptake through photosynthesis, with more in deep shade (10 %) than in full or partial sun (3–4 %) (Fig. 5.3). Label was detected in receiver plant shoots in all transfer experiments indicating movement of transferred C out of fungal and into plant tissues. In the field, AM western redcedar seedlings absorbed 18 % of isotope transferred between paper birch and Douglas-fir, suggesting that most C was transferred through the EM network, but that some was also transferred through alternate soil pathways (Simard and Durall 2004).

Later, Philip (2006) used dual labelling in the field to show that the direction of net C transfer between Douglas-fir and paper birch switched twice over the growing season: (1) from rapidly growing Douglas-fir to bud-bursting paper birch in spring, (2) then reversing, from nutrient and photosynthate-enriched paper birch to stressed understory Douglas-fir in summer; and (3) switching yet again, from still-photosynthesizing Douglas-fir to senescent paper birch in the fall. The C moved bidirectionally between

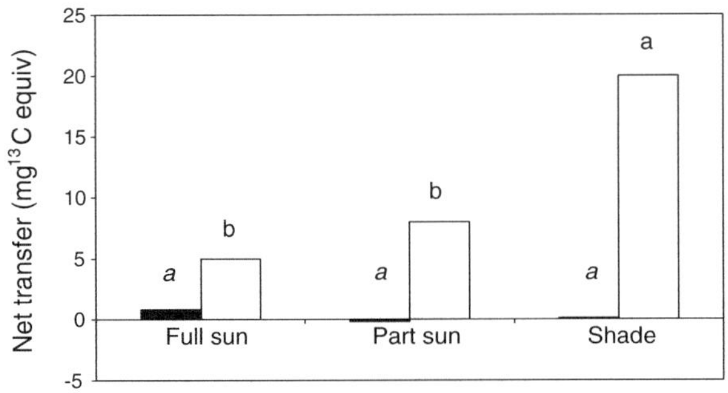

Fig. 5.3 Mean net C transfer from paper birch to Douglas-fir in deep shade, partial shade and full ambient sunlight in the first (two-year-old seedlings (*filled bars*)) and second (three-year-old seedlings, *open bars*) experiment years. Within years, means denoted by the same letter (*a* or *b*) do not differ significantly ($P < 0.05$). Net transfer was greater than zero in all treatments in year three but not year two ($P < 0.05$). Reproduced, with permission, from Simard et al. (1997a, b)

paper birch and Douglas-fir through multiple belowground pathways, including MNs, soils and a discontinuous hyphal pathway. These results suggest a dynamic interplay between paper birch, Douglas-fir and the interconnecting fungi, with C and nutrients moving in the direction of greater need over the growing season.

The question still remains, however, as to whether C transfer through networks affects productivity of the plant and fungal species in nature over the long-term (see Hoeksema, Chap. 9, this volume). In a long term field experiment in Douglas-fir–paper birch mixtures, we found net photosynthetic rates were higher in mid-growing season where mycorrhizal networks that linked the two tree species were left intact than where they had been isolated using trenching and barriers five years earlier. After 21 years, paper birch with intact mycorrhizal networks with neighboring Douglas-fir (i.e., not trenched) maintained higher annual growth rates than birch trees that were isolated by trenching, and appeared to be due to neighbor subsidies at no cost to growth rates of Douglas-fir (Louw 2015). Simard et al. (1997a) similarly found in deep forests that understory interior Douglas-fir seedlings had higher net photosynthetic rates, height growth rates and foliar nutrients where they had access to the EM networks of century-old paper birch and Douglas-fir trees than where the connections were severed. In this study, the EM diversity was higher where seedling roots accessed those of the older trees (Simard et al. 1997a). These results suggest that networks have the potential for longer-term benefits to the productivity and health of mixed species forests, but whether this is the result of C or nutrient transfers remains elusive.

Low Arctic Tussock Tundra

Arctic warming has exceeded global mean warming by up to 3 times over the past 150 years, as positive feedbacks facilitate even more warming (Sturm et al. 2001; Anisimov et al. 2007). In low Arctic tussock tundra, these changes are associated with enhanced competition, growth and spread of the rhizomatous EM shrub *Betula nana* (Bret-Harte et al. 2008). *Betula nana* has been shown to increasingly dominate tussock tundra in long-term warming and fertilization experiments (Chapin et al. 1995; Shaver et al. 2001; Mack et al. 2004), and we hypothesized that EM networks may be involved in its expansion. Deslippe and Simard (2011) found that EM networks facilitated interplant C transfer among *Betula nana* individuals, but not between or within other tundra species examined (Fig. 5.4). Total C transfer among conspecific *B. nana* pairs was 10.7 % of photosynthesis, with the majority of C transferred through rhizomes or root grafts (5.2 %) and the EM network pathway (4.1 %), and very little through the soil (1.4 %). There was greater C transfer in June than in August, suggesting that leaf expansion drove C demand in these receivers. Targeting rhizospheres of the most highly enriched receiver plants, and using a combined stable isotope probing of DNA (DNA-SIP) pyrosequencing approach, Deslippe et al. (2015) found that a

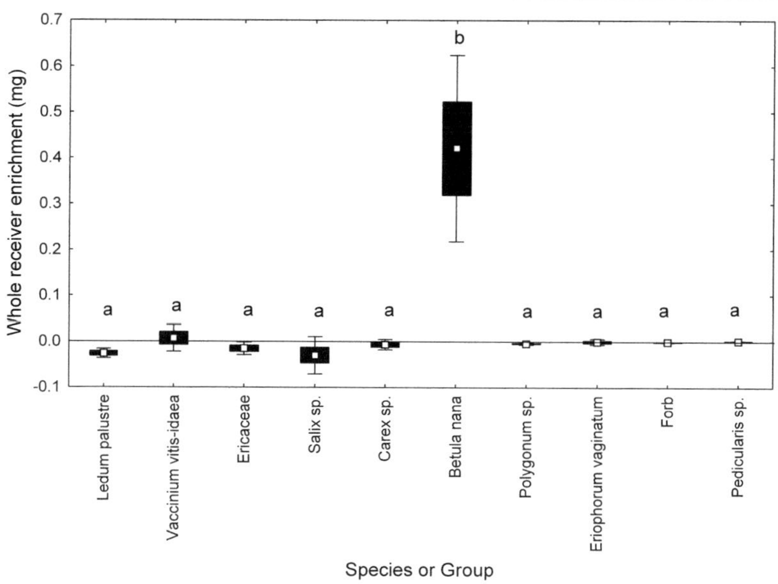

Fig. 5.4 Whole-receiver enrichment calculated as the sum of leaf, stem and rhizome enrichments in mg of excess ^{12}C-equivalent, for all receiver plants regardless of donor species (*Betula nana* or *Ledum palustre* donors). Inner spread denotes the standard error of the mean; whiskers denote the 99 % confidence interval about the mean, lower-case letters denote significant difference ($P < 0.05$) among means as determined by a Tukey's post hoc test. Reproduced, with permission, from Deslippe and Simard (2011)

member of the genus *Cortinarius,* closely affiliated with *C. fennoscandicus,* was unique in being highly enriched in ^{13}C. These data suggest a role for this *Cortinarius* sp. in C-transfer among *B. nana* plants. *C. fennoscandicus* is considered to be specific to *B. nana,* consistent with our observation that *B. nana* was unique among members of this Arctic tundra plant community in its ability to transfer C through MNs. In a related study, Deslippe et al. (2011) showed that long-term experimental warming of this plant community increased the abundance of *Cortinarius spp*. *Cortinarius* have high-biomass growth forms, medium-distance exploration types, and enhanced capacities to degrade complex organic matter, thus securing access to limiting N to grow. At the same time, it reduced the prevalence of the short-distance explorer *Russula* spp., some of which have high affinities for labile N. The increased populations of EM fungi with long-distance exploration types may increase connectivity among *B. nana* individuals, thus facilitating greater C transfer through mycorrhizal networks and increased competitive ability of this expanding shrub species. Warming-induced increases in the population of networking fungal genets may facilitate the expansion of *B. nana* with global change (Deslippe et al. 2011) and *C. fennoscandicus* may therefore be considered a keystone species in the regime shift of Arctic tundra to shrub tundra as climate warms (Deslippe et al. 2015).

These three case studies highlight that light availability, phenology and stress associated with shading, season, drought, warming, disturbance and herbivory can be important factors in C, N or H_2O transfer through networking fungi. The amount of C transferred through EM networks increased with shading of receiver plants (Simard et al. 1997a, b) and appeared to peak during leaf expansion or high photosynthetic activity (Philip 2006; Deslippe and Simard 2011). Carbon transfer also increased with soil disturbance and defoliation (Teste et al. 2010; Song et al. 2015), but water transfer and not C transfer increased with climatic aridity in interior Douglas-fir, resulting in increased seedling establishment (Bingham and Simard 2012a). In Bingham and Simard (2012a), EM network facilitation may act to extend the niche breadth of interior Douglas-fir seedlings in a very dry climate.

Ectomycorrhizal facilitation of seedling survival and growth represents a positive feedback for forest regeneration, but as seedling density increases, competition and density-dependent mortality provide additional negative feedbacks that may stabilize the forest community (Simard 2009). Based on these observations, we expect that facilitation via MNs will remain important to the stability of interior Douglas-fir ecosystems as climate warming increases the severity and duration of drought and associated disturbance stress. In the Arctic tundra, however, C transfer among *B. nana* may serve to destabilize the ecosystem by promoting the dominance of *B. nana* and the conversion of tundra landscapes to shrub-lands as climate warms (Deslippe and Simard 2011). Paleoclimate data for Arctic Alaska suggest a strong control of vegetation composition on fire return intervals, with shrub communities supporting more frequent fires (Higuera et al. 2008, 2009). Increases in fire frequency and extent in Arctic Alaska were recently observed (Hu et al. 2010) and can led to unprecedented losses of C from the Arctic tundra biome (Mack et al. 2011). Thus C-transfer through EM network of *Cortinarius sp.* and *B. nana* appears to be an avenue for cross-scale interactions affecting higher order processes, in this case potentially altering local disturbance regimes. This represents a positive feedback that serves to destabilize the current plant community and amplify ecosystem change.

5.3.2 Nitrogen

Interplant N transfer from N-rich to N-poor plants has been studied for decades because of the implications for improved production in agriculture and forest systems. Early studies focused on transfers of N-rich decomposition products from N_2-fixing to non-N_2-fixing plants without distinguishing transfer pathways, but the benefits of direct transfers through mycorrhizal networks have become a focus of recent work. These studies have employed ^{15}N external labelling and ^{15}N natural abundance techniques to trace the direction and magnitude of N transfer from N_2-fixing, N-fertilized or N-enriched source plants to non-N_2 fixing, unfertilized or N-depleted sink plants (for previous reviews of N transfer studies and methods, see Simard et al. 2002; He et al. 2003, 2009; Selosse et al. 2006). While some of the more recent tracer experiments are briefly summarized here, we also describe new efforts to elucidate interplant N transfer based on species patterns in natural abundance of ^{15}N.

Studies dating back to the 1980s established that NH_4^+, NO_3^- or amino acids can transfer along a N gradient from *Rhizobium*- or *Frankia*-infected donors either through AM networks (Haystead et al. 1988; Bethlenfalvay et al. 1991; Frey and Schuepp 1993; Johansen and Jensen 1996; Moyer-Henry et al. 2006) or EM networks (Arnebrant et al. 1993; Ekblad and Huss-Danell 1995; He et al. 2004, 2005, 2006). In the EM studies, N has been shown to transfer from *Frankia*-N_2-fixing *Alnus glutinosa* or *A. incana* to non-N_2-fixing *Pinus contorta* or *P. sylvestris* (Arnebrant et al. 1993; Ekblad and Huss-Danell 1995), and from *Frankia*-N_2-fixing *Casuarina cunninghamiana* to non-N_2-fixing *Eucalyptus maculata* (the latter two species form both EM and AM networks; He et al. 2004, 2005). It should be noted, however, that these studies likely included some degree of 'incompatible' EMF species due to high specificity for one of the host plants (Pölme et al. 2013; Roy et al. 2013; Horton et al. 2013). Nevertheless, the amount of N transferred from N-enriched to N-depleted plants through EM networks has ranged from 0 to 40 %, but <5 % between non-N_2-fixing plants (Table 5.2). In all studies, the transferred N has been found in both roots and shoots of the receiver plants. In some cases, N

Table 5.2 Summary of published studies examining N transfer through ectomycorrhizal networks (adapted from He et al. 2009)

Donor plant	Receiver plant	Networking fungus	Element transferred	N % transferred	Source-sink	Reference
Alnus glutinosa	*Pinus contorta*	*Paxillus involutus*	$^{15}NH_4^+$	15.0	N_2-fixer to non-N_2-fixer	Arnebrant et al. (1993)
Alnus incana	*Pinus sylvestris*	*P. involutus*	$^{15}NH_4^+$	49.0	N_2-fixer to non-N_2-fixer	Ekblad and Huss-Danell (1995)
Betula pendula	*Picea abies*	*Scleroderma citrinum*	$^{15}NH_4^+$	<0.3 % to *Picea*	Mycelium to coniferous	Ek et al. (1996)
Picea abies	*B. pendula*	*S. citrinum*	$^{15}NH_4^+$	5.3 % to *Betula*	Mycelium to deciduous	Ek et al. (1996)
Casuarina cunninghamiana	*C. cunninghamiana*	*Pisolithus tinctorius*	$^{14}NH_4^+$	30.5	N_2-fixer to non-N_2-fixer	He et al. (2003)
C. cunninghamiana	*Eucalyptus maculata*	*P. tinctorius*	$^{14}NH_4^+$	32.8	N_2-fixer to non-N_2-fixer	He et al. (2003)
C. cunninghamiana	*E. maculata*	*P. tinctorius*	$^{15}NH_4^+$	10.1	N_2-fixer to non-N_2-fixer	He et al. (2004, 2005)
C. cunninghamiana	*E. maculata*	*P. tinctorius*	$^{15}NO_3^-$	5.3	N_2-fixer to non-N_2-fixer	He et al. (2004, 2005)
E. maculata	*C. cunninghamiana*	*P. tinctorius*	$^{14}NH_4^+$	26.7	Non-N_2-fixer to N_2-fixer	He et al. (2003)
E. maculate	*C. cunninghamiana*	*P. tinctorius*	$^{15}NH_4^+$	39.1	Non-N_2-fixer to N_2-fixer	He et al. (2004, 2005)
E. maculate	*C. cunninghamiana*	*P. tinctorius*	$^{15}NO_3^-$	23.6	Non-N_2-fixer to N_2-fixer	He et al. (2003)
E. maculate	*E. maculate*	*P. tinctorius*	$^{14}NH_4^+$	10.0	Non-N_2-fixer to N_2-fixer	He et al. (2003)
Pinus sabiniana	*P. sabiniana*	Field soil	$^{15}NO_3^-$	<1 %	Plant size	He et al. (2006)
P. sabiniana	*Quercus douglasii*	Field soil	$^{15}NO_3^-$	<1 %	Plant size	He et al. (2006)
Pseudotsuga menziesii	*P. menziesii*	Field soil	$^{15}NH_4^+$ $^{15}NO_3$	<1 %	Plant size	Teste et al. (2009)

transfer has been shown to improve productivity not only of the N-poor-receivers but also the N-rich donor plants.

Nitrogen transfer between non-N_2-fixing donors and receivers has also been examined in temperate forests (He et al. 2006; Teste et al. 2009), which is useful not only for understanding N dynamics and distribution in N-limited ecosystems, but also for improving our understanding of the role of N in regulating N and C transfers through networks. In these systems, one-way transfer from conifer trees to EM conspecifics or other EM or AM heterospecifics has generally been very low (<5 % of N pulse) compared to transfer from N_2-fixing to non-N_2-fixing plants. Teste et al. (2009) found that N-rich *Pseudotsuga menziesii* saplings transferred both N and C to N-poor conspecific germinants through mycorrhizal networks and that this corresponded with greater two-year seedling survival. The relative amounts of N (0.0018 %) and C (0.0063 % of photo-assimilate) transferred suggest that they moved together in amino acids (a stoichiometry of 2N:7C; which is similar to glutamine 2N:5C; alanine, cysteine and lysine 2N:6C; threonine and aspartic acid 2N:8C), but the compounds were never identified. He et al. (2006) found that EM *Pinus sabiniana* rapidly transferred N to other *P. sabiniana*, AM-EM *Quercus douglasii*, AM *Ceanothus cuneatus* and AM herbs, suggesting transfer occurred through multiple belowground pathways. In a recent study, Teste et al. (2015) found that non-N2-fixing donor plants transferred about 4 % of their N to receivers, with greater enrichment in receivers that formed ectomycorrhizas or cluster roots than AM receivers.

Differences in natural abundance $\delta^{15}N$ signatures between fungi, plants and soils, or N_2-fixing and non-N_2-fixing plants, or non-mycorrhizal AM, EM and ERM mycorrhizal fungi and plants, potentially provide a novel non-destructive approach for examining N transfer in intact ecosystems (Hobbie and Hobbie 2008; He et al. 2009). These differences exist because soil organisms and mycorrhizal fungi discriminate against ^{15}N when they decompose organic matter and formulate and transfer compounds to plants; hence, host plants are depleted in ^{15}N, whereas mycorrhizal fungi are enriched in ^{15}N (Hobbie and Hobbie 2008). Based on these principles, Hobbie and Hobbie (2008) estimated that up to 20 % of net primary production is used to support mycorrhizal fungi in N-limited forests and arctic tundra ecosystems. Differences in natural abundance of $\delta^{13}C$ signatures have been successfully applied to quantify C transfer through networks between AM C_3 and C_4 plants (Watkins et al. 1996), but to our knowledge few have used $\delta^{15}N$ signatures to elucidate N transfer through EM networks. In a sub-boreal spruce forest, Kranabetter and MacKenzie (2010) inferred that understory partial MH plants, *Orthillia secunda* and *Pyrola ascarifolia* (Tedersoo et al. 2007), acquired their N from mycorrhizal networks. This is because their foliar $\delta^{15}N$ signatures were as high as nearby EM sporocarps. This contrasted with EM *Abies lasiocarpa* and ERM *Vaccinium membranaceum*, whose $\delta^{15}N$ signatures increased with both N concentration and soil N supply, suggesting that their N was supplied via mycorrhizal uptake of inorganic and/or organic N from soils.

5.3.3 Phosphorus

Phosphorus has been shown to transfer between plants much less freely than N, and this is thought to be due to its lower soil mobility (Lambers et al. 2010). Other reasons in some ecosystems may include lower plant demand or less leakage out of roots (Eissenstat and Newman 1990; Newman et al. 1992; Johansen and Jensen 1996). Transfer of P has been little studied in EM networks, contrasting with AM networks. In AM networks, P transfer between plants has been considered too small (0.1–3 %) or too slow to be of much relevance to living plants, except possibly under highly P deficient conditions (Newman and Eason 1989a, b; Eason and Newman 1990a, b; Ikram et al. 1994; Johansen and Jensen 1996; Yao et al. 2003; Wilson et al. 2006). For example, Newman and Eason (1989a, b) found that P transfer between AM plants was slower than the life-span of leaves or tillers, while transfer between tillers within a plant was faster than their life-span. This contrasts with senescing, stressed or grazed AM donor plants, however, which have been shown to deliver larger quantities of P or N to living AM neighbors, even into their shoots (Eason and Newman 1990a, b; Hamel et al. 1991; Ikram et al. 1994; Bethlenfalvay et al. 1991; Johansen and Jensen 1996; Tuffen et al. 2002). In this case, the dying plant roots appear to exude large amounts of P or N into their mycorrhizas, and this is then available for transfer to connected plants of the same species. Nutrients captured from dying roots have been shown to increase several-fold if the dying and living roots shared the same mycorrhizal fungi, but the involvement of direct hyphal links has sometimes been inconclusive (Ikram et al. 1994; Johansen and Jensen 1996).

Evidence for P transfer in EM networks is considerably more equivocal, and none of the studies appear to clearly demonstrate that P can move from one EM plant to another via EM networks. In the first study to investigate P transfer, Finlay and Read (1986) fed ^{32}P to the rhizomorphs of *Suillus bovinus* connecting *Pinus contorta* and *Pinus sylvestris*. They found that the isotope was transferred up to 40 cm one-way by symplastic flow through the rhizomorphs into both of the host plants and throughout the mycelium. This appeared related to plant size and mycorrhizal colonization, but not to transpiration rates. The isotope did not move in the other direction, however, namely from the roots of the plants back into the mycelium, indicating one-way flow (Finlay and Read 1986). In a later study, Finlay (1989) showed that ^{32}P fed to the mycelium of *Larix*-specific *Suillus bovinus* (Basionym *Boletinus cavipes*) moved preferentially to *Larix eurolepis* compared to *Pinus sylvestris* linked together in an EM network, and this was related to the considerably greater fungal colonization of *L. eurolepis*. There was no evidence that the P moved from one plant to another in this study either. Moreover, this result must be interpreted cautiously given that the fungus may have low compatibility or even parasitic interactions with *Pinus sylvestris,* even under laboratory conditions.

5.3.4 Water

Hydraulic redistribution via mycorrhizal networks is a potentially important mechanism for recharging depleted surface soil layers during drought, delaying or reducing the loss of root and hyphal conductivity, and reducing root or hyphal turnover (Warren et al. 2008). It could allow roots and hyphae to recover quickly from drought, and even small amounts of hydraulically redistributed water could aid in the transport of nutrients through mycorrhizal networks. Plants in a mycorrhizal network that are the source of hydraulically redistributed water benefit from rehydration and increased longevity of their mycorrhizal root tips (Querejeta et al. 2003). Connected receiver plants have a more secure source of water and nutrients (Schoonmaker et al. 2007). Thus, the daily rehydration of roots and hyphae via hydraulic redistribution should enhance both long-term resource acquisition by plants connected in a mycorrhizal network as well as nutrient cycling by the soil microbial community (Querejeta et al. 2007; Egerton-Warburton et al. 2008; Prieto et al. 2012). Lilleskov et al. (2009) found that hydraulic redistribution through mycorrhizal pathways was great enough to push sporocarps through crusts, rock and cement fractures during extremely dry seasons.

Hydraulic redistribution between plants via EM has been demonstrated in the lab (Querejeta et al. 2003, 2012; Egerton-Warburton et al. 2007; Plamboeck et al. 2007; Bingham and Simard 2011) and the field (Schoonmaker et al. 2007; Warren et al. 2008) (Table 5.3). Querejeta et al. (2003) used laboratory microcosms to show that EM fungi provided a pathway for movement of ^2H_2O and fluorescent dye from donor EM plants to soil through hydraulic redistribution. Brooks et al. (2006) observed that ^2H_2O applied to the soil moved horizontally through the soil matrix in an asymmetric pattern, presumably due to hydraulic redistribution by both interior Douglas-fir and EM fungi. Field experiments have shown that movement of $^2H_20$ applied directly to soil varies horizontally depending on the potential to form an EM network, soil moisture conditions and the functional form of the plants present (Brooks et al. 2006; Egerton-Warburton et al. 2007; Schoonmaker et al. 2007). Recent studies have shown that water hydraulically lifted via taproots is transferred directly to EM fungal symbionts and subsequently translocated from plant to plant via the EM network independently of soil pathways (Egerton-Warburton et al. 2007; Bingham and Simard 2011). In a field study, EM networks were shown to provide a direct pathway for hydraulic redistribution of ^2H_2O from old stumps to nearby ponderosa pine seedlings (Warren et al. 2008).

Hydraulic redistribution is thought to be an especially important process in ecosystems that experience annual or seasonal drought because it can help maintain the integrity of the plant-fungal-soil system, limiting the possibility of embolism-induced catastrophic hydraulic failure of plants (Warren et al. 2008) and leading to a more diverse community of plants and fungi. Hydraulic redistribution has occurred in soils as wet as −0.05 MPa soil water potential, but it appears to be most important in drier soils at −0.4 to −0.8 MPa, or lower (Querejeta et al. 2007). In support of this, Schoonmaker et al. (2007) found greater hydraulic redistribution

Table 5.3 Summary of published studies examining water transfer or hydraulic redistribution involving mycorrhizas or ectomycorrhizal networks

Donor plant	Receiver plant	Networking fungus	Element transferred	% of seedling water that was transferred	Water potential gradient	Other factors tested	Reference
Pinus sylvestris	*Pinus sylvestris*	*Suillus bovinus*	$^{3}H_2O$	Not known, but transferred at rate of 27 cm h^{-1}	Cup of labelled water placed near rhizomorph system	Not necessarily tested for interplant transfer, but autoradiographs suggest so	Duddridge et al. (1980)
Pinus sylvestris	*Pinus sylvestris*	*Suillus bovinus*	$^{14}CO_2$	Not known, but needle water potential of receivers reduced by half and seedlings died one week after severing	Severed mycelia strands, creating water potential gradient detected in shoots of receiver plant	^{14}C transferred from donor to roots of receiver seedling	Brownlee et al. (1983)
Artemisia tridentata	*Artemisia tridentata* in mixture with *Agropyron spicatum* and *A. desertorum*	Conducted in the field. Mycorrhizal networks not tested but considered important in process	Examined Ψ_s	Not known, but Ψ_s of soil layers increased by approx. 1 l m^{-2} of ground area per night	Diel gradient in Ψ_s, manipulated by reducing transpiration of *Artemisia* plants	All AM plants considered beneficiaries of increased Ψ_s	Richards and Caldwell (1987)

(continued)

Table 5.3 (continued)

Donor plant	Receiver plant	Networking fungus	Element transferred	% of seedling water that was transferred	Water potential gradient	Other factors tested	Reference
Quercus agrifolia	None, *Quercus agrifolia* in laboratory microcosms	*Cortinarius collinitus, Cenococcum geophilum*, Glomalean spp.	Fluorescent tracers (Lucifer Yellow carbohydrazide, Cascade Blue hydrazide, Sulforhodamine G)	Not known, but Ψ_s increased dramatically overnight, most mycorrhizas were labelled after night-time labelling of roots. HR occurred via mycorrhizas	Induced drought by withholding irrigation from upper chambers	Labeling of mycorrhizas declined with continuing drought and age, but mycorrhizal fungi persisted at $\Psi_s < -20$ MPa, their integrity possibly maintained by HR	Querejeta et al. (2003)
Pseudotsuga menziesii	*Pseudotsuga menziesii* and soil	Field soil. Mycorrhizal networks not tested but considered important in process	Examined Ψ_s changes, D_2O water	50 % of water in Douglas-fir from HR, <1 % in other plants from HR. HR was 0.17 mm day^{-1} when Ψ_s −1.0 MPa in August, replenishing 50 % of water depleted daily	Diel gradient in Ψ_s, seasonal variation in Ψ_s, increased source water potential through irrigation. Most HR occurred at night and in the latter part of the summer	D_2O water appeared in plants 5 m from source, including EM *Pseudotsuga menziesii*, EM *Tsuga heterophylla*, EM *Mahonia berberis*, but little in ERM *Vaccinium*	Brooks et al. (2006)

(continued)

Table 5.3 (continued)

Donor plant	Receiver plant	Networking fungus	Element transferred	% of seedling water that was transferred	Water potential gradient	Other factors tested	Reference
Quercus agrifolia (coastal live oak) seedlings	EM *Quercus agrifolia* and AM *Q. Agrifolia* and *Salvia mellifera,* but no transfer to AM *Eriogonum Fasciculatum* or *Keckiella antirrhinoides*	Mixture of EM suspensions in mesocosms: *Cenococcum geophilum, Cortinarius evernius, Boletus amygdalinus, Boletus dryophilus, Pisolithus tinctorius*	D_2O, Lucifer yellow carbohydrazide dye	Unknown, transferred to xylem of mycorrhizal roots, hyphae and soil	Oaks with taproot accessing in water source to depleted EM and AM seedlings	More dye transferred to plant species with greater mycorrhizal dependency, varied with mycorrhizal species	Egerton-Warburton et al. (2007)
Arctostaphylos viscida (manzanita) shrub	*Pinus lambertiana* (sugar pine) and *Pseudotsuga menziesii* (Douglas-fir) seedlings	Field soil and root fragments	D_2O, Lucifer yellow carbohydrazide dye	0.01–0.04 % of D_2O to Douglas-fir roots only; dye transferred to both species, but more to sugar pine	Water with tracers added to Manzanita compartment	Amount of dye transferred varied with mycorrhizal species	Plamboeck et al. (2007)
Pseudotsuga menziesii mature tree	*Pseudotsuga menziesii* establishing seedling	Field soil, no difference in transfer through MN versus soil	D_2O	1.4 % transferred, representing 21.6 % of seedling water	Mature tree with taproot to seedlings	Decreased with distance	Schoonmaker et al. (2007)

(continued)

Table 5.3 (continued)

Donor plant	Receiver plant	Networking fungus	Element transferred	% of seedling water that was transferred	Water potential gradient	Other factors tested	Reference
Pinus ponderosa cut stump	*Pinus ponderosa* establishing seedling	Field soil, transfer through MN	D_2O, acid fuchsin dye	25 % more transferred through MN than soil; 15 % of total site water use	From cut mature stump to seedlings	Increased with cut tree age	Warren et al. (2007, 2008)
Pseudotsuga menziesii seedling	*Pseudotsuga menziesii* seedling	Field soil, transfer through MN under ambient (CO_2)	D_2O	Not known	Established seedling with taproot in water source to depleted seedling	Decreased with (CO_2)	Bingham and Simard (2011)
Quercus agrifolia	*Quercus agrifolia*	*Pisolithus tinctorius*, *Scleroderma* sp., plus field inoculum applied to laboratory microcosms; fungicide added to half of the microcosms as a control	D_2O	Not known, but >10 ‰ enrichment in receiver seedlings	Irrigation withdrawal from donor chamber, Ψ_s as low as −1.2 MPa; unexpectedly greater Ψ_s gradient resulted from fungicide application	Increased transfer with fungicide application, likely due to reduced water retention resulting from death of hyphae	Querejeta et al. (2012)

Ψ_s = soil water potential. HR = hydraulic redistribution from trees growing in drier compared to wetter site conditions, regardless of the redistribution pathway. The importance of EM networks versus other pathways for hydraulically redistributed water likely increases with aridity, where resistance of water movement through an EM network should be less than that from dry soils to roots (Ishikawa and Bledsoe 2000). Indeed, Bingham and Simard (2011) found that hydraulic redistribution through EM networks increased with declining CO_2 concentration, and may have been a mechanism for increased survival of water-stressed seedlings. While some EM fungi may be less active under drier conditions, others are known to persist under dry soil conditions, particularly those forming rhizomorphs (Parke et al. 1983; Molina et al. 1992; Nardini et al. 2000). In this case, the long-distance rhizomorphs allow these fungi to access water elsewhere, such as moist microsites nearby.

The physiological relevance of water transfer depends on whether the amount of hydraulically redistributed water affects water relations of the plants, their mycorrhizal hyphae, or the soil in which the hyphae and plant roots are growing. Hydraulic redistribution from *Acer saccharum* trees accounted for 3–60 % of water uptake by understory plants (Dawson 1993). In a dry interior Douglas-fir forest, Schoonmaker et al. (2007) found that 22 % of seedling water was supplied via soil and EM network pathways from nearby conspecific mature trees. In a coastal Douglas-fir forest, Brooks et al. (2006) found up to 50 % of water in Douglas-fir seedlings was acquired by hydraulic redistribution through EM networks. Warren et al. (2007) found that the amount of hydraulically redistributed water through EM networks in a ponderosa pine forest amounted to <0.15 mm day^{-1} or up to 15 % of total site water use. The benefits to receiver plants of hydraulically redistributed water via EM networks, however, must be weighed against the costs of competition of the source tree for the same redistributed water or surrounding soil water (Schoonmaker et al. 2007).

The relevance of hydraulic redistribution through mycorrhizal networks is also related to the speed and distance over which water is transferred relative to other pathways. The speed at which water transports through mycorrhizal networks has ranged from 0.13 to 2.2 m day^{-1} (Duddridge et al. 1980; Brownlee et al. 1983; Warren et al. 2008), which is similar to transport rates in xylem but generally faster than the <0.4 m day^{-1} transferred through soils. The distance over which water has been hydraulically redistributed has also varied with time. Warren et al. (2008) found that 90 year-old source trees released hydraulically redistributed water up to 7 m away over 21 days, and the zone of influence increased with time and the age of the source trees. Schoonmaker et al. (2007) found that hydraulically distributed water in receiver seedlings tended to decrease with distance from source trees, but that it occurred up to 5 m away, within the range of earlier studies (Dawson 1993; Ludwig et al. 2003).

Like other resource transfers, hydraulic redistribution is a patchy process, and is a function of the variation in water potential gradients among soils, plants and the mycorrhizal network. Any factor that affects the spatial or temporal patchiness of water potential gradients in soils, plants and the mycorrhizal network should thus influence the magnitude of hydraulic redistribution. These factors can include weather, topography, soil properties, phenology or plant-fungal community characteristics. In addition, any factor that reduces plant transpiration, such as drought, herbivory or disturbance should enhance hydraulic redistribution via mycorrhizal networks, provided the network remains functional. The environmental controls on hydraulic redistribution via mycorrhizal networks have been addressed in only a handful of studies, leaving many questions waiting to be answered.

5.4 Experimental Designs for Mycorrhizal Network Studies in the Field

Since the classic studies of Björkman (1960) and Reid and Woods (1969), very few manipulated-field experiments focused on mycorrhizal networks have been conducted (see Hoeksema, Chap. 9, this volume). Feasibility and financial costs still remain considerable barriers to field studies addressing mycorrhizal network ecophysiology. The lack of feasibility is mostly due to the delicate nature of the hyphal networks and their potentially delayed establishment after experimental setup. Like many ecological field studies, rigorous ‘controls’ in experimental design are challenging. In mycorrhizal network studies, controls such as AM host plants (in an EM system) and fine mesh barriers to prevent hyphal movement have been used successfully (Simard et al. 1997a, b; Teste et al. 2009; Bingham and Simard 2011, 2012a, b; Deslippe and Simard 2011). In biodiverse ecosystems with abundant AM and EM plants, non-mycorrhizal plants (e.g., Proteaceae species) could be used with or without mesh barriers (Teste et al. 2014, 2015). Similarly, in studies examining interspecific transfer, using plants that form mycorrhizas only with fungal species that do not form a network with other plant species, would be an ideal control since differences would not be due to lack of mycorrhizal formation. This setup would resemble the fine mesh barrier approach used by Booth (2004) and Teste et al. (2009), with the added feature of comparing the effect of mycorrhizal roots intermingling with mycorrhizal roots that do or do not form mycorrhizal networks.

Booth (2004) used both EM and AM species and a novel combination of trenching and physical barriers (“networking cylinders”) that allowed him to include almost all possible combinations of mycorrhizal networking and direct root interactions. Seedlings were planted into four root/mycorrhizal network treatments in the field surrounded by overstorey trees (Booth 2004; Booth and Hoeksema 2010). Seedlings were planted (i) directly into the soil (root and mycorrhizal networks allowed); (ii) into circular slit trenches, preventing interactions with overstorey roots and mycorrhizal mycelia (no roots and no networks); (iii) into “networking cylinders”, to prevent direct root competition with overstorey trees, but enable mycorrhizal mycelia to invade and potentially form mycorrhizal networks (no roots with mycorrhizal network); or (iv) into “trenched networking cylinders” where “networking cylinders” were installed and also trenched to detect soil moisture or increased mineralization effects of the installation (no roots and no mycorrhizal network with disturbance effects).

The experimental design of Booth (2004) is quite robust, and we recommend it for mycorrhizal network studies in forests or in soils that permit slit trenches. In more sandy soils or easily manipulated agriculture soil, the ‘rotated-core method’ introduced by Johnson et al. (2002) may be best. Here the non-mycorrhizal network treatment is achieved by periodic rotation of the cores; analogous to manual severing of roots and hyphae accomplished in rhizoboxes in the laboratory (Philip et al. 2010). Teasing apart belowground interplant transfer pathways with cores, mesh barriers, physically severing links, or small slit trenches around plants are most commonly used for ecological field experiments. However, all of these approaches have unintended effects that are either minimized or accounted for with secondary control treatments, such as the combined use of trenching and "networking cylinders" described by Booth (2004).

The use of C_3 and C_4 plants in conjunction with a clever microcosm and stable isotope tracing can be a powerful alternative to using cores and mesh, especially when studying the functioning of mycorrhizal networks (Walder et al. 2012). The existence of a strong difference in isotope fractionation during C_3 and C_4 photosynthesis gives these types of plants a distinct C isotope ratio, sufficiently different (>15 ‰ in many cases) to allow the tracing of C in mycorrhizal network studies (Walder et al. 2012). Using the same concept, myco-heterotrophic plants permit an even more robust model system if they can be grown in microcosms or included in manipulated field experiments. Hynson et al. (2012) conducted a full factorial field manipulation experiment that quantified the contribution of mycorrhizal networks in the C economy of forest understory plants. They implemented a mycorrhizal network treatment with trenching around small colonies of Pyroleae plants (no access to the networks), in conjunction with the well-established stable isotope and mixed modeling methodology (Preiss and Gebauer 2008). In the field, MH plants could also be used in conjunction with cores or mesh to sever links to the mycorrhizal network and determine the biological importance of such links for survival, growth, and C and nutrient status. Similar to differences in natural abundance $\delta^{13}C$ signatures between C_3 and C_4 plants, $\delta^{15}N$ signatures can also be used to elucidate N transfer through EM networks, as discussed above. This is more difficult, however, due to the narrower $\delta^{15}N$ signature differences that normally occur between plants under natural conditions.

5.5 Resource Transfers and Complexity Models

The flow of resources through EM networks is compatible with complex adaptive systems (CAS) theory that postulates ecosystems can be represented by energy, resource and information flows among parts (Levin 2005). In CAS, ecosystems are modelled as adaptive dynamic networks of interacting parts where feedbacks and cross-scale interactions lead to self-organization and emergent properties (Bascompte 2009; Parrott 2010). The spatial and temporal patterns in ecosystems have commonly been modeled as networks, and usually have been characterized as complex with non-linear, scale-free (or power law) topology and behaviour (Sole et al. 2002). Thus, the scale-free topology of EM networks described by Beiler et al. (2010) is consistent with self-organization in CAS, where mycorrhizal colonization and nutrient fluxes through the MN provide feedbacks (positive or negative) to plants that can influence the stability of the ecosystem. Mycorrhizal networks can thus be considered fundamental agents of self-organization in ecosystems because they provide direct avenues for feedbacks and cross-scale interactions via resource transfers that are large enough to affect plant establishment and growth, and hence ecosystem structure and function (Simard et al. 2012).

A fundamental property of MNs as agents of CAS is that the parts (e.g., plant and fungal species) are subject to selective pressures through localized interactions with each other, other parts and processes, leading to local adaptation and influence on the functioning of the network (Sole et al. 2002). The local, bottom-up, iterative development of nodes and links through differential growth, strengthening and weakening (e.g., self-thinning or pruning of plants or fungi) that is characteristic of MNs (Boddy and Jones 2007; Heaton et al. 2012) is also a fundamental feature of CAS. In theory, mycorrhizal networks have high adaptive capacity because the mutualism is reacting, adapting and evolving to the constant change brought by shifts in local interactions among symbionts and with their environment, and because members of the network together comprise high genetic diversity.

These behaviours, adaptive capacities and interactions of the network parts can influence the whole ecosystem (Gorzelak et al. 2015). In CAS, it is increasingly understood that the global system is comprised of many local interacting and overlapping complex systems, where local state-shifts can rapidly propagate to cause state-shifts of the whole system. In other words, a forcing at one scale can cause a critical transition to occur on another scale; hence it is important to understand local scale processes such as nutrient fluxes through EM networks so that we can predict how they may propagate upward to affect higher-level processes such as global change (Barnosky et al. 2012). For example, local extinction of strong networking fungal species and mortality of tree seedlings dependent on mycorrhizal networks for establishment could reduce forest regeneration, causing a reduction in forest cover that cascades upward to affect climate. Thus, modelling the dynamic interactions and selection pressures in networks will help us understand, predict and manage the dynamics and resilience of ecosystems under changing environments.

5.6 Conclusions and Future Directions

In this chapter, we have shown that resource transfers through EM networks are highly variable, and the magnitude, direction and fate depends on the resource being transferred, the pattern and strength of energy potential gradients across plants, fungi and soils within the system, and the degree and type of environmental stress that the system is under. Some studies provide support that C, N and H_2O transfers can be sufficiently large to improve plant establishment and growth, particularly in MH plants or plants that are nutrient- or water-deficient due to environmental limitations. For example, C fluxes through EM networks have been shown to supply 0–10 % of autotrophic, up to 85 % of partial MH, and 100 % of fully MH plant C. This C supply has been associated with increased survival and growth of autotrophic plants, and appears essential for survival of MH plants.

Nitrogen fluxes between N_2-fixing and non-N_2-fixing plants have supplied up to 40 % of receiver N and this has been associated with increased plant productivity. Hydraulic redistribution involving EM fungi has supplied up to 50 % of plant water, and in some cases this has been shown as essential for plant survival, but how much of this water transfers through EM networks remains uncertain. Phosphorus transfer through EM networks has not been adequately demonstrated nor related to plant performance, and needs to be better studied. Indeed, the majority studies on EM networks do not adequately relate resource transfers to plant performance.

Our findings support the idea that EM networks are avenues, and transferred compounds are messengers, for cross-scale interactions and feedbacks within communities, consistent with complex adaptive system theory. Our experiments in temperate forests and Arctic tussock tundra demonstrate that these feedbacks through EM networks could either stabilize or destabilize plant communities as climate changes.

Considerably more research is needed to better understand how resource transfers through EM networks affect ecosystem structure and function. The following eight areas form a partial list of research needs. First, better methods are needed for long-term tracer experiments to determine how nutrient transfers affect long-term plant performance, interspecific interactions and community succession. This research needs to be carried out in nature, not just labs, so that we can more realistically understand the importance of EM networks in ecosystems (Simard and Durall 2004; van der Heijden and Horton 2009). Natural experiments that examine spatial and temporal patterns in natural abundance of stable isotopes, combined with molecular biology, show promise in this area. Second, the resource compounds that are transferred, including hormones, defence signals and allelopathic chemicals, and the genes that regulate the transfer and receipt of these compounds, need to be identified (Song et al. 2010; Barto et al. 2011; Witzany 2012). Multi-factorial experiments that investigate interactions among different transferred compounds and the different genes involved should also be conducted. With this understanding, better approaches for conservation management of ecosystems that are resilient to stress and disturbance can be developed. Third, the stoichiometry of transferred compounds should be better studied as a method to develop cost-benefit models of trade predicting plant community development under different conditions (Johnson 2009). This should involve quantification of isotope signatures of the component plant and fungal species and soil C pools (Hobbie and Hobbie 2008). Ultimately, models should be expanded to include transfer of signals, hormones and allelochemicals, not just C, nutrients and H_2O. Fourth, the roles of EM networks, other associated soil organisms and resource transfers in the sequestration, storage and stabilization of C, and the cycling of nutrients and H_2O in ecosystems, is urgently needed as habitat conversion and global change threaten to destabilize these pools (Treseder 2004; Treseder et al. 2007; Cairney 2012). This research should include the role of networks in stabilizing existing plant communities to mitigate large pulses of C to the atmosphere as pressures from drought, insects and diseases increase. Fifth, how resource transfer through EM networks interact with resource cycling in other biotic networks, including microbial, plant and animal networks, is needed so that conservation efforts better target key structures and functions to maintain ecosystem stability (Simard et al. 2013). Sixth, the role mycorrhizal networks and nutrient fluxes play in facilitating or inhibiting migration of plant and fungal species, including invasion of unwanted weeds, is needed. This knowledge can help with restoring ecosystems, enhancing their adaptive capacity, or with designing assisted migrations strategies that are being widely considered world-wide to smooth ecosystem transitions as climate changes (Aitken et al. 2008; Pickles et al. 2011). Seventh, the role mycorrhizal networks play in selection, and how this informs theories of evolution and ecology should be explored using experimentation and modeling. This can build on the important work of Dudley and File (2008) and Whitham et al. (2006). Finally, it is critical that mycorrhizal researchers become increasingly involved in inter-disciplinary research to help address the highly complex problems we face with global change. Only through collaborative research and communication will the role of mycorrhizas in ecology and evolution be appreciated, understood and incorporated in global conservation strategies. For example, analysis and modelling nutrient fluxes through interacting meta-networks within the framework of CAS theory could be useful for conservation management if the long-term objectives are to maintain resilient ecosystems or assist in re-organization of novel ones that are productive, adaptive and resilient to global change.

References

Abuzinadah RA, Read DJ (1989) Carbon transfer associated with assimilation of organic nitrogen sources by silver birch (*Betula pendula* Roth.). Trees 3:17–23

Agerer R (2001) Exploration types of ectomycorrhizae. A proposal to classify ectomycorrhizal mycelial systems according to their patterns of differentiation and putative ecological importance. Mycorrhiza 11:107–114

Agerer R (2006) Fungal relationships and structural identity of their ectomycorrhizae. Mycol Prog 5:67–107

Aitken SN, Yeaman S, Holliday JA, Wang T, Curtis-McLane S (2008) Adaptation, migration or extirpation: climate change outcomes for tree populations. Evol Appl 1:95–111

Allen MF (2007) Mycorrhizal fungi: highways for water and nutrients in arid soils. Vadose Zone J 6:291–297

Allen MF (2009) Bidirectional water flows through the soil–fungal–plant mycorrhizal continuum. New Phytol 182:290–293

Anand M, Gonzalez A, Guichard F, Kolasa J, Parrot L (2010) Ecological systems as complex systems: challenges for an emerging science. Diversity 2:395–410

Anisimov OA, Vaughan DG, Callaghan TV, Furgal C, Marchant C, Prowse TD, Vilhjalmsson H, Walsh JE (2007) Polar regions (Arctic and Antarctic). In: Parry ML, Canziani OF, Palutikof JP, van der Linden PJ, Hanson CE (eds) Climate Change 2007: impacts, adaptation and vulnerability, Contribution of Working Group II to the fourth assessment report of the Intergovernmental Panel on Climate Change. Cambridge University Press, Cambridge, pp 653–685

Arnebrant K, Ek H, Finlay RD, Söderström B (1993) Nitrogen translocation between *Alnus glutinosa* (L.) Gaertn. Seedlings inoculated with *Frankia* sp. and *Pinus contorta* Dougl. ex Loud seedlings connected by a common ectomycorrhizal mycelium. New Phytol 24:231–242

Asay A (2013) Mycorrhizal facilitation of kin recognition in interior Douglas-fir (*Pseudotsuga menziesii* var. *glauca*). Master's thesis, University of British Columbia, Vancouver, Canada

Augé RM, Toler HD, Saxton AM (2014) Arbuscular mycorrhizal symbiosis alters stomatal conductance of host plants more under drought than under amply watered conditions: a meta-analysis. Mycorrhiza 25:13–24

Babikova Z, Gilbert L, Bruce TJA, Birkett M, Caulfield JC, Woodcock C, Pickett JA, Johnson D (2013) Underground signals carried through common mycelial networks warn neighbouring plants of aphid attack. Ecol Lett 16:835–843

Bago B, Zipfel W, Williams RM, Jun J, Arreola R, Lammers PJ, Pfeffer PE, Shachar-Hill Y (2002) Translocation and utilization of fungal storage lipid in the arbuscular mycorrhizal symbiosis. Plant Physiol 128:109–124

Bahram M, Põlme S, Kõljalg U, Tedersoo L (2011) A single European aspen (*Populus tremula*) tree individual may potentially harbour dozens of *Cenococcum geophilum* ITS genotypes and hundreds of species of ectomycorrhizal fungi. FEMS Microbiol Ecol 75:313–320

Barnosky AD, Hadley EA, Bascompte J et al (2012) Approaching a state shift in Earth's biosphere. Nature 486:52–58

Barto EK, Hilker M, Muller F, Mohney BK, Weidenhamer JD, Rillig MC (2011) The fungal fast lane: common mycorrhizal networks extend bioactive zones of allelochemicals in soils. PLoS ONE 6(11):e27195. doi:10.1371/journal.pone.0027195

Bascompte J (2009) Mutualistic networks. Front Ecol Env 7(4):29–436

Bascompte J, Jordano P, Melian CJ, Olesen JM (2003) The nested assembly of plant-animal mutualistic networks. PNAS 100:9382–9387

Beiler KJ, Durall DM, Simard SW, Maxwell SA, Kretzer AM (2010) Mapping the wood-wide web: mycorrhizal networks link multiple Douglas-fir cohorts. New Phytol 185:543–553

Beiler KJ, Simard SW, Durall DM (2015) Topology of tree-mycorrhizal fungus interaction networks in xeric and mesic Douglas-fir forests. J Ecol. doi:10.1111/1365-2745.12387

Bertness MD, Callaway R (1994) Positive interactions in communities. Trends Ecol Evol 9:191–193

Bethlenfalvay GJ, Reyes-Solis MG, Camel SB et al (1991) Nutrient transfer between the root zones of soybean and maize plants connected by a common mycorrhizal mycelium. Physiol Plant 82:423–432

Bidartondo MI (2005) The evolutionary ecology of myco-heterotrophy. New Phytol 167:335–352

Biedrzycki ML, Jilany TA, Dudley SA, Bais HP (2010) Root exudates mediate kin recognition in plants. Commun Integr Biol 3:28–35

Bingham MA, Simard SW (2011) Do mycorrhizal network benefits to survival and growth of interior Douglas-fir seedlings increase with soil moisture stress? Ecol Evol 306–316

Bingham MA, Simard SW (2012a) Ectomycorrhizal networks of old *Pseudotsuga menziesii* var. *glauca* trees facilitate establishment of conspecific seedlings under drought. Ecosystems 15:188–199

Bingham MA, Simard SW (2012b) Mycorrhizal networks affect ectomycorrhizal fungal community similarity between conspecific trees and seedlings. Mycorrhiza 22:317–326

Bingham MA, Simard SW (2013) Seedling genetics and life history outweigh mycorrhizal network potential to improve conifer regeneration under drought. For Ecol Manage 287:132–139

Björkman E (1960) *Monotropa hypopitys* L.—An epiparasite on tree roots. Physiol Plant 13:308–327

Bledsoe C, Allen MF, Southworth D (2014) Beyond mutualism: complex mycorrhizal interaction. In: Lüttge U et al (eds) Progress in botany, vol 75. Doi:10.1007/978-3-642-38797-5_10. © Springer, Berlin

Boddy L, Jones TH (2007) Mycelial responses in heterogeneous environments: parallels with macroorganisms. In Gadd G, Watkinson SC, Dyer P (eds) Fungi in the environment. Cambridge University Press, Cambridge, pp 112–158

Booth MG (2004) Mycorrhizal networks mediate overstorey–understorey competition in a temperate forest. Ecol Lett 7:538–546

Booth MG, Hoeksema JD (2010) Ectomycorrhizal networks counteract competitive effects of canopy trees on seedling survival. Ecology 91:2294–2302

Bornyasz MA, Graham RC, Allen MF (2005) Ectomycorrhizae in a soil-weathered granitic bedrock regolith: linking matrix resources to plants. Geoderma 126:141–160
Bougoure JJ, Brundrett MC, Grierson PF (2010) Carbon and nitrogen supply to the underground orchid, *Rhizanthella gardneri*. New Phytol 186:947–956
Bret-Harte MS, Mack MC, Goldsmith GR et al (2008) Plant functional types do not predict biomass responses to removal and fertilization in Alaskan tussock tundra. J Ecol 96:713–726
Brooks R, Meinzer FC, Warren JM, Domec J-C, Coulombe R (2006) Hydraulic redistribution in a Douglas-fir forest: lessons from system manipulations. Plant Cell Env 29:138–150
Brownlee C, Duddridge JA, Malibari A, Read DJ (1983) The structure and function of mycelial systems of ectomycorrhizal roots with special reference to their role in forming inter-plant connections and providing pathways for assimilate and water transport. Plant Soil 71:433–443
Büking H, Heyser W (2001) Microautoradiographic localization of phosphate and carbohydrates in mycorrhizal roots of *Populus tremula* × *Populus alba* and the implications for transfer processes in ectomycorrhizal associations. Tree Physiol 21:101–107
Butterfield BJ (2009) Effects of facilitation on community stability and dynamics: synthesis and future directions. J Ecol 97:1192–1201
Cairney JWG (2005) Basidiomycete mycelia in forest soils: dimensions, dynamics and roles in nutrient distribution. Mycol Res 109:7–20
Cairney JWG (2012) Extramatrical mycelia of ectomycorrhizal fungi as moderators of carbon dynamics in forest soil. Soil Biol Biochem 37:198–208
Cameron DD, Preiss K, Gebauer G, Read DJ (2009) The chlorophyll-containing orchid *Corallorhiza trifida* derives little carbon through photosynthesis. New Phytol 183:358–364
Cameron DD, Johnson I, Leake JR, Read DJ (2010) Giving and receiving: measuring the carbon cost of mycorrhizas in the green orchid, *Goodyera repens*. New Phytol 180:176–184
Chapin FS III, Shaver GR, Giblin AE, Nadelhoffer KJ, Laundre JA (1995) Responses of Arctic tundra to experimental and observed changes in climate. Ecology 76:694–711
Chiarello NJ, Hickman C, Mooney HA (1982) Endomycorrhizal role in interspecific transfer of phosphorus in a community of annual plants. Science 217:941–943
Connell JH, Slatyer RO (1977) Mechanisms of succession in natural communities and their role in community stability and organization. Am Nat 111:1119–1144
Dawson TE (1993) Hydraulic lift and water use by plants: implications for water balance, performance and plant–plant interactions. Oecologia 95:565–574
Deslippe JR, Simard SW (2011) Below-ground carbon transfer among *Betula nana* may increase with warming in Arctic tundra. New Phytol 192:689–698
Deslippe JR, Hartmann M, Mohn WW, Simard SW (2011) Long-term experimental manipulation of climate alters the ectomycorrhizal community of *Betula nana* in Arctic tundra. Global Change Biol 17:1625–1636
Deslippe J, Hartmann M, Grayston SJ, Simard SW, Mohn WW (2015) Stable isotope probing implicates Cortinarius collinitus in carbon transfer through ectomycorrhizal mycelial networks in the field. New Phytologist (submitted)
Deveau A, Kohler A, Frey-Klett P, Martin F (2008) The major pathways of carbohydrate metabolism in the ectomycorrhizal basidiomycete *Laccaria bicolor* S238N. New Phytol 180:379–390
Dickie IA, Guza RC, Krazewski SE, Reich PB (2004) Shared ectomycorrhizal fungi between a herbaceous perennial (*Helianthemum bicknellii*) and oak (*Quercus*) seedlings. New Phytol 164:375–382
Duddridge JA, Malibari A, Read DJ (1980) Structure and function of mycorrhizal rhizomorphs with special reference to their role in water transport. Nature 287:834–836
Dudley SA, File AL (2008) Kin recognition in an annual plant. Biol Lett 3:435–438
Eason WR, Newman EI (1990a) Rapid cycling of nitrogen and phosphorus from dying roots of *Lolium perenne*. Oecologia 82:432–436
Eason WR, Newman EI (1990b) Rapid loss of phosphorus from dying ryegrass roots: the chemical components involved. Oecologia 84:359–361

Egerton-Warburton LM, Graham RC, Hubbert KR (2003) Spatial variability in mycorrhizal hyphae and nutrient and water availability in a soil-weathered bedrock profile. Plant Soil 249:331–342
Egerton-Warburton LM, Querejeta JI, Allen MF (2007) Common mycorrhizal networks provide a potential pathway for the transfer of hydraulically lifted water between plants. J Exp Botany 58:1473–1483
Egerton-Warburton LM, Querejeta JI, Allen MF (2008) Efflux of hydraulically lifted water from mycorrhizal fungal hyphae during imposed drought. Plant Signal Behav 3:68–71
Eissenstat DM, Newman EI (1990) A comparison of phosphorus and nitrogen transfer between plants of different phosphorus status. Oecologia 82:342–347
Ek H, Andersson S, Soderstrom B (1996) Carbon and nitrogen flow in silver birch and Norway spruce connected by a common mycorrhizal mycelium. Mycorrhiza 6:465–467
Ekblad A, Huss-Danell K (1995) Nitrogen fixation by *Alnus incana* and nitrogen transfer from *A. incana* to *Pinus sylvestris* influenced by macronutrient and ectomycorrhiza. New Phytol 131:453–459
File AL, Murphy GP, Dudley SA (2011) Fitness consequences of plants growing with siblings: reconciling kin selection, niche partitioning and competitive ability. Proc R Soc B. doi:10.1098/rspb.2011.1995
Finlay RD (1989) Functional aspects of phosphorus uptake and carbon translocation in incompatible ectomycorrhizal associations between *Pinus sylvestris* and *Suillus grevillei* and *Boletinus cavipes*. New Phytol 112:185–192
Finlay RD, Read DJ (1986) The structure and function of the vegetative mycelium of ectomycorrhizal plants. II. The uptake and distribution of phosphorus by mycelial strands interconnecting host plants. New Phytol 103:157–165
Fitter AH, Hodge A, Daniell TJ, Robinson D (1999) Resource sharing in plant–fungus communities: did the carbon move for you? Trends Ecol Evol 14:70–71
Francis R, Read DJ (1984) Direct transfer of carbon between plants connected by vesicular-arbuscular mycorrhizal mycelium. Nature 307:53–56
Fraser EC, Lieffers VJ, Landhäusser SM (2006) Carbohydrate transfer through root grafts to support shaded trees. Tree Physiol 26:1019–1023
Frey B, Schuepp H (1993) A role of vesicular-arbuscular (VA) mycorrhizal fungi in facilitating interplant nitrogen transfer. Soil Biol Biochem 25:651–658
Fricker MD, Boddy L, Bebber DP (2007) Network organisation of mycelial fungi. In: Howard RJ, Gow NAR (eds) The Mycota. Springer, Berlin, pp 309–330
Gai J, Christie P, Cai X, Fan J, Zhang J, Feng G, Li X (2009) Occurrence and distribution of arbuscular mycorrhizal fungal species in three types of grassland community of the Tibetan Plateau. Ecol Res 24:1345–1350
Giovannetti M, Sbrana C, Avio L, Stranil P (2004) Patterns of belowground plant interconnections established by means of arbuscular mycorrhizal networks. New Phytol 164:175–181
Giovannetti M, Avio L, Fortuna P, Pellegrino E, Sbrana C, Strani P (2005) At the root of the Wood Wide Web: self recognition and non-self incompatibility in mycorrhizal networks. Plant Signal Behav 1:1–5
Girlanda M, Segreto R, Cafasso D, Liebel HT, Rodda M, Ercole E, Perotto S (2011) Photosynthetic Mediterranean meadow orchids feature partial mycoheterotrophy and specific mycorrhizal associations. Am J Bot 98:1148–1163
Gorzelak MA, Asay AK, Pickles BJ, Simard SW (2015) Inter-plant communication through mycorrhizal networks mediates complex adaptive behaviour in plant communities. Ann Bot Plants (in press)
Graham B, Bormann F (1966) Natural root grafts. Bot Rev 32:255–292
Gyuricza V, Thiry Y, Wannijn J, Declerck S, de Boulois HD (2010) Radiocesium transfer between *Medicago truncatula* plants via a common mycorrhizal network. Environ Microbiol 12:2180–2189
Hamel C, Barrantes-Cartin U, Furlan V et al (1991) Endomycorrhizal fungi in nitrogen transfer from soybean to maize. Plant Soil 138:33–40

Haystead A, Malajczuk N, Grove TS (1988) Underground transfer of nitrogen between pasture plants infected with vesicular-arbuscular mycorrhizal fungi. New Phytol 108:417–423
He X-H, Critchley C, Bledsoe C (2003) Nitrogen transfer within and between plants through common mycorrhizal networks (CMNs). Crit Rev Plant Sci 22:531–567
He XH, Critchley C, Ng H, Bledsoe C (2004) Reciprocal N ($^{15}NH_4^+$ or $^{15}NO_3^-$) transfer between non-N_2-fixing *Eucalyptus maculata* and N_2-fixing *Casuarina cunninghamiana* linked by the ectomycorrhizal fungus *Pisolithus* sp. New Phytol 163:629–640
He XH, Critchley C, Ng H, Bledsoe C (2005) Nodulated N_2-fixing *Casuarina cunninghamiana* is the sink for net N transfer from non-N_2-fixing *Eucalyptus maculata* via an ectomycorrhizal fungus *Pisolithus* sp. supplied as ammonium nitrate. New Phytol 167:897–912
He XH, Bledsoe CS, Zasoski RJ, Southworth D, Horwath WR (2006) Rapid nitrogen transfer from ectomycorrhizal pines to adjacent ectomycorrhizal and arbuscular mycorrhizal plants in a California oak woodland. New Phytol 170:143–151
He XH, Xu M, Qiu GY, Zhou J (2009) Use of ^{15}Nstable isotope to quantify nitrogen transfer between mycorrhizal plants. J Plant Ecol 2:107–118
Heaton L, Obara B, Grau V, Jones N, Nakagaki T, Boddy L, Fricker MD (2012) Imaging and analysis of microbial networks. Fung Biol Rev 26:12–29
Higuera PE, Brubaker LB, Anderson PM, Brown TA, Kennedy AT, Hu FS (2008) Frequent fires in ancient shrub tundra: implications of paleorecords for arctic environmental change. PLoS One 3(3)
Higuera PE, Brubaker LB, Anderson PM, Hu FS, Brown TA (2009) Vegetation mediated the impacts of postglacial climate change on fire regimes in the south-central Brooks Range, Alaska. Ecol Monogr 79:201–219
Hobbie EA, Agerer R (2010) Nitrogen isotopes in ectomycorrhizal sporocarps correspond to belowground exploration types. Plant Soil 32:71–83
Hobbie EA, Hobbie JE (2008) Natural abundance of ^{15}N in nitrogen-limited forests and tundra can estimate nitrogen cycling through mycorrhizal fungi: a review. Ecosystems 11:815–830
Horton TR, Bruns TD (2001) The molecular revolution in ectomycorrhizal ecology: peeking into the black-box. Mol Ecol 10:1855–1871
Horton TR, Hayward J, Tourtellot SG, Taylor DL (2013) Uncommon ectomycorrhizal networks: richness and distribution of *Alnus*-associating ectomycorrhizal fungal communities. New Phytol 198:978–980
Hu FS, Higuera PE, Walsh JE, Chapman WL, Duffy PA, Brubaker LB, Chipman ML (2010) Tundra burning in Alaska: linkages to climatic change and sea ice retreat. J Geophys Res Biogeosci 115
Hynson NA, Bruns TD (2010) Fungal hosts for mycoheterotrophic plants: a nonexclusive, but highly selective club. New Phytol 185:598–601
Hynson NA, Mambelli S, Amend A, Dawson T (2012) Measuring carbon gains from fungal networks in understory plants from the tribe Pyroleae (Ericaceae): a field manipulation and stable isotope approach. Oecologia 169:307–317
Ikram A, Jensen ES, Jakobsen I (1994) No significant transfer of N and P from *Pueraria phaseoloides* to *Hevea brasiliensis* via hyphal links of arbuscular mycorrhiza. Soil Biol Biochem 26:1541–1547
Ishikawa CM, Bledsoe CS (2000) Seasonal and diurnal patterns of soil water potential in the rhizosphere of blue oaks: evidence for hydraulic lift. Oecologia 125:459–465
Jedd G (2011) Fungal evodevo: organelles and multicellular complexity. Trends Cell Biol 21:12–19
Johansen A, Jensen ES (1996) Transfer of N and P from intact or decomposing roots of pea to barley interconnected by an arbuscular mycorrhizal fungus. Soil Biol Biochem 28:73–81
Johnson NC (2009) Resource stoichiometry elucidates the structure and function of arbuscular mycorrhizas across scales. New Phytol 185:631–647
Johnson D, Leake JR, Read DJ (2002) Novel in-growth core system enables functional studies of grassland mycorrhizal mycelial networks. New Phytol 152:555–562

Kiers ET, Franken O, Hart MM, Duhamel M, Verbruggen E, Bago A, Vandenkoornhuyse P, Beesetty Y, Fellbaum CR, Palmer TM, Jansa J, Mensah JA, Kowalchuk GA, West SA, Bücking H (2011) Reciprocal rewards stabilize cooperation in the mycorrhizal symbiosis. Science 333:880–882

Kranabetter JM, MacKenzie WM (2010) Contrasts among mycorrhizal plant guilds in foliar nitrogen concentration and $\delta^{15}N$ along productivity gradients of a boreal forest. Ecosystems 13:108–117

Lambers H, Brundrett MC, Raven JA, Hopper SD (2010) Plant mineral nutrition in ancient landscapes: high plant species diversity on infertile soils is linked to functional diversity for nutritional strategies. Plant Soil 334:11–31

Leake JR, Cameron DD (2010) Physiological ecology of mycoheterotrophy. New Phytol 185:601–605

Leake J, Johnson D, Donnelly D, Muckle G, Boddy L, Read D (2004) Networks of power and influence: the role of mycorrhizal mycelium in controlling plant communities and agroecosystem functioning. Can J Bot 82:1016–1045

Leffler AJ, Peek MS, Ryel RJ, Ivans CY, Caldwell MM (2005) Hydraulic redistribution through the root systems of senesced plants. Ecology 86:633–642

Lerat S, Gauci R, Catford J, Vierheilig H, Piché Y, Lapointe L (2002) ^{14}C transfer between the spring ephemeral *Erythronium americanum* and sugar maple saplings via arbuscular mycorrhizal fungi in natural stands. Oecologia 132:181–187

Levin SA (2005) Self-organization and the emergence of complexity in ecological systems. Bioscience 55:1075–1079

Lilleskov EA, Bruns TD, Dawson TE, Camacho FJ (2009) Water sources and controls on water loss rates of epigeous ectomycorrhizal fungal sporocarps during summer drought. New Phytol 182:483–494

Lilleskov EA, Hobbie EA, Horton TR (2011) Conservation of ectomycorrhizal fungi: exploring the linkages between functional and taxonomic responses to anthropogenic N deposition. Fungal Ecol 4:174–183

Louw D (2015) Interspecific interactions in mixed stands of paper birch (*Betula papyrifera*) and interior Douglas-fir (*Pseudotsuga menziesii* var. *glauca*). Masters thesis, University of British Columbia, Vancouver, Canada

Lucic E, Fourrey C, Kohler A, Martin F, Chalot M, Brun-Jacob A (2008) A gene repertoire for nitrogen transporters in *Laccaria bicolor*. New Phytol 180:343–364

Ludwig F, Dawson TE, Kroon H, Berendse F, Prins HHT (2003) Hydraulic lift in *Acacia tortilis* trees on an East African savanna. Oecologia 134:293–300

Mack MC, Schuur EAG, Bret-Harte MS, Shaver GR, Chapin FS III (2004) Ecosystem C storage in Arctic tundra reduced by long-term nutrient fertilization. Nature 431:440–443

Mack MC, Bret-Harte MS, Hollingsworth TN, Jandt RR, Schuur EAG, Shaver GR, Verbyla DL (2011) Carbon loss from an unprecedented Arctic tundra wildfire. Nature 475(7357):489–492

Maestre FT, Callaway RM, Valladares F, Lortie CJ (2009) Refining the stress-gradient hypothesis for competition and facilitation in plant communities. J Ecol 97:199–205

Mangan SA, Herre EA, Bever JD (2010) Specificity between Neotropical tree seedlings and their fungal mutualists leads to plant–soil feedback. Ecology 91:2594–2603

Marschner H, Bell D (1994) Nutrient uptake in mycorrhizal symbiosis. Plant Soil 159:89–102

Martin F, Stewart GR, Genetet I, Le Tacon F (1986) Assimilation of $^{15}NH_4$ by beech (*Fagus sylvatica* L.) ectomycorrhizas. New Phytol 102:85–94

Martin F, Aerts A, Ahren D et al (2008) The genome of *Laccaria bicolor* provides insights into mycorrhizal symbiosis. Nature 452:88–93

Martos F, Munoz F, Pailler T, Kottke I, Gonneau C, Selosse M-A (2012) The role of epiphytism in architecture and evolutionary constraint within mycorrhizal networks of tropical orchids. Mol Ecol 21:5098–5109

McKendrick SL, Leake JR, Read DJ (2000) Symbiotic germination and development of myco-heterotrophic plants in nature: transfer of carbon from ectomycorrhizal *Salix repens* and

Betula pendula to the orchid *Corallorhiza trifida* through shared hyphal connections. New Phytol 145:539–548

Meding SM, Zasoski RJ (2008) Hyphal-mediated transfer of nitrate, arsenic, cesium, rubidium, and strontium between arbuscular mycorrhizal forbs and grasses from a California oak woodland. Soil Biol Biochem 40:126–134

Merckx V, Bidartondo MI, Hynson NA (2009) Myco-heterotrophy: when fungi host plants. Ann Bot 104:1255–1261

Molina R, Massicotte H, Trappe JM (1992) Specificity phenomenon in mycorrhizal symbiosis: community-ecological consequences and practical implications. In: Allen MF (ed) Mycorrhizal functioning: an integrative plant-fungal process. Chapman Hall, New York, pp 357–423

Motomura H, Selosse M-A, Martos F, Kagawa A, Yukawa T (2010) Mycoheterotrophy evolved from mixotrophic ancestors: evidence in *Cymbidium* (Orchidaceae). Ann Bot 106:573–581

Moyer-Henry KA, Burton JW, Israel D et al (2006) Nitrogen transfer between plants: a ^{15}N natural abundance study with crop and weed species. Plant Soil 282:7–20

Nara K (2006) Ectomycorrhizal networks and seedling establishment during early primary succession. New Phytol 169:169–178

Nardini A, Salleo S, Tyree MT, Vertovec M (2000) Influence of the ectomycorrhizas formed by *Tuber melanosporum* Vitt. on hydraulic conductance and water relations of *Quercus ilex* L. seedlings. Ann For Sci 57:305–312

Nehls U, Grunze B, Willmann M, Reich M, Küster H (2007) Sugar for my honey: carbohydrate partitioning in ectomycorrhizal symbiosis. Phytochemistry 68:82–91

Neumann RB, Cardon ZG (2012) The magnitude of hydraulic redistribution by plant roots: a review and synthesis of empirical and modeling studies. New Phytol 194:337–352

Newman EI (1988) Mycorrhizal links between plants: their functioning and ecological significance. Adv Ecol Res 18:243–270

Newman EI, Eason W (1989a) Cycling of nutrients from dying roots to living plants, including the role of mycorrhizas. Plant Soil 115:211–215

Newman EI, Eason W (1989b) Rates of phosphorus transfer within and between ryegrass (*Lolium perenne*) plants. Funct Ecol 7:242–248

Newman EI, Eason W, Eissenstat DM, Ramos MIRF (1992) Interactions between plants: the role of mycorrhizae. Mycorrhiza 1:47–53

Onguene NA, Kuyper TW (2002) Importance of the ectomycorrhizal network for seedling survival and ectomycorrhiza formation in rain forests of south Cameroon. Mycorrhiza 12:13–17

Parke JL, Lindermann RG, Black CH (1983) The role of ectomycorrhizas in drought tolerance of Douglas fir seedlings. New Phytol 95:83–95

Parrott L (2010) Measuring ecological complexity. Ecol Indic 10:1069–1076

Perry DA (1995) Self-organizing systems across scales. Trends Ecol Evol 19:241–244

Perry DA (1998) A moveable feast: the evolution of resource sharing in plant-fungal communities. Trends Ecol Evol 13:432–434

Perry DA, Margolis H, Choquette C, Molina R, Trappe JM (1989a) Ectomycorrhizal mediation of competition between coniferous tree species. New Phytol 112:501–511

Perry DA, Amaranthus MP, Borchers JG, Borchers SL, Brainerd RE (1989b) Bootstrapping in ecosystems. Bioscience 39:230–237

Philip LJ (2006) Carbon transfer between ectomycorrhizal paper birch (*Betula papyrifera*) and Douglas-fir (*Pseudotsuga menziesii*). PhD thesis, UBC, Vancouver, Canada

Philip LJ, Simard SW, Jones MD (2010) Pathways for below-ground carbon transfer between paper birch and Douglas-fir seedlings. Plant Ecol Divers 3:221–233

Pickles BJ, Egger KN, Massicotte HB, Green SG (2011) Mycorrhizas and climate change. Fungal Ecol 5:75–83

Pickles BJ, Wilhelm R, Asay AK, Hahn AS, Simard SS, Mohn WW Carbon assimilation by soil communities: mycorrhizae dominate the uptake and sub-surface transfer of carbon between paired Douglas-fir seedlings. ISME J (submitted)

Plamboeck AH, Dawson TE, Egerton-Warburton LM, North M, Bruns TD, Querejeta JI (2007) Water transfer via ectomycorrhizal fungal hyphae to conifer seedlings. Mycorrhiza 17:439–447

Pölme S, Bahram M, Yamanaka T, Nara K, Dai YC, Grebenc T, Kraigher H, Toivonen M, Wang P-H, Matsuda Y, Naadel T, Kennedy PG, Kõljalg U, Tedersoo L (2013) Biogeography of ectomycorrhizal fungi associated with alders (*Alnus* spp.) in relation to biotic and abiotic variables at the global scale. New Phytol 198:1239–1249

Preiss K, Gebauer G (2008) A methodological approach to improve estimates of nutrient gains by partially myco-heterotrophic plants. Isot Environ Health Stud 44:393–401

Preiss K, Adam IKU, Gebauer G (2010) Irradiance governs exploitation of fungi: fine-tuning of carbon gain by two partially myco-heterotrophic orchids. Proc R Soc B 277:1333–1336

Prieto I, Armas C, Pugnaire FI (2012) Water release through plant roots: new insights on its consequences at the plant and ecosystem level. New Phytol 193:830–841

Pringle A (2009) Mycorrhizal networks. Current Biol 19:R838

Querejeta JI, Egerton-Warburton LM, Allen MF (2003) Direct nocturnal water transfer from oaks to their mycorrhizal symbionts during severe soil drying. Oecologia 134:55–64

Querejeta JI, Egerton-Warburton LM, Allen MF (2007) Hydraulic lift may buffer rhizosphere hyphae against the negative effects of severe soil drying in a California oak savanna. Soil Biol Biochem 39:409–417

Querejeta JI, Egerton-Warburton LM, Prieto I, Vargas R, Allen MF (2012) Changes in soil hyphal abundance and viability can alter the patterns of hydraulic redistribution by plant roots. Plant Soil 355:63–73

Read DJ, Francis R, Finlay RD (1985) Mycorrhizal mycelia and nutrient cycling in plant communities. In: Fitter AH (ed) Ecological interactions in soil. Blackwell Scientific Publications, Oxford, pp 193–217

Reid CPP, Woods FW (1969) Translocation of ^{14}C-labeled compounds in mycorrhizae and its implications in interplant nutrient cycling. Ecology 50:179–187

Richard F, Millot S, Gardes M, Selosse M-A (2005) Diversity and specificity of ectomycorrhizal fungi retrieved from an old-growth Mediterranean forest dominated by *Quercus ilex*. New Phytol 166:1011–1023

Richards JH, Caldwell MM (1987) Hydraulic lift: substantial nocturnal water transport between soil layers by *Artemisia tridentata* roots. Oecologia 73:486–489

Rillig MC, Mummey DL (2006) Mycorrhizas and soil structure. New Phytol 171:41–53

Roy M, Dubois M-P, Proffit M, Vincenot L, Desmarais E, Selosse M-A (2008) Evidence from population genetics that the ectomycorrhizal basidiomycete *Laccaria amethystina* is an actual multihost symbiont. Mol Ecol 17:2825–2838

Roy M, Rochet J, Manzi S, Jargeat P, Gryta H, Moreau P-A, Gardes M (2013) What determines *Alnus*-associated ectomycorrhizal community diversity and specificity? A comparison of host and habitat effects at a regional scale. New Phytol 198:1228–1238

Schimel JP, Bennett J (2004) Nitrogen mineralization: challenges of a changing paradigm. Ecology 85:591–602

Schoonmaker AL, Teste FP, Simard SW, Guy RD (2007) Tree proximity, soil pathways and common mycorrhizal networks: their influence on utilization of redistributed water by understory seedlings. Oecologia 154:455–466

Schweitzer JA, Bailey JK, Fischer DG, LeRoy CJ, Lonsdorf EV, Whitham TG, Hart SC (2008) Plant-soil-microorganism interactions: heritable relationship between plant genotype and associated soil microorganisms. Ecology 89:773–781

Selosse M-A, Roy M (2009) Green plants that feed on fungi: facts and questions about mixotrophy. Trends Plant Sci 14:64–70

Selosse M-A, Richard F, He X, Simard SW (2006) Mycorrhizal network: des liaisons dangereuses? Trends Ecol Evol 21:621–628

Shaver GR, Bret-Harte MS, Jones MH, Johnstone J, Gough L, Laundre JA, Chapin FS III (2001) Species changes interact with fertilizer addition to control 15 years of change in tundra. Ecology 82:3163–3181

Simard SW (2009) The foundational role of mycorrhizal networks in the self-organization of interior Douglas-fir forests. For Ecol Manage 258S:S95–S107

Simard SW, Durall DM (2004) Mycorrhizal networks: a review of their extent, function, and importance. Can J Bot 82:1140–1165

Simard SW, Perry DA, Smith JE, Molina R (1997a) Effects of soil trenching on occurrence of ectomycorrhizae on *Pseudotsuga menziesii* seedlings grown in mature forests of *Betula papyrifera* and *Pseudotsuga menziesii*. New Phytol 136:327–340

Simard SW, Perry DA, Jones MD, Myrold DD, Durall DM, Molina R (1997b) Net transfer of carbon between tree species with shared ectomycorrhizal fungi. Nature 388:579–582

Simard SW, Jones MD, Durall DM (2002) Carbon and nutrient fluxes within and between mycorrhizal plants. In: van der Heijden M, Sanders I (eds) Mycorrhizal ecology. Ecological studies, vol 157. Springer, Berlin, pp 33–61

Simard SW, Beiler KJ, Bingham MA, Deslippe JR, Philip LJ, Teste FP (2012) Mycorrhizal networks: mechanisms, ecology and modelling. Fungal Biol Rev 26:39–60

Simard SW, Martin K, Vyse A, Larson B (2013) Meta-networks of fungi, fauna and flora as agents of complex adaptive systems, Chap 7, pp 133–164. In: Puettmann K, Messier C, Coates KD (eds) Managing world forests as complex adaptive systems: building resilience to the challenge of global change. Routledge, New York, 369 pp. ISBN 978-0-415-51977

Smith SE, Read DL (2008) Mycorrhizal symbiosis, 3rd edn. Academic Press, London

Smith SE, Smith FA (1990) Structure and function of the interfaces in biotrophic symbioses as they relate to nutrient transport. New Phytol 114:1–38

Sole RV, Ferrer-Cancho R, Montoya JM, Valverde S (2002) Selection, tinkering and emergence in complex networks. Complexity 8:20–33

Song YY, Zeng RS, Xu JF, Li J, Yihdego WG (2010) Interplant communication of tomato plants through underground common mycorrhizal networks. PLoS ONE 5(10):e13324. doi:10.1371/journal.pone.0013324

Song YY, Simard SW, Carroll A, Mohn WW, Zheng RS (2015) Defoliation of interior Douglas-fir elicits carbon transfer and defense signalling to ponderosa pine neighbors through ectomycorrhizal networks. Sci Rep 5:8495

Sturm M, Racine CR, Tape K (2001) Increasing shrub abundance in the Arctic. Nature 411:546–547

Taylor AFS (2002) Fungal diversity in ectomycorrhizal communities: sampling effort and species detection. Plant Soil 244:19–28

Taylor AFS, Gebauer G, Read DJ (2004) Uptake of nitrogen and carbon from double-labelled (15N and 13C) glycine by mycorrhizal pine seedlings. New Phytol 164:383–388

Tedersoo L, Pellet P, Koljalg U, Selosse M-A (2007) Parallel evolutionary paths to mycoheterotrophy in understorey Ericaceae and Orchidaceae: ecological evidence for mixotrophy in Pyroleae. Oecologia 151:206–217

Tedersoo L, Jairus T, Horton BM, Abarenkov K, Suvi T, Saar I, Kõljalg U (2008) Strong host preference of ectomycorrhizal fungi in a Tasmanian wet sclerophyll forest as revealed by DNA barcoding and taxon-specific primers. New Phytol 180:479–490

Teste FP, Simard SW, Durall DM, Guy RD, Jones MD (2009) Access to mycorrhizal networks and roots of trees: importance for seedling survival and resource transfer. Ecology 90:2808–2822

Teste FP, Simard SW, Durall DM, Guy RD, Berch SM (2010) Net carbon transfer between *Pseudotsuga menziesii* var. *glauca* seedlings in the field is influenced by soil disturbance. J Ecol 98:429–439

Teste FP, Veneklaas EJ, Dixon KW, Lambers H (2014) Complementary plant nutrient-acquisition strategies promote growth of neighbour species. Funct Ecol 28:819–828

Teste FP, Veneklaas EJ, Dixon KW, Lambers H (2015) Is nitrogen transfer among plants enhanced by contrasting nutrient-acquisition strategies? Plant Cell Environ 38:50–60

Tlalka M, Bebber DP, Darrah PR, Watkinson SC, Fricker MD (2008) Quantifying dynamic resource allocation illuminates foraging strategy in *Phanerochaete velutina*. Fungal Genet Biol 45:1111–1121

Treseder KK (2004) A meta-analysis of mycorrhizal responses to nitrogen, phosphorus and atmospheric CO_2 in field studies. New Phytol 164:347–355

Treseder KK, Turner KM, Mack MC (2007) Mycorrhizal responses to nitrogen fertilization in boreal ecosystems: potential consequences for soil carbon storage. Global Change Biol 13:78–88

Tuffen F, Eason WR, Scullion J (2002) The effect of earthworms and arbuscular mycorrhizal fungi on growth of and ^{32}P transfer between *Allium porrum* plants. Soil Biol Biochem 34:1027–1036

Twieg BD, Durall DM, Simard SW (2007) Ectomycorrhizal fungal succession in mixed temperate forests. New Phytol 176:437–447

van der Heijden MGA, Horton TR (2009) Socialism in soil? The importance of mycorrhizal fungal networks for facilitation in natural ecosystems. J Ecol 97:1139–1150

van der Putten WH (2009) A multitrophic perspective on functioning and evolution of facilitation in plant communities. J Ecol 97:1131–1138

Walder F, Niemann H, Natarajan M, Lehmann MF, Boller T, Wiemken A (2012) Mycorrhizal networks: common goods of plants shared under unequal terms of trade. Plant Physiol 159:789–797

Warren JM, Meinzer FC, Brooks JR, Domec JC, Coulombe R (2007) Hydraulic redistribution of soil water in two old-growth coniferous forests: quantifying patterns and controls. New Phytol 173:753–765

Warren JM, Brooks JR, Meinzer FC, Eberhart J (2008) Hydraulic redistribution of water from *Pinus ponderosa* trees to seedlings: evidence for an ectomycorrhizal pathway. New Phytol 178:382–394

Waters JR, Borowicz VA (1994) Effect of clipping, benomyl, and genet on ^{14}C transfer between mycorrhizal plants. Oikos 71:246–252

Watkins NK, Fitter AH, Graves JD, Robinson D (1996) Carbon transfer between C_3 and C_4 plants linked by a common mycorrhizal network, quantified using stable carbon isotopes. Soil Biol Biochem 28:471–477

Whitham TG, Bailey JK, Schweitzer JA, Shuster SM, Bangert RK, LeRoy CJ, Lonsdorf EV, Allan GJ, DiFazio SP, Potts PM, Fischer DG, Gehring KA, Lindroth RL, Marks JC, Hart SC, Wimp GM, Wooley SC (2006) A framework for community and ecosystem genetics: from genes to ecosystems. Nature Rev Genet 7:510–523

Wilkinson DM (1998) The evolutionary ecology of mycorrhizal networks. Oikos 82:407–410

Wilson GWT, Hartnett DC, Rice CW (2006) Mycorrhizal-mediated phosphorus transfer between tallgrass prairie plants *Sorghastrum nutans* and *Artemisia ludoviciana*. Funct Ecol 20:427–435

Witzany G (2012) Biocommunication of fungi. Springer, Berlin

Woods AJ, Heppner D, Kope HH, Burleigh J, Maclauchlan L (2010) Forest health and climate change: A British Columbia perspective. For Chron 86:412–422

Wu, Nara K, Hogetsu T (2001) Can ^{14} C-labeled photosynthetic products move between *Pinus densiflora* seedlings linked by ectomycorrhizal mycelia? New Phytol 149:137–146

Xu H, Kemppainen M, El Kayal W, Hee Lee S, Pardo AG, Cooke JEK, Zwiazek JJ (2015) Overexpression of *Laccaria bicolor* aquaporin *JQ585595* alters root water transport properties in ectomycorrhizal white spruce (*Picea glauca*) seedlings. New Phytol 205:757–770

Yang H, Bognor M, Steinhoff Y-D, Ludewig U (2010) H^+-independent glutamine transport in plant root tips. PLoS ONE 5(1):e8917. doi:10.1371/journal.pone.0008917

Yao Q, Li XL, Ai WD, Christie P (2003) Bi-directional transfer of phosphorus between red clover and perennial ryegrass via arbuscular mycorrhizal hyphal links. Eur J Soil Biol 39:47–54

Zimmer K, Hynson NA, Gebauer G, Allen EB, Allen MF, Read DJ (2007) Wide geographical and ecological distribution of nitrogen and carbon gains from fungi in pyroloids and monotropoids (Ericaceae) and in orchids. New Phytol 175:166–175

Chapter 6
The Role of Ectomycorrhizal Networks in Seedling Establishment and Primary Succession

Kazuhide Nara

Abstract In developed forests and secondary successional sites, host plants can readily access ectomycorrhizal (ECM) fungi because of the ubiquitous ECM mycelia and spores in soil, but this is not the case in some primary successional sites. In a volcanic desert on Mt Fuji, Japan, most of the area is non-mycorrhizal habitat and has poorly developed soil spore-banks. ECM habitat, i.e. pioneer willow shrubs and a small surrounding area containing ECM mycelia, are quite sparsely distributed, accounting for about 1 % of the ground surface in total. Such unique conditions provide us an interesting opportunity to explore the magnitude and role of direct mycelial connections between plants, i.e. ECM networks, in the field. It is difficult to observe individual ECM mycelial spread in soil, but the distribution of sporocarps and ECM roots having the same genotype indicate the spread of a mycelium in soil. We applied microsatellite markers to genotype sporocarps and ECM tips, and found that genets of two pioneer *Laccaria* species were small in size (mostly <1 m) and ephemeral. In contrast, genets of *Scleroderma* included some long-lived large genets (>10 m). These results indicate that ECM networks could vary considerably in size and longevity, even in the same site and associated with the same host species. Field transplanting experiment revealed that current-year willow seedlings rarely formed ECM associations in most habitats of the desert and showed poor growth. ECM infection from spores did not significantly improve seedling growth, indicating a small isolated mycelium on a tiny seeding may not be enough to acquire sufficient nutrients from extremely nutrient poor scoria. In contrast, seedlings transplanted near the pre-established willow shrubs, where ECM networks are available, readily formed ECM associations and grew well. Moreover, artificially reproduced ECM networks in previously non-mycorrhizal habitats significantly improved the growth of connected seedlings in 10 of 11 ECM fungal species in this desert. Therefore, ECM networks appear to be mostly positive and could be critical to seedling establishment, at least in this primary successional

---

K. Nara (✉)
Department of Natural Environmental Studies, Graduate School of Frontier Sciences, The University of Tokyo, 5-1-5 Kashiwanoha, Kashiwa, Chiba 277-8563, Japan
e-mail: nara@k.u-tokyo.ac.jp

T.R. Horton (ed.), *Mycorrhizal Networks*,
Ecological Studies 224, DOI 10.1007/978-94-017-7395-9_6

setting. Some previously proposed mechanisms may be less relevant to the observed positive effect of ECM networks on seedling establishment. For example, plant-to-plant carbon transfer through ECM networks might work for seedlings in dark forest floor, but not in primary successional settings characterized by strong sun light. More relevant mechanisms should include rapid ECM colonization with low costs, larger absorbing surface area than a solitary mycelium, and nutrient translocation within a network from nutrient rich soil patches to most demanding parts, often seedlings.

Keywords Ectomycorrhizal fungi · Primary succession · Seedlings establishment · Pioneer plants · Succession · Nutrient transfer

6.1 Introduction

Ectomycorrhizal (ECM) trees dominate many forest ecosystems from the tropics to subpolar zones. ECM colonization is usually found in most of the fine roots of trees, sometimes in up to nearly 100 % of the root tips of the host. Thus, ECM root tips and their mycelia are ubiquitous in forest soil, especially in surface organic soil. ECM fungal diversity usually exceeds the diversity of trees in temperate and boreal forests (Horton and Bruns 2001). Dozens or even up to a few hundred ECM fungal species inhabit a single hectare of forest, where the number of host tree species can range from one (e.g., conifer plantations) to >30 (e.g., mixed dipterocarp forests in Southeast Asia). Although increasing evidence indicates that many ECM fungi exhibit host preferences in mixed forests (Ishida et al. 2007; Tedersoo et al. 2008; Kennedy et al. Chap. 8, this Volume), most ECM fungi do not have strict host specificities and can colonize different host species (Molina et al. 1992; Horton and Bruns 1998; Selosse et al. 2006; Murata et al. 2013; Molina and Horton, Chap. 1, this Volume). Because the same ECM fungal species can occur on different host species, some researchers refer to the existence of common mycorrhizal networks (CMNs), or more cautiously, "potential" CMNs (e.g., Kennedy et al. 2003). The word "networks" is also used in ecological network theory to describe plant–fungal associations at the species level (e.g., Chagnon et al. 2012). In these contexts, networks do not necessarily mean direct mycelial connections between plants; however, this chapter focuses on these direct connections.

In nature, each ECM fungal species exists as a group of genetically different individuals (genets), which collectively form an ECM fungal population. The size of a genet, which is often defined as the largest distance between sporocarps of the same genotype, ranges from <1 to >10 m depending on the ECM fungal species and the environment (Douhan et al. 2011 and references therein). While some studies have examined belowground genets (e.g., Zhou et al. 2001; Kretzer et al. 2004; Lian et al. 2006; Wadud et al. 2007, 2008; Beiler et al. 2010), the sampling scales were relatively small. In fact, examining all belowground genets within a forest is impossible or impractical (Bahram et al. 2011). Thus, the total number of genets of all ECM fungal species in a forest remains uncertain. However, the number of ECM associations defined at the tree-fungus individual level in a forest far clearly exceeds the number of host–fungus associations defined at the species level.

Genetically different vegetative mycelia of an ECM fungus (dikaryotic mycelia for Basidiomycetes) do not fuse. In fact, nutrients and carbon are quickly transferred between fused ECM fungal mycelia belonging to the same *Pisolithus* genotype, but such transfers do not occur between different mycelial genets (Wu et al. 2012). Thus, direct physiological and nutritional interactions are only possible within a system that includes a genetically identical ECM fungal mycelium and its colonizing hosts. Different ECM fungal genets compete with each other for soil space, nutrients, water, and host photosynthates, and may therefore engage in some indirect ecological interactions. Yet, the focus of this chapter is on direct ECM fungal mycelial links and their potential roles in seedling establishment. Therefore, I will only use the term "ectomycorrhizal (ECM) network" for such a direct link between plants on an ECM fungal mycelium. Although confirming the existence of such ECM networks in the field is difficult, several pioneer studies have used both host and fungal simple sequence repeat (SSR) markers to simultaneously identify fungal and host individuals and thus demonstrated the existence of complex ECM networks; a single ECM fungal genet is associated with multiple hosts, while a single host is associated with multiple fungal genets (Lian et al. 2006; Beiler et al. 2010).

Because ECM hosts depend largely on ECM fungi for soil nutrients, host plants need ECM fungal colonization to grow normally and survive in the field (Smith and Read 2008). Two major infection pathways exist: spores (and sclerotia) and ECM networks. In developed forests, both types of inoculum are ubiquitous. Thus, the lack of ECM fungal inocula is not a limiting factor for seedling establishment in these forests.

However, in severely disturbed areas, especially primary successional sites, the inoculum potential is reduced to a level that critically limits seedling establishment. Research at such sites could potentially pinpoint the ecological roles of ECM networks. Here, I summarize the data related to ECM fungi and host seedling performance during early primary succession, particularly in the volcanic desert on Mt. Fuji, where a great deal of strong evidence for the ecological roles of ECM networks has been documented. For more information on the more general roles of ECM fungi and ECM networks, I recommend other chapters in this book, and previous reviews (Newman 1988; Simard et al. 2002; Simard and Durall 2004; Smith and Read 2008; van der Heijden and Horton 2009) and references therein.

6.1.1 Primary Successional Sites for ECM Network Research

To quantify the effect of ECM networks in seedling establishment, including a non-mycorrhizal control treatment in which no ECM fungi exist under natural conditions is ideal, as in vitro experiments. However, non-mycorrhizal controls are very difficult to attain in forests because of the ubiquitous distribution of ECM networks, spores, and sclerotia, by which seedlings are readily colonized. Given the lack of non-mycorrhizal controls and the presence of mixed ECM infection from different sources, isolating the effect of ECM networks on seedlings is very difficult. For example, seedlings are readily colonized by ECM fungi and grow well in forests and secondary successional sites, even after clear-cut logging (Jones et al. 2003; Twieg et al. 2007) and wildfires (Horton et al. 1998; Baar et al. 1999; Stendell et al. 1999); however, no studies have quantified the relative contribution of ECM networks to seedling establishment under such conditions.

Primary successional sites are characterized by the complete absence of ECM fungal inoculum at the initial stage. Disturbances that create bare ground for primary succession include volcanic activity, glacier retreat, sand dune movement, mining and so on. Among these, glacier retreat and sand dune movement are regarded as gradual and relatively mild disturbances. At these primary successional sites, ECM propagules become available over a relatively short time frame. For example, in a recently stabilized sand dune in Oregon, USA, Ashkannejhad and Horton (2006) found that the majority of soil samples contained infective ECM fungal spores, especially those of suilloids species (e.g., Suillus and Rhizopogon). At a glacier forefront site, tree seedlings were readily colonized by ECM fungi, even during the early stages of primary succession (Helm et al. 1996; Cazares et al. 2005, but see Fujiyoshi et al. 2011). ECM inoculum potential in volcanic sites appears to be much lower, even long after the eruption (Allen et al. 1992; Nara and Hogetsu 2004; Obase et al. 2007, see below). Although the reason soil inoculum potential is so low at volcanic sites compared to other primary successional sites remains uncertain, completely sterile substrates created by volcanic activities may not be suitable for the survival of ECM fungal spores.

Mount Fuji, the highest mountain in Japan, erupted in 1707. Scoria deposits covered the area along the southeast slope up to 10 m deep. Although vegetation has been gradually recovering for about 300 years since this last eruption, vegetation coverage remains at only about 5 % of the ground surface at 1450–1600 m above sea level. Vegetation is now patchily distributed, forming isolated vegetation islands of various sizes (i.e., various developmental stages) in a sea of scoria (Nara et al. 2003a). Relatively slow vegetation recovery in this desert is not due to limited precipitation, as the average annual precipitation is nearly 5000 mm. Unstable and extremely nutrient-poor scoria substrates are likely the main limiting factor for vegetation recovery. In fact, total nitrogen (N) concentration in bare-ground scoria is below 0.002 % in this desert. Low N concentration in soil has also been observed in the Mt. St. Helens volcano systems (about 0.01 %; Halvorson and Smith 2009).

Read (1989) proposed a successional model of mycorrhizal types for sand dune ecosystem: where primary succession begins with non-mycorrhizal plants, followed by arbuscular mycorrhizal (AM) dominance, a subsequent shift to ECM trees, and lastly a climax of ericoid-dominant communities. Although the empirical evidence does not always support this model (Dickie et al. 2013), vegetation succession in the volcanic desert on Mt. Fuji partially follows it. The formation of vegetation islands is initiated by a non-mycorrhizal perennial species *Polygonum cuspidatum*, which vegetatively enlarges and creates a stable habitat in the unstable scoria desert (Hirose and Tateno 1984; Zhou et al. 2003). The stabilized habitat enables further establishment of other plant species, including some AM perennials and/or the pioneer ECM plant *Salix reinii* (hereafter *Salix*, unless otherwise specified). This *Salix* species is a ground-covering dwarf willow species, usually <1 m in plant height. Because *Salix* seedlings are very small and light-demanding, they cannot establish within the interior of vegetation islands. Thus, their seedling establishment is restricted to the periphery of a vegetation island. About one-third of vegetation islands contain *Salix*. The total coverage of *Salix* is about 1 % of the ground surface in the Mt. Fuji desert (Nara et al. 2003a). Other EM plants include *Larix kaempferi* and *Betula ermanii* (hereafter *Larix* and *Betula*, respectively), but they are still rare (0.003 % of the total *Salix* coverage). Because the study area belongs to the mixed forest zone dominated by ECM trees such as *Quercus*, *Fagus*, *Carpinus*, and *Abies*, eventually, the area likely will become a mixed forest of these tree species after long successional processes. What is most important here is that ECM islands in a largely non-mycorrhizal habitat are sparsely distributed in a potentially ECM forest zone, providing an ideal opportunity to examine the functions of ECM networks.

6.1.2 Species and Size of ECM Networks in Primary Succession

ECM fungal sporocarps were found under all established *Salix* shrubs in the volcanic desert on Mt. Fuji. These ECM fungi exhibit a clear successional pattern (Nara et al. 2003a, b). Pioneer ECM fungal species that colonize newly established *Salix* in a previously non-mycorrhizal vegetation island include *Laccaria laccata*, *Laccaria amethystina*, and *Inocybe lacera*. With growth of *Salix* patches, two additional pioneer species, *Scleroderma bovista* and *Laccaria murina*, appear mainly outside of each vegetation island. Late-seral colonizers, such as *Hebeloma*, *Cortinarius*, *Russula*, and *Tomentella* (resupinate), are only found inside of larger *Salix* islands, i.e., where soil organic matter has accumulated (Nara et al. 2003a). Although belowground ECM fungal communities include some additional fungal species, the species composition and successional patterns are quite similar to those of sporocarps (Nara et al. 2003b).

In relation to the size of ECM networks, the genet sizes of four ECM fungi in the Mt. Fuji desert are shown in Fig. 6.1. These genets were identified using multiple microsatellite markers (or SSR markers), which are highly polymorphic codominant genetic markers (Wadud et al. 2006a, b). When the genet size is defined as the largest distance between the sporocarps, all 56 *L. laccata* genets identified from 2004 to 2006 were smaller than 1.4 m, with a mean size of 0.28 m (Wadud et al. 2014). Similarly, all 47 *L. amethystina* genets were smaller than 1.2 m, with a mean of 0.38 m (Wadud et al. 2014). Note that sporocarps of the same genotype are always clustered (Fig. 6.1a, b), indicating that they are not fragmented. Belowground genets of both *Laccaria* species were also examined by combining fine-scale soil core sampling and SSR genotyping (Wadud et al. 2007, 2008). The mean belowground genet areas of *L. laccata* and *L. amethystina* were 0.20 ± 0.02 and 0.25 ± 0.04 m^2, respectively, during the fruiting season (Fig. 6.1e). These values were reduced to 0.08 ± 0.01 and 0.06 ± 0.01 m^2, respectively, during the following spring (Fig. 6.1f), due to the disappearance of some larger sporocarp-produced genets and the production of many new small genets (potentially offspring genets, sharing at least one of two alleles at every locus with the previous-year sporocarp). Genets of both *Laccaria* species exhibit strong turnover every year (Wadud et al. 2014), and the majority of genets disappeared after sporocarp formation (Wadud et al. 2007, 2008).

In contrast to the *Laccaria* species, *S. bovista* genets were relatively large (Nakaya et al. unpublished data). Of the 56 genets identified in 2000 and 2001, 37 were smaller than 4.0 m, although some large genets did occur (Fig. 6.1c), the largest of which was 18.4 m in size. Genet composition was nearly the same between the 2 years, indicating less turnover of genets. *S. bovista* produces well-developed rhizomorphs and preferentially colonizes from the periphery to the outside of each vegetation island, where the root density is small and environmental conditions are relatively harsh due to strong sun radiation. Well-developed rhizomorphs and heat-tolerance abilities (Ingleby et al. 1985) support *S. bovista* vegetative growth year after year, allowing the establishment of a large belowground ECM network in the harsh habitat. Although estimating the exact age of a large *S. bovista* genet is difficult, it would likely be much older than *Laccaria* genets, even up to dozens of years old or more.

The cosmopolitan species *Cenococcum geophilum* is a minor component in the volcanic desert on Mt. Fuji (Nara et al. 2003b). Defining the size of ECM networks through SSR genotyping may be difficult for this species, as *C. geophilum* produces sclerotia that are potentially dispersive propagules and genetically identical to the parent mycelium. Indeed, the same genotype was observed in distantly separated vegetation islands (Wu et al. 2005; Fig. 6.1d). However, considering that the sporadic occurrence of the same genotypes (the same genotypes were not detected in neighboring soil cores sampled at 1–2-m intervals), the size of ECM networks for this fungus may not be >2 m.

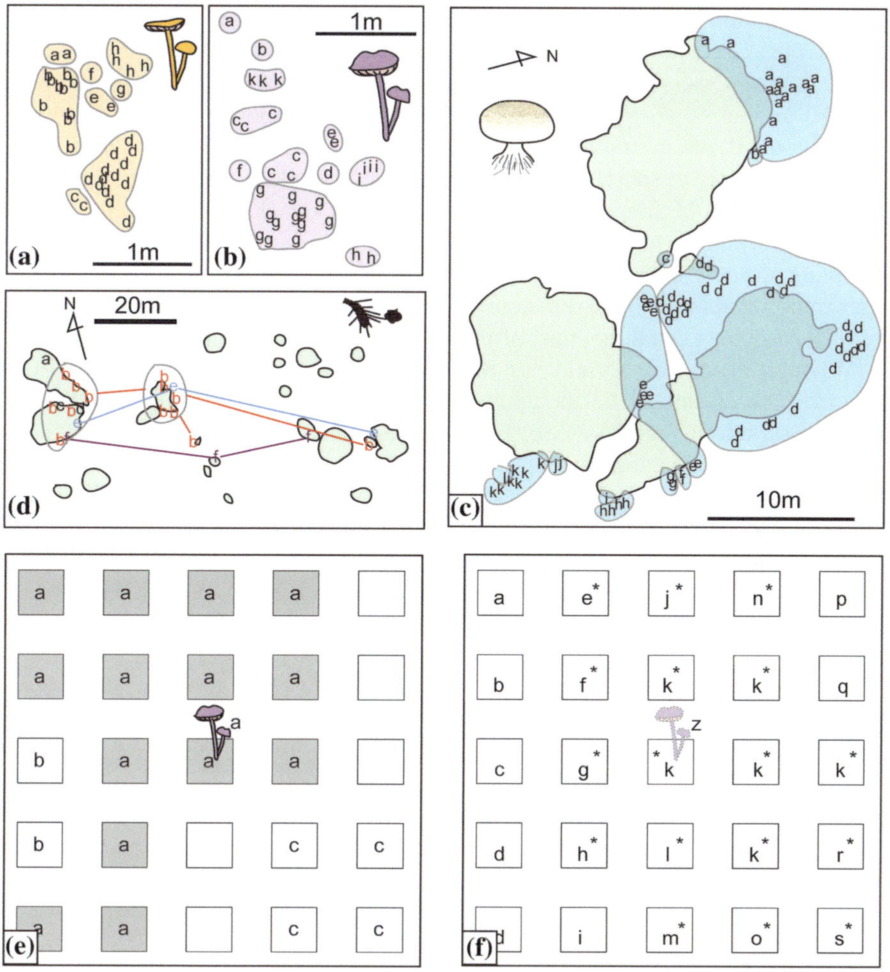

Fig. 6.1 Examples of genet distribution of ECM fungi in the volcanic desert on Mt. Fuji, Japan: **a** sporocarps of *Laccaria laccata* in 2004; **b** sporocarps of *Laccaria amethystina* in 2004; **c** sporocarps of *Scleroderma bovista* in 2000; **d** Sclerotia and ECM tips of *Cenococcum geophilum* in 2002; **e** ECM tips of *L. amethystina* at the fruiting in 2005 and **f** ECM tips of *L. amethystina* 9 months after the fruiting. Different *alphabets* indicate different genets identified by microsatellite markers. Each panel has a different scale bar. *Green areas* surrounded by solid line in (**c**) and (**d**) are vegetation islands. In (**e**) or (**f**), each of which is a representative plot among 10 replicate 1 m × 1 m square plots, 25 soil cubes (5 cm^3) were collected at every 20 cm distance, where a focal sporocarp was located at the center of the plot. ECM tips having the same genotype as the focal sporocarp are shown in *grey squares* (**e**). *Asterisks* in (**f**) indicate potential offspring genets generated by the focal sporocarp in the previous year. See text for details. Data from Wadud et al. 2007, 2008, 2014; Wu et al. 2005; and Nakaya et al. (unpublished)

To date, the genet sizes of other ECM fungal species in this desert are unavailable. Gryta et al. (1997) demonstrated that the genet size of *Hebeloma cylindrosporum* was <3.6 m. Thus, *Hebeloma* genets in the Mt. Fuji desert may be as small as that of *H. cylindrosporum*. In fact, basidiospores of three *Hebeloma* species in the desert germinate very well in the presence of *Salix* roots (Ishida et al. 2008), indicating a spore-dependent regeneration pattern. Similarly, *I. lacera*, another major ECM fungal component in the desert, lacks a developed mycelial system and exhibits high rates of basidiospore germination (Ishida et al. 2008), indicating a spore-dependent reproduction strategy. Given that spore-dependent regeneration is related to the dominance of small genets, the genet sizes of *I. lacera* are likely to be as small as those of the two *Laccaria* species described above.

Although genet size should not necessarily be the same as the size of ECM networks, the clustered and mutually exclusive distribution patterns of individual genets observed in Fig. 6.1a–c, e, f indicate that each identified genet is not fragmented and could function as a nutritional unit or an ECM network. Note also that the size and magnitude of ECM networks could substantially vary among ECM fungal species, even when they are associated with the same host species at the same site. Differences in genet turnover patterns also indicate that the longevity of ECM networks could substantially differ among fungal species (i.e., ECM networks of both *Laccaria* species are rapidly replaced, whereas those of *S. bovista* are relatively stable for many years).

Most *Salix* species appear as pioneer ECM hosts at many primary successional sites in northern cool–temperate regions. Similar to the Mt. Fuji site, species belonging to *Laccaria*, *Inocybe*, *Hebeloma*, and *Cortinarius* are also major components at other volcanic sites (Obase et al. 2005, 2007), retreating glacier sites (Helm et al. 1996; Jumpponen et al. 1999, 2002), and in dune ecosystems (van der Heijden et al. 1999), although the sizes of ECM networks in these primary systems are not available. Given that phylogenetically close species share similar genet sizes (Vincenot et al. 2012), the sizes of ECM genets or networks found on Mt. Fuji may be applicable to other *Salix*-dominated primary sites.

At some other primary successional sites in the warm–temperate zone, *Alnus* and *Pinus* could serve as the pioneer ECM plants, on which *Alpova* (Yamanaka and Okabe 2006) and suilloid fungi (Ashkannejhad and Horton 2006), respectively, become major ECM fungal symbionts. Unfortunately, the genet structures of these ECM fungi have not been examined in primary successional sites; thus, the size of these ECM networks remains uncertain. However, because suilloid fungi in forest areas are usually composed of some large genets (up to >10 m; e.g., Dahlberg and Stenlid 1994; Dahlberg 1997; Hirose et al. 2004), the size of ECM networks in *Pinus*-dominated successional sites may potentially be larger than those in *Salix*-dominated sites.

6.1.3 ECM Networks and Conspecific Seedling Establishment During Primary Succession

Virtually all *Salix* seedlings in the volcanic desert on Mt. Fuji are only found in close proximity to established *Salix* shrubs (Fig. 6.2). While non-*Salix* vegetation (i.e., non-ECM vegetation) is far more common and provides similar physical (e.g., shading, wind protection, stable soil surface, soil water content) and chemical conditions (e.g., soil nutrients), *Salix* seedlings are almost absent in these habitats. Using chloroplast and nuclear SSR markers, Lian et al. (2003) demonstrated that a large *Salix* patch in this desert was composed of many genetically different *Salix* individuals (Fig. 6.2). Moreover, larger *Salix* genets were found inside to upper areas of the patch, whereas the genets became smaller toward the lower periphery along the slope (Fig. 6.2). Therefore, *Salix* seedlings have been repeatedly establishing in the vicinity of previously established shrubs during the long successional period on Mt. Fuji.

Small areas surrounding the established *Salix* are the only places in which ECM networks are available in this desert (Fig. 6.2). The area covered by these ECM networks covers approximately 1 % of the ground surface. Considering the exclusive establishment of *Salix* seedlings in this limited habitat, one would expect some mechanisms of ECM networks that facilitate conspecific seedling establishment. In fact, the pattern resembles a greenhouse experiment using Japanese red pine, where seedling growth was improved by ECM infection through the ECM

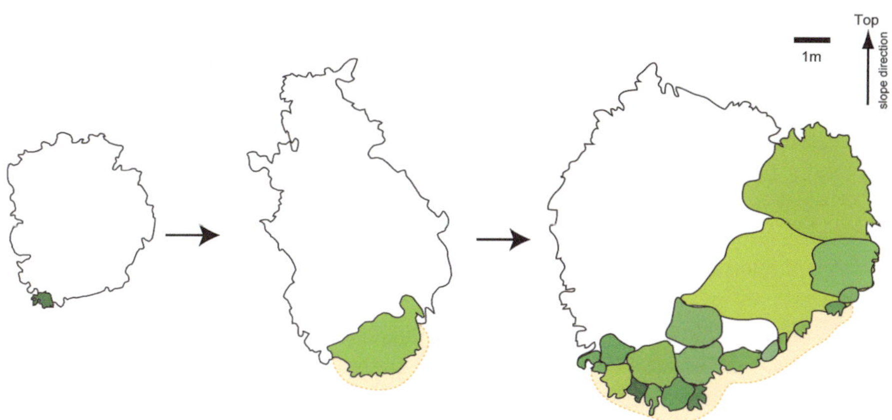

Fig. 6.2 Initial colonization of *Salix reinii* and subsequent development of *Salix* islands in the volcanid desert on Mt. Fuji, Japan. (*left*) *Salix* usually colonize at the periphery of a vegetation island, shown in *green* color in the upper panel. (*middle*) The established *Salix* grows vegetatively to several square meters. (*right*) The *Salix* island further develops to a large coverage area by recruiting new *Salix* establishments. Different green color areas indicate *Salix* genets defined by chloroplast and nuclear microsatellite markers (illustrated from the data in Lian et al. 2006). The *orange area* enclosed by broken line is the place where naturally established *Salix* seedlings are observed

Fig. 6.3 A greenhouse experiment that shows the positive effect of ECM networks, specifically rapid colonization through the network, on seedling growth. The left end tree is 2 year-old ECM tree (*Pinus densiflora*), which was planted in autoclaved nursery soil with a mesh. At the same time, seeds of *P. densiflora* were placed in three rows with different distances (15, 30, 45 cm) from the ECM tree. ECM mycelia spread from the left ECM mother tree and infected seedlings sequentially along the distances. Earlier infection resulted in better growth because of longer support from the ECM networks (Nara 1998)

network provided by an ECM mother tree (Fig. 6.3). No ECM fungal propagules existed in the sterilized soil; thus, the ECM network was the only available source of fungal symbionts in this experiment. Given that host plants need ECM fungal symbionts for nutrient acquisition and growth (Smith and Read 2008; Fig. 6.4), ECM networks could be the primary factor affecting seedling establishment in ECM propagule-poor soil.

Transplanted non-mycorrhizal *Salix* seedlings remained largely non-mycorrhizal in bare ground and in non-ECM vegetation islands (one of 31 surviving seedlings was ECM), indicating the scarcity of ECM fungal propagules in these predominant habitats (Nara and Hogetsu 2004). In contrast, seedlings near the established *Salix* were readily colonized (27 of 33 seedlings) by ECM fungi, most of which were the same species as those on the established *Salix*. Because ECM infection was directly related to the improved nutrient status and growth of the seedlings, the performance of *Salix* seedlings was better near the established shrubs (Fig. 6.5). Also, seedling performance near actively photosynthesizing shrubs differed from that near shrubs with lower photosynthetic activity (Fig. 6.5). Because the amount of photosynthate from the host can directly affect belowground ECM networks, this observed difference in seedling performance might be related to the activity of ECM networks.

Fig. 6.4 *Salix reinii* seedlings inoculated with a variety of ectomycorrhizal fungi in sterile nursery soil

The improved seedling performance near the established *Salix* may have been due to other soil factors (e.g., better soil conditions) or nonnetwork ECM colonization, i.e., from spores or sclerotia (see below). To isolate the effect of ECM networks from these potentially confounding factors, Nara (2006a) conducted a field experiment in which all biotic and abiotic conditions were identical except for the presence of ECM networks. First, most ECM fungal species in the desert, including rarely studied *Inocybe* and *Russula* species, were isolated from sporocarps collected in the desert. These isolated strains were used to prepare "ECM mother trees" of *Salix*. After a 1-year growth period, ECM mother trees were transplanted to non-*Salix* islands to artificially create ECM networks in non-ECM habitats in the field. Current-year *Salix* seedlings were also planted near the ECM mother tree to evaluate the effect of individual ECM networks on seedling performance. ECM infection through ECM networks was observed in all of the 11 ECM fungal species examined, whereas other ECM infection (e.g., from spores) remained undetectable. Compared to seedlings in the non-mycorrhizal treatment, in which a non-mycorrhizal mother tree was used as the nurse plant, the growth of seedlings connected to the ECM networks was significantly improved (Fig. 6.5). N and phosphorus contents of the seedlings in ECM networks were considerably higher than values for seedlings in the absence of ECM networks (Nara 2006a). These findings unequivocally indicate that most ECM networks, although not all, alleviate competition with larger conspecific plants for soil nutrients and improve seedling performance (but see Jakobsen and Hammer, Chap. 4, this Volume). Therefore, ECM networks alone could account for the observed facilitation of seedling establishment near established conspecific shrubs in this volcanic desert

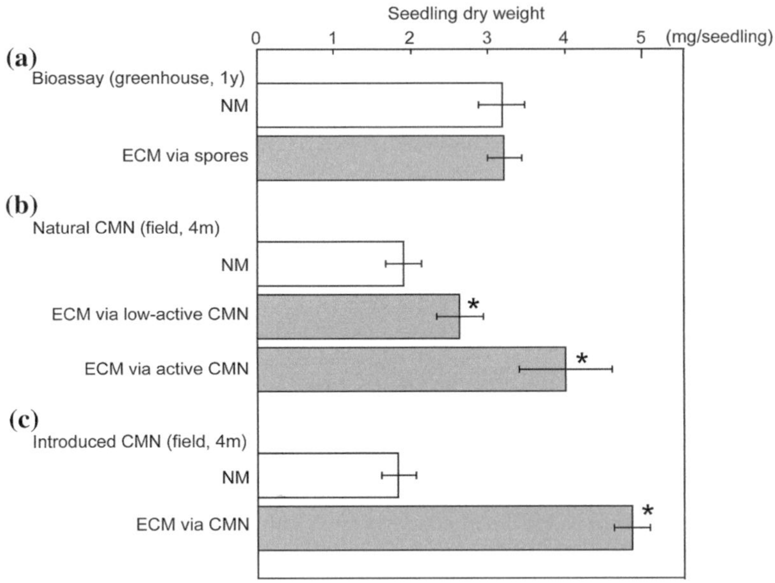

Fig. 6.5 The growth of *Salix* seedlings in response to ECM fungal infection through **a** soil spore banks (Nara et al. unpublished), **b** natural ECM networks (Nara and Hogetsu 2004), and **c** introduced ECM networks (Nara 2006a). Mean ± SE. *Asterisks* indicates statistically significant difference from NM seedlings in each treatment ($P < 0.05$). To compare the data among similar soil nutrient conditions, seedlings in vegetation-patch scoria are shown while excluding seedlings in bare ground scoria. In the experiment (**b**), the possibility of spore infection could not be excluded. The difference between low-active common mycorrhizal networks (CMN) and active CMN is based on the photosynthetic activity of host shrubs (see Nara et al. 2003a; Nara and Hogetsu 2004). Note that experiment (**a**) was conducted in a greenhouse for a longer period than the field experiments (**b**) and (**c**). Thus, the direct comparison between experiments might be difficult, though the comparison between NM and ECM seedlings within an experiment is rigorous. ECM infection from spores in field scoria has no effect on the growth of solitary seedlings (**a**), while the growth is significantly improved by ECM infection in more nutrient rich nursery soil (Fig. 6.4)

(Fig. 6.2). Moreover, Nara (2006a) also observed that the extent of the benefit of ECM networks varied considerably among ECM fungal species, although most effects were positive.

6.1.4 ECM Networks Mediate Primary Succession of Trees

In the primary successional sere on Mt. Fuji, the establishment of *Larix* and *Betula* follows *Salix* establishment. Unlike dwarf *Salix* shrubs, both of these tree species become tall and could form initial forests. Thus, the establishment of *Larix* and *Betula* is a critical step in succession toward forests, although the establishment of both species is still limited in the volcanic desert. Nara (2006b) surveyed established *Larix* and *Betula* individuals, mostly young saplings, in a 21-ha area of the volcanic desert on Mt. Fuji. All of the 26 naturally established *Larix* and 39 naturally established *Betula* individuals were found in the vicinity of established *Salix* shrubs, where ECM networks are available. ECM networks very likely facilitated their establishment, as in the case of conspecific *Salix* seedling establishment. Corroborating this hypothesis, ECM fungal communities observed in naturally established *Larix* and *Betula* saplings (Nara 2006b) and transplanted *Larix* and *Betula* seedlings near established *Salix* shrubs (Nara and Hogetsu 2004) were dominated by ECM fungi common to *Salix*.

Bioassay experiments have revealed that ECM inoculum in the soils (scoria) was less available for *L. kaempferi* (Pinaceae) than for *Salix*, where ECM infection (mostly by *Suillus grevillei* and *Suillus laricinus*) was found on <20 % of *Larix* bioassay seedlings (Nara et al. unpublished data). Growth rates did not significantly differ between infected and uninfected seedlings in this *Larix* bioassay experiment, as in the *Salix* bioassay described below, indicating that ECM infection from spores did not improve the growth of this secondary colonizing tree species in the volcanic desert.

6.1.5 Network Versus Nonnetwork ECM Fungal Infection for Seedling Establishment in Primary Succession

Salix bioassay experiments using field scoria collected from the Mt. Fuji desert have revealed that spore banks are poorly developed in bare ground and under non-ECM vegetation (Nara et al. unpublished data). In contrast, infective spore banks were found more frequently under small *Salix* shrubs (9 of 20 scoria samples) and large *Salix* shrubs (33 of 50). The number of ECM sporocarps produced increases with *Salix* size (Nara et al. 2003a), and their spores should be deposited at short distances from the sporocarps (Galante et al. 2011). Thus, the increased frequency of spore banks may be related to the amount of spore deposits. Species belonging to *Laccaria*, *Inocybe*, and *Scleroderma* were dominant in the spore bank community.

To evaluate the importance of ECM networks in seedling establishment, quantifying the effect of other ECM infection pathways (e.g., spores) would be useful. While all existing ECM networks should originate from spores initially, for discussion, I would like to differentiate network from nonnetwork infection specifically for "a seedling," based on whether a colonizing ECM mycelium is solely supported by the seedling alone (non-network) or supported jointly by other established hosts, often larger and preexisting ones (network). In the above bioassay experiment, we can evaluate the potential effect of nonnetwork ECM infection. The dry weight (DW) of infected bioassay seedlings did not differ from that of uninfected bioassay seedlings (Fig. 6.5). This result was unexpected, as ECM colonization is a prerequisite for hosts to grow normally (Smith and Read 2008) and seedling growth was actually improved by mycelial inoculation in an in vitro experiment using nursery soil (Fig. 6.4). Growth improvement over an order of magnitude is also observed in other inoculation experiments (e.g., Abuzinadah and Read 1986, 1989). Regardless, ECM fungal infection itself clearly has no significant effect on either seedling growth in the scoria substrate, or in all likelihood, on seedling establishment in the volcanic desert.

The relative importance of ECM networks and spore infection in seedling establishment is expected to be context-dependent. In secondary successional sites and some primary successional sites where soil spore banks have already developed and available soil nutrients are greater than in volcanic deserts, ECM infection from spores plays a critical role in initial seedling establishment (e.g., Baar et al. 1999; Ashkannejhad and Horton 2006). In these cases, a small nonnetwork ECM mycelium supported by a single seedling appears to be effective in absorbing enough soil nutrients to improve host growth. In primary successional sites like volcanic deserts, available soil nutrients, especially N, are quite deficient. Thus, a tiny *Salix* seedling, often ≪10 mg in DW after the first growing season in this desert, could not develop a large enough mycelium to explore soil nutrients within extremely nutrient-poor scoria to the extent needed to improve host growth in addition to its own demands. Established ECM networks, however, could provide seedlings with access to a larger nutrient pool (Fig. 6.6). Therefore, the importance of ECM networks, despite being quite sparsely distributed, could be more pronounced in primary successional sites.

Because the initial establishment of ECM plants in previously non-mycorrhizal habitats has been rare in the volcanic desert on Mt. Fuji (<30 events/ha during 300 years after eruption), little information is available on how initial ECM colonization occurs.

However, considering the species composition of pioneer ECM fungi, wind-dispersed spores were likely responsible for the initial establishment. In British heathlands, Collier and Bidartondo (2009) demonstrated that host seedlings can remain non-mycorrhizal and wait for ECM spores for a substantial period with almost no growth, and then after ECM infection occurs, the seedlings begin to grow. Similar events may be possible in some nutrient-rich and moderate environmental microhabitats in the volcanic desert, for example, near animal feces in the shelter of existing non-mycorrhizal plants, although this remains to be substantiated. Moreover, considering the rapid turnover of ECM genets in some fungal species such as *Laccaria*, colonization from spores should still be a widespread and critical process for building new ECM networks on established hosts, some of which can extend to seedlings and facilitate their establishment.

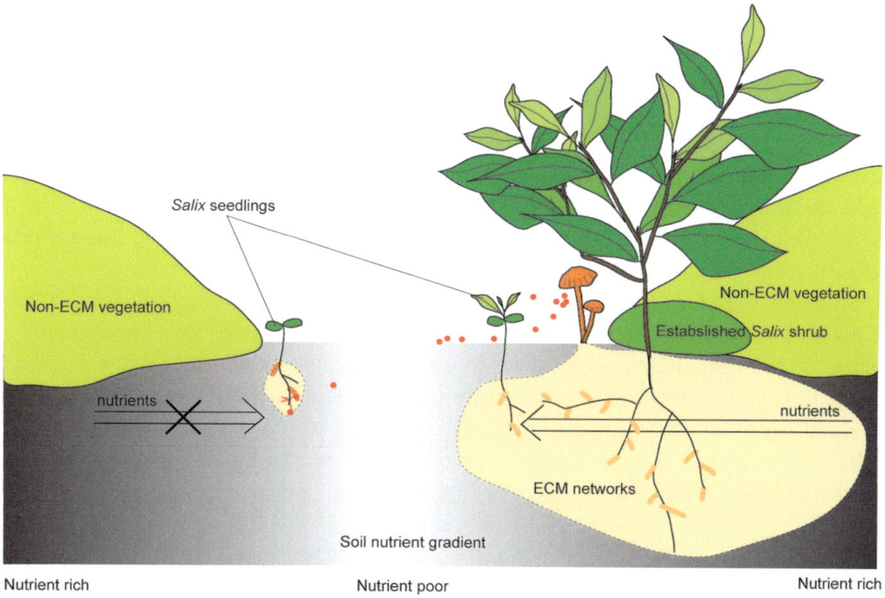

Fig. 6.6 A schematic diagram showing the facilitation of seedling establishment mediated by ECM networks in the volcanic desert on Mt. Fuji. Soil nutrients are relatively rich inside the vegetation. A tiny *Salix* seedling germinated beside a vegetation island has no access to the rich nutrient pool. In contrast, established *Salix* shrubs provide sufficient photosynthate to develop ECM networks from inside to the periphery of a vegetation island. *Salix* seedlings connected to the ECM networks can use the rich nutrient pool through nutrient translocations within the network

6.1.6 Mechanisms of Facilitated Seedling Establishment via ECM Networks in Primary Succession: Verification of Previous and New Models

Previous studies and reviews have put forth many potential mechanisms or models of how ECM networks could facilitate seedling establishment, most of which are not supported by unequivocal evidence or are even refutable due to contradictory evidence (Newman 1988; Perry et al. 1989; Dickie et al. 2002; Simard et al. 2002; Booth 2004; van der Heijden and Horton 2009; Hoeksema Chap. 9, this Volume). Some of the proposed mechanisms are also expected to be less relevant in primary successional sites. For example, plant-to-plant carbon transfer through common mycorrhizal networks has attracted much attention and is still rather controversial (see Simard et al. 1997; Robinson and Fitter 1999; Simard and Durall 2004; Simard et al., Chap. 5, this Volume, and references therein). Such carbon transfer could only be ecologically significant under light-limited conditions, such as in the shade of a closed forest canopy, similar to mycoheterotrophic and mixotrophic forest floor plants (Selosse et al. 2006). In primary successional sites, strong sunlight is available in most areas because of low vegetation coverage, and thus carbon is unlikely to be a limiting factor for seedlings. If a direct carbon supply from large established shrubs to seedlings could have aided seedling establishment in the volcanic desert, *Salix* seedlings should also occur inside the vegetation islands, where soil nutrient conditions are better but where light conditions are limited under the canopy of *Salix* shrubs or other perennial herbs. However, this is apparently not the case, i.e., no *Salix* seedlings occur under light-limiting conditions (Fig. 6.2), indicating that direct carbon transfer between plants is not responsible for the observed facilitation near *Salix* shrubs.

Nutrient transfer (e.g., N) between plants has also been suggested repeatedly (e.g., Newman 1988; Simard et al. 2002; Selosse et al. 2006) as a mechanism of facilitation, although unequivocal evidence that isolates direct ECM pathways from soil pathways is scarce. As discussed by van der Heijden and Horton (2009), plant productivity is usually limited by N, especially in primary succession, making it unlikely that plants give it away for free to other plants through ECM networks. Most evidence of plant-to-plant N transfer has been documented in studies involving N-fixing plants (e.g., Arnebrant et al. 1993; Ekblad and Huss-Danell 1995; He et al. 2005); however, none of the aforementioned ECM seedlings and shrubs in the volcanic desert on Mt. Fuji are N-fixing plants. Thus, at least in this desert, nutrient transfer between plants would be less relevant to the observed seedling establishment via ECM networks.

A greater diversity of ECM fungi provided by networks has also been proposed as a potential mechanism of facilitated seedling establishment (e.g., Newman 1988; Simard et al. 2002). In fact, because spores of "late successional ECM fungi" such as *Russula* and *Cortinarius* rarely germinate even in the presence of host roots (Ishida et al. 2008), ECM networks could provide more diverse ECM fungal symbionts to seedlings than what is attainable by spores alone. However, this mechanism is less relevant in the volcanic desert on Mt. Fuji. In Nara and Hogetsu (2004), transplanted non-mycorrhizal seedlings of ECM hosts near the established shrubs were primarily colonized by pioneer species, which would be readily attainable by spores; furthermore, ECM fungal diversity did not significantly affect seedling growth.

Cost–benefit relationships could determine host performance in mycorrhizal networks (van der Heijden and Horton 2009). If soil nutrient supply from the network is proportional to carbon investment from individual hosts, larger established hosts would outcompete seedlings, as shown in some studies of AM networks (e.g., Kytoviita et al. 2003; Jakobsen and Hammer, Chap. 4, this Volume). In a recent study using in vitro root organ culture systems, Kiers et al. (2011) demonstrated that AM fungi and hosts can distinguish cooperative partners from less-cooperative ones and will preferentially reward cooperative ones in relation to the obtained benefits. Although similar mechanisms might be possible in ECM networks, little evidence exists, and the mechanism may be less relevant to seedling establishment in the volcanic desert on Mt. Fuji. A typical *Salix* shrub of several square meters in coverage area usually harbors $\gg$1 kg dry weight of leaf biomass (Nara et al. 2003a), which is incomparably greater than the leaf biomass of a current year seedling ($\ll$10 mg even at the end of the growing season). In addition, *Salix* shrubs maintain higher N concentration in leaves and higher photosynthetic activity per unit leaf during the growing season (Nara et al. 2003a; Nara and Hogetsu 2004). Thus, the photosynthate supply to ECM fungi from a solitary seedling may be far less than, or even negligible, compared to that supplied from the established *Salix* shrub. If the cost–benefit relationships between ECM networks and individual hosts are proportional, tiny seedlings would be unable to compete with established shrubs. However, available evidence from the volcanic desert indicates that the benefit from ECM networks is not perfectly proportional to their carbon investment and is rather advantageous to seedlings.

Factors that appear to be relevant to the observed facilitation of seedling establishment via ECM networks in the volcanic desert include (1) rapid ECM fungal colonization and (2) larger ECM mycelia enabling access to larger amounts of soil nutrients. Both mechanisms could be provided to seedlings with less cost because incomparably larger hosts connected to the ECM network are supporting or maintaining the network. These mechanisms have been proposed repeatedly in other ecological settings (e.g., Newman 1988; Simard et al. 2002 and references therein).

1. Because ECM hosts largely depend on ECM fungi for soil nutrients, rapid ECM colonization provides seedlings with access to soil nutrients at earlier stages in the growing season, resulting in better seedling growth (as shown in Fig. 6.3). The same mechanism likely partially accounts for the observed facilitation of seedling establishment. *Salix* shrubs on Mt. Fuji flower and leaf in the spring and produce seeds in early summer. Dispersed seeds germinate immediately after landing in areas with sufficient soil water conditions. As in most tree species, the production of new roots on established shrubs begins well before bud bursting. Thus, ECM networks, even current-year belowground genets like *Laccaria* (Fig. 6.1f), are readily available for *Salix* seedlings at germination. In contrast, ECM colonization on germinated seedlings in the absence of ECM networks would be inevitably delayed to late summer or fall when most ECM fungal spores are dispersed. Even in the presence of soil spore banks (although infrequent), some time is required for seedlings to stimulate spore germination, to be infected by ECM fungal hyphae, and to invest carbon (or even N at the initial stage) to hyphae to develop functional ECM mycelial systems. Given the short growth period on Mt. Fuji, earlier access to soil nutrients through ECM fungi would be an important prerequisite for survival.
2. As noted above, a small ECM mycelium supported by a tiny *Salix* seedling is not effective for improving seedling growth, likely because of insufficient nutrients provided from the mycobiont. Moreover, the same ECM fungal species significantly improved seedling growth in nursery soil that contained much higher concentrations of N and other soil nutrients compared to the scoria (Fig. 6.4). Given the extremely nutrient-poor scoria in the volcanic desert on Mt. Fuji, ECM mycelia would need to be of larger sizes to acquire an adequate amount of soil nutrients to improve host growth to the extent that could be

expected in more nutrient-rich soil. The nutrient pool accumulated in the existing ECM mycelia would also be relevant.

Mechanisms that have rarely been proposed in the literature but are likely to be relevant in the volcanic desert on Mt. Fuji include (3) nutrient translocation from areas of nutrient-rich soil to seedlings and (4) nutrient gradients between ECM fungi and individual hosts, especially in relation to the difference in nutrient status between mature trees and seedlings. This latter mechanism fundamentally differs from the source–sink gradients "between plants" that have been frequently discussed in the literature, particularly plant-to-plant carbon transfer (e.g., Simard and Durall 2004; See Simard et al. Chap. 5).

3. As in many primary successional sites, the soil nutrient distribution is quite heterogeneous in the volcanic desert on Mt. Fuji. While the bare-ground scoria contains very low levels of N, the interior of vegetation patches contains relatively higher levels of N because of accumulated organic matter and associated microbial activities (Hirose and Tateno 1984). In fact, the soil N level inside a vegetation island is about 10 times higher than that outside the island. Given that the seedling establishment of ECM plants is restricted to the periphery of a vegetation island because of their light-demanding properties, a solitary seedling with a tiny root system and a small ECM mycelium would have no access to such a nutrient pool (Fig. 6.6). On the other hand, a larger ECM mycelium supported by mature *Salix* shrubs would have an enhanced ability to explore the relatively rich nutrient resources inside of the vegetation island (Fig. 6.6). Given that nutrients are quickly translocated within a mycelium (e.g., Finlay and Read 1986b; Wu et al. 2012), this mechanism could facilitate seedling establishment near the established ECM shrubs. Several previous studies have suggested nutrient transfer from dying or senescent parts (roots/hyphae) to active parts within an ECM network (e.g., Newman 1988). Although such transfer could also be important in optimizing nutrient usage even in primary successional settings, the mechanism proposed here is not restricted to such plant or fungal tissues. Given that ECM mycelia preferentially explore nutrient-rich patches in soil (Bending and Read 1995; Read and Perez-Moreno 2003) and that nutrients quickly move within the mycelia or even to connecting plants (Finlay and Read 1986b), this process could be a mechanism facilitating seedling establishment in nutrient-poor habitats.
4. Fungi contain far higher concentrations of N compared to plants: N in fungi ranges from 4 to 6 % in sporocarps (e.g., Vogt et al. 1981), whereas N in plants is usually <3 % even in leaves. In contrast, plants contain higher concentrations of sugars compared to fungi. Therefore, exchanging N (or other nutrients) and sugars along the nutrient gradients in ECM symbioses is expected to be easier than exchanging against the gradient. In fact, some studies have demonstrated that carbon transfer from ECM fungi to hosts does not occur (Wu et al. 2001; cf. Simard et al. 2002). Although actual concentrations at the fungus–plant interface may potentially vary (sometimes enough to generate reverse gradients) and transporters may enable substrate exchanges against the normal gradients, no

unequivocal evidence has demonstrated ecologically relevant net transfer against the gradients (i.e., against the dominant directions) between ECM symbionts (except mycoheterotrophs). Here, I summarize the data for N concentrations of ECM fungi and host shrubs/seedlings to address the hypothesis that nutrient exchanges along the gradient could, by itself, potentially account for facilitated seedling establishment in primary successional sites (Fig. 6.7).

As in most deciduous tree species, *Salix* shrubs retrieve nutrients (especially N) from leaves (and fine roots) before leaf-fall and store them in dormant parts such as buds, stems, and roots during winter. The stored nutrients are then used for leafing/rooting in spring. During the growing season, the shrubs also absorb nutrients from soil. Because this process is repeated every year, the established shrubs have accumulated a large amount of N since the initial establishment. In fact, leaves of established *Salix* shrubs contain relatively higher N concentrations than seedlings of conspecific *Salix* or other species (*Larix* and *Betula*), indicating a better

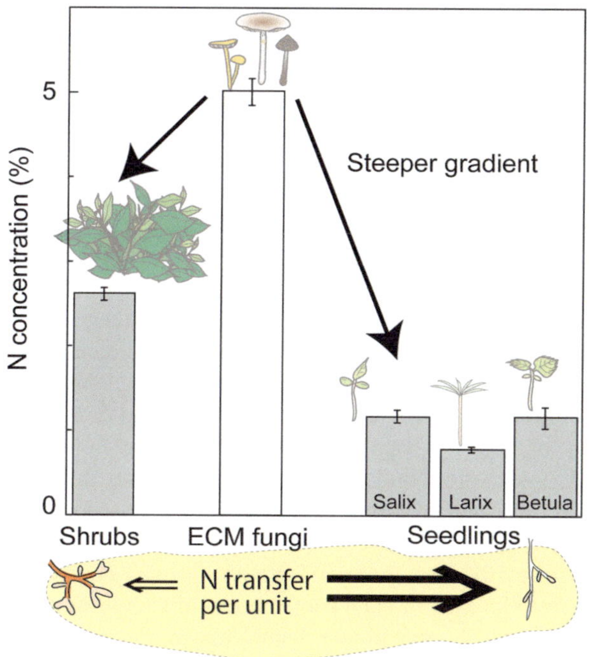

Fig. 6.7 A hypothetical model of nutrient transfer between an ectomycorrhizal (ECM) fungus and hosts in relation to the strength of nutrient concentration gradients. Empirical data of leaf N concentration of established *Salix* shrubs (Nara et al. 2003a), current year seedlings (Nara and Hogetsu 2004), and ECM fungal sporocarps (Nara et al. unpublished) on Mt Fuji are shown. Because of the woody lifestyle of hosts and accumulated nutrients over many years, nutrient status of the larger established plants is better than current-year seedlings. The difference in nutrient gradients in an ECM network could drive favorable nutrient movement to the seedlings. See text for details

N status in established shrubs than in seedlings (Fig. 6.7). N concentrations of ECM fungi are much higher than in any type of host, generating potential gradients for fungus-to-plant N transfer; however, the gradients are much steeper in fungus-to-seedling transfers than in fungus-to-shrub transfers (Fig. 6.7). Thus, N in the common ECM mycelium pool could be transferred to seedlings more easily than to established shrubs, just by following the strength of the gradient.

Plants usually allocate more sugar to roots when in nutrient starvation. Thus, although the photosynthetic activity per unit leaf is much higher in the established shrubs with higher N concentrations, sugar concentrations in roots could be comparable or even higher in seedlings than in shrubs. Regardless, given that ECM fungal growth is primarily limited by carbon supply, no need would exist for ECM fungi to refuse additional carbon sources. In fact, seedling roots are quickly colonized by ECM fungi when the network mycelia come into contact with them (Wu et al. 1999; Fig. 6.3), even with an initial expense of carbon to establish new ECM colonization on seedlings (Finlay and Read 1986a). Once ECM colonization establishes, carbon would be transferred between symbionts, following the carbon gradients between them.

In this model, hosts and ECM fungi are independently seeking the most efficient pathways, seeking deficient substances to enhance their respective performances. For example, both established shrubs and seedlings are pursuing the optimal strategies for nutrient and carbon allocation within their own structures under their current nutrient status, while the carbon-deficient ECM fungus is exploring all potential carbon sources (i.e., compatible roots), allocating its resources to the most demanding parts within the mycelium, such as near the newly contacted roots. At the interface of the hosts and fungus in an ECM network, both carbon and nutrients move in relation to the strength of their gradients in this simplified model. As a result of the woody habit of ECM hosts, the strength of the gradients (and possibly carbon) within an ECM network could differ between the seedlings and larger trees and function as a mechanism of favorable transfer to the seedlings in a primary successional setting. Under this model, the magnitude and direction of transfer, and thus the effects of ECM networks on seedlings, are context-dependent. For example, if seedlings cannot conduct photosynthesis under the deep shade of trees in a closed forest, N concentrations could be higher in seedlings than in mature trees. In this case, N could be preferentially transferred to the large trees within an ECM network, following the gradient. Whether sugar concentrations could be higher in fungal mycelia than in plant roots under some environmental conditions is not certain, but if so, carbon movement from fungi to plants and thus plant-to-plant carbon transfer (as in Simard et al. 1997) could also be explained by this model. Inconsistent results between AM networks (many negative results) and ECM networks (mostly positive results) in terms of seedling performance (van der Heijden and Horton 2009) may be due in part to the difference in host habit [i.e., herbaceous (AM) and woody (ECM)] and eventually to the different nutrient statuses of larger competing plants within a network. Because nutrient/sugar concentrations in plant roots and fungal mycelia should also vary between combinations of host and fungal species (although little has been documented), various outcomes of networks in seedling establishment are possible under this model.

By examining the unique conditions of a primary succession site, positive effects of ECM networks upon seedling establishment have been unequivocally confirmed. Although the unique setting has also enabled the verification of several previous models of ECM network functions and has generated additional potential mechanisms, none of these models or mechanisms has been conclusively addressed to date. Because of the relatively simple study environment, primary successional sites could prove the most promising locations for further research on ECM networks and seedling establishment.

Acknowledgments I thank T. Horton for inviting me to contribute to this volume and constructive comments on a draft of this chapter. This work was supported by JSPS KAKENHI Grants 16780111, 19380083, 21658054, and 22380083.

References

Abuzinadah RA, Read DJ (1986) The role of proteins in the nitrogen nutrition of ectomycorrhizal plants III: protein utilization by *Betula*, *Picea* and *Pinus* in mycorrhizal association with *Hebeloma crustuliniforme*. New Phytol 103:507–514

Abuzinadah RA, Read DJ (1989) The role of proteins in the nitrogen nutrition of ectomycorrhizal plants IV: the utilization of peptides by Birch (*Betula pendula* L) infected with different mycorrhizal Fungi. New Phytol 112:55–60

Allen MF, Crisafulli C, Friese CF, Jeakins SL (1992) Re-formation of mycorrhizal symbioses on Mount St Helens, 1980-1990: interactions of rodents and mycorrhizal fungi. Mycol Res 96:447–453

Arnebrant K, Ek H, Finlay RD, Soderstrom B (1993) Nitrogen translocation between *Alnus glutinosa* (L) Gaertn seedlings inoculated with *Frankia* sp and *Pinus contorta* Doug Ex-Loud seedlings connected by a common ectomycorrhizal mycelium. New Phytol 124:231–242

Ashkannejhad S, Horton TR (2006) Ectomycorrhizal ecology under primary succession on coastal sand dunes: interactions involving *Pinus contorta*, suilloid fungi and deer. New Phytol 169:345–354

Baar J, Horton TR, Kretzer AM, Bruns TD (1999) Mycorrhizal colonization of *Pinus muricata* from resistant propagules after a stand-replacing wildfire. New Phytol 143:409–418

Bahram M, Põlme S, Kõljalg U, Tedersoo L (2011) A single European aspen (*Populus tremula*) tree individual may potentially harbour dozens of *Cenococcum geophilum* ITS genotypes and hundreds of species of ectomycorrhizal fungi. FEMS Microbiol Ecol 75:313–320

Beiler KJ, Durall DM, Simard SW, Maxwell SA, Kretzer AM (2010) Architecture of the wood-wide web: *Rhizopogon* spp. genets link multiple Douglas-fir cohorts. New Phytol 185:543–553

Bending GD, Read DJ (1995) The structure and function of the vegetative mycelium of ectomycorrhizal plants V: foraging behavior and translocation of nutrients from exploited litter. New Phytol 130:401–409

Booth MG (2004) Mycorrhizal networks mediate overstorey-understorey competition in a temperate forest. Ecol Lett 7:538–546

Cazares E, Trappe JM, Jumpponen A (2005) Mycorrhiza-plant colonization patterns on a subalpine glacier forefront as a model system of primary succession. Mycorrhiza 15:405–416

Chagnon PL, Bradley RL, Klironomos JN (2012) Using ecological network theory to evaluate the causes and consequences of arbuscular mycorrhizal community structure. New Phytol 194:307–312

Collier FA, Bidartondo MI (2009) Waiting for fungi: the ectomycorrhizal invasion of lowland heathlands. J Ecol 97:950–963

Dahlberg A (1997) Population ecology of *Suillus variegatus* in old Swedish Scots pine forests. Mycol Res 101:47–54

Dahlberg A, Stenlid J (1994) Size, distribution and biomass of genets in populations of *Suillus bovinus* (L, Fr) Roussel revealed by somatic incompatibility. New Phytol 128:225–234

Dickie IA, Koide RT, Steiner KC (2002) Influences of established trees on mycorrhizas, nutrition, and growth of *Quercus rubra* seedlings. Ecol Monogr 72:505–521

Dickie IA, Martinez-Garcia LB, Koele N, Grelet GA, Tylianakis JM, Peltzer DA, Richardson SJ (2013) Mycorrhizas and mycorrhizal fungal communities throughout ecosystem development. Plant Soil 367:11–39

Douhan GW, Vincenot L, Gryta H, Selosse MA (2011) Population genetics of ectomycorrhizal fungi: from current knowledge to emerging directions. Fungal Biol 115:569–597

Ekblad A, Huss-Danell K (1995) Nitrogen fixation by *Alnus incana* and nitrogen transfer from *A. incana* to *Pinus sylvestris* influenced by macronutrients and ectomycorrhiza. New Phytol 131:453–459

Finlay RD, Read DJ (1986a) The structure and function of the vegetative mycelium of ectomycorrhizal plants I: translocation of C14-labeled carbon between plants interconnected by a common mycelium. New Phytol 103:143–156

Finlay RD, Read DJ (1986b) The structure and function of the vegetative mycelium of ectomycorrhizal plants II: the uptake and distribution of phosphorus by mycelial strands interconnecting host plants. New Phytol 103:157–165

Fujiyoshi M, Yoshitake S, Watanabe K, Murota K, Tsuchiya Y, Uchida M, Nakatsubo T (2011) Successional changes in ectomycorrhizal fungi associated with the polar willow *Salix polaris* in a deglaciated area in the High Arctic, Svalbard. Polar Biol 34:667–673

Galante TE, Horton TR, Swaney DP (2011) 95% of basidiospores fall within 1 m of the cap: a field- and modeling-based study. Mycologia 103:1175–1183

Gryta H, Debaud JC, Effosse A, Gay G, Marmeisse R (1997) Fine-scale structure of populations of the ectomycorrhizal fungus *Hebeloma cylindrosporum* in coastal sand dune forest ecosystems. Mol Ecol 6:353–364

Halvorson JJ, Smith JL (2009) Carbon and nitrogen accumulation and microbial activity in Mount St. Helens pyroclastic substrates after 25 years. Plant Soil 315:211–228

He XH, Critchley C, Ng H, Bledsoe C (2005) Nodulated N(2)-fixing *Casuarina cunninghamiana* is the sink for net N transfer from non-N(2)-fixing *Eucalyptus maculata* via an ectomycorrhizal fungus *Pisolithus* sp using (15)NH(4)(+) or (15)NO(3)(-) supplied as ammonium nitrate. New Phytol 167:897–912

Helm DJ, Allen EB, Trappe JM (1996) Mycorrhizal chronosequence near Exit Glacier, Alaska. Can J Bot 74:1496–1506

Hirose D, Kikuchi J, Kanzaki N, Futai K (2004) Genet distribution of sporocarps and ectomycorrhizas of *Suillus pictus* in a Japanese white pine plantation. New Phytol 164:527–541

Hirose T, Tateno M (1984) Soil-nitrogen patterns Induced by colonization of *Polygonum cuspidatum* on Mt Fuji. Oecologia 61:218–223

Horton TR, Bruns TD (1998) Multiple-host fungi are the most frequent and abundant ectomycorrhizal types in a mixed stand of Douglas fir (*Pseudotsuga menziesii*) and bishop pine (*Pinus muricata*). New Phytol 139:331–339

Horton TR, Bruns TD (2001) The molecular revolution in ectomycorrhizal ecology: peeking into the black-box. Mol Ecol 10:1855–1871

Horton TR, Cazares E, Bruns TD (1998) Ectomycorrhizal, vesicular-arbuscular and dark septate fungal colonization of bishop pine (*Pinus muricata*) seedlings in the first 5 months of growth after wildfire. Mycorrhiza 8:11–18

Ingleby K, Last FT, Mason PA (1985) Vertical-distribution and temperature relations of sheathing mycorrhizas of *Betula* spp. growing on coal spoil. Forest Ecol Manag 12:279–285

Ishida TA, Nara K, Hogetsu T (2007) Host effects on ectomycorrhizal fungal communities: insight from eight host species in mixed conifer-broadleaf forests. New Phytol 174:430–440

Ishida TA, Nara K, Tanaka M, Kinoshita A, Hogetsu T (2008) Germination and infectivity of ectomycorrhizal fungal spores in relation to their ecological traits during primary succession. New Phytol 180:491–500

Jones MD, Durall DM, Cairney JWG (2003) Ectomycorrhizal fungal communities in young forest stands regenerating after clearcut logging. New Phytol 157:399–422

Jumpponen A, Trappe JM, Cazares E (1999) Ectomycorrhizal fungi in Lyman Lake Basin: a comparison between primary and secondary successional sites. Mycologia 91:575–582

Jumpponen A, Trappe JM, Cazares E (2002) Occurrence of ectomycorrhizal fungi on the forefront of retreating Lyman Glacier (Washington, USA) in relation to time since deglaciation. Mycorrhiza 12:43–49

Kennedy PG, Izzo AD, Bruns TD (2003) There is high potential for the formation of common mycorrhizal networks between understorey and canopy trees in a mixed evergreen forest. J Ecol 91:1071–1080

Kiers ET, Duhamel M, Beesetty Y, Mensah JA, Franken O, Verbruggen E, Fellbaum CR, Kowalchuk GA, Hart MM, Bago A, Palmer TM, West SA, Vandenkoornhuyse P, Jansa J, Bucking H (2011) Reciprocal rewards stabilize cooperation in the mycorrhizal symbiosis. Science 333:880–882

Kretzer AM, Dunham S, Molina R, Spatafora JW (2004) Microsatellite markers reveal the below ground distribution of genets in two species of *Rhizopogon* forming tuberculate ectomycorrhizas on Douglas fir. New Phytol 161:313–320

Kytoviita MM, Vestberg M, Tuom J (2003) A test of mutual aid in common mycorrhizal networks: established vegetation negates benefit in seedlings. Ecology 84:898–906

Lian CL, Narimatsu M, Nara K, Hogetsu T (2006) *Tricholoma matsutake* in a natural *Pinus densiflora* forest: correspondence between above- and below-ground genets, association with multiple host trees and alteration of existing ectomycorrhizal communities. New Phytol 171:825–836

Lian CL, Oishi R, Miyashita N, Nara K, Nakaya H, Wu BY, Zhou ZH, Hogetsu T (2003) Genetic structure and reproduction dynamics of *Salix reinii* during primary succession on Mount Fuji, as revealed by nuclear and chloroplast microsatellite analysis. Mol Ecol 12:609–618

Molina R, Massicotte H, Trappe JM (1992) Specificity phenomena in mycorrhizal symbiosis: community-ecological consequences and practical implications. In: Allen MJ (ed) Mycorrhizal functioning: an integrative plant-fungal process. Chapman & Hall, London, pp 357–423

Murata M, Kinoshita A, Nara K (2013) Revisiting the host effect on ectomycorrhizal fungal communities: implications from host-fungal associations in relict *Pseudotsuga japonica* forests. Mycorrhiza 23:641–653

Nara K (1998) Composition of symbiotic fungi and their ecological roles in beech forests. In Kaneko S, Sahashi N (eds) Fungi in beech forests. Bun-ichi Sogo Shuppan, Japan, pp 79–149 (in Japanese)

Nara K (2006a) Ectomycorrhizal networks and seedling establishment during early primary succession. New Phytol 169:169–178

Nara K (2006b) Pioneer dwarf willow may facilitate tree succession by providing late colonizers with compatible ectomycorrhizal fungi in a primary successional volcanic desert. New Phytol 171:187–198

Nara K, Hogetsu T (2004) Ectomycorrhizal fungi on established shrubs facilitate subsequent seedling establishment of successional plant species. Ecology 85:1700–1707

Nara K, Nakaya H, Hogetsu T (2003a) Ectomycorrhizal sporocarp succession and production during early primary succession on Mount Fuji. New Phytol 158:193–206

Nara K, Nakaya H, Wu BY, Zhou ZH, Hogetsu T (2003b) Underground primary succession of ectomycorrhizal fungi in a volcanic desert on Mount Fuji. New Phytol 159:743–756

Newman EI (1988) Mycorrhizal links between plants: their functioning and ecological significance. Adv Ecol Res 18:243–270
Obase K, Tamai Y, Miyamoto T, Yajima T (2005) Macrofungal flora on the volcano Usu, deforested by 2000 eruptions. Eur J Forest Res 8:65–70
Obase K, Tamai Y, Yajima T, Miyamoto T (2007) Mycorrhizal associations in woody plant species at the Mt. Usu volcano, Japan. Mycorrhiza 17:209–215
Perry DA, Margolis H, Choquette C, Molina R, Trappe JM (1989) Ectomycorrhizal mediation of competition between coniferous tree species. New Phytol 112:501–511
Read DJ (1989) Mycorrhizas and nutrient cycling in sand dune ecosystems. Proc Roy Soc Edin 96B:89–110
Read DJ, Perez-Moreno J (2003) Mycorrhizas and nutrient cycling in ecosystems—a journey towards relevance? New Phytol 157:475–492
Robinson D, Fitter A (1999) The magnitude and control of carbon transfer between plants linked by a common mycorrhizal network. J Exp Bot 50:9–13
Selosse MA, Richard F, He XH, Simard SW (2006) Mycorrhizal networks: des liaisons dangereuses? Trends Ecol Evol 21:621–628
Simard SW, Durall DM (2004) Mycorrhizal networks: a review of their extent, function, and importance. Can J Bot 82:1140–1165
Simard SW, Durall DM, Jones MD (2002) Carbon and nutrient fluxes within and between mycorrhizal plants. In: van der Heijden MGA, Sanders IR (eds) Mycorrhizal Ecology. Springer, Berlin, pp 33–74
Simard SW, Perry DA, Jones MD, Myrold DD, Durall DM, Molina R (1997) Net transfer of carbon between ectomycorrhizal tree species in the field. Nature 388:579–582
Smith SE, Read DJ (2008) Mycorrhizal symbiosis, 3rd edn. Academic Press, London
Stendell ER, Horton TR, Bruns TD (1999) Early effects of prescribed fire on the structure of the ectomycorrhizal fungus community in a Sierra Nevada ponderosa pine forest. Mycol Res 103:1353–1359
Tedersoo L, Jairus T, Horton BM, Abarenkov K, Suvi T, Saar I, Koljalg U (2008) Strong host preference of ectomycorrhizal fungi in a Tasmanian wet sclerophyll forest as revealed by DNA barcoding and taxon-specific primers. New Phytol 180:479–490
Twieg BD, Durall DM, Simard SW (2007) Ectomycorrhizal fungal succession in mixed temperate forests. New Phytol 176:437–447
van der Heijden MGA, Vries FW, Kuyper TW (1999) Mycorrhizal associations of *Salix repens* L. communities in succession of dune ecosystems I: above-ground and below-ground views of ectomycorrhizal fungi in relation to soil chemistry. Can J Bot 77:1821–1832
van der Heijden MGA, Horton TR (2009) Socialism in soil? The importance of mycorrhizal fungal networks for facilitation in natural ecosystems. J Ecol 97:1139–1150
Vincenot L, Nara K, Sthultz C, Labbe J, Dubois MP, Tedersoo L, Martin F, Selosse MA (2012) Extensive gene flow over Europe and possible speciation over Eurasia in the ectomycorrhizal basidiomycete *Laccaria amethystina* complex. Mol Ecol 21:281–299
Vogt KA, Edmonds RL, Grier CC (1981) Biomass and nutrient concentrations of sporocarps produced by mycorrhizal and decomposer fungi in *Abies amabilis* stands. Oecologia 50:170–175
Wadud MA, Lian CL, Nara K, Hogetsu T (2006a) Isolation and characterization of five microsatellite loci in an ectomycorrhizal fungus *Laccaria laccata*. Mol Ecol Notes 6:700–702
Wadud MA, Lian CL, Nara K, Ishida TA, Hogetsu T (2006b) Development of microsatellite markers from an ectomycorrhizal fungus, *Laccaria amethystina*, by a dual-suppression-PCR technique. Mol Ecol Notes 6:130–132
Wadud MA, Lian CL, Nara K, Ishida TA, Hogetsu T (2007) Below-ground genet structure of an ectomycorrhizal fungus *Laccaria amethystina* in a volcanic desert on Mount Fuji. J Agrofor Environ 1:157–162
Wadud MA, Lian CL, Nara K, Reza MS, Hogetsu T (2008) Below-ground genet differences of an ectomycorrhizal fungus *Laccaria laccata* infecting *Salix* stands in primary successional stage. J Agrofor Environ 2:1–6

Wadud MA, Nara K, Lian CL, Ishida TA, Hogetsu T (2014) Genet dynamics and ecological functions of the pioneer ectomycorrhizal fungi *Laccaria amethystine* and *Laccaria laccata* in a volcanic desert on Mount Fuji. Mycorrhiza 24:551–563

Wu B, Nara K, Hogetsu T (1999) Competition between ectomycorrhizal fungi colonizing *Pinus densiflora*. Mycorrhiza 9:151–159

Wu B, Maruyama H, Teramoto M, Hogetsu T (2012) Structural and functional interactions between extraradical mycelia of ectomycorrhizal *Pisolithus* isolates. New Phytol 194:1070–1078

Wu B, Nara K, Hogetsu T (2001) Can C14-labeled photosynthetic products move between *Pinus densiflora* seedlings linked by ectomycorrhizal mycelia? New Phytol 149:137–146

Wu B, Nara K, Hogetsu T (2005) Genetic structure of *Cenococcum geophilum* populations in primary successional volcanic deserts on Mount Fuji as revealed by microsatellite markers. New Phytol 165:285–293

Yamanaka T, Okabe H (2006) Distribution of *Frankia*, ectomycorrhizal fungi, and bacteria in soil after the eruption of Miyake-Jima (Izu Islands, Japan) in 2000. J Forest Res 11:21–26

Zhou ZH, Miwa M, Matsuda Y, Hogetsu T (2001) Spatial distribution of the subterranean mycelia and ectomycorrhizae of *Suillus grevillei* genets. J Plant Res 114:179–185

Zhou ZH, Miwa M, Nara K, Wu BY, Nakaya H, Lian CL, Miyashita N, Oishi R, Maruta E, Hogetsu T (2003) Patch establishment and development of a clonal plant, *Polygonum cuspidatum*, on Mount Fuji. Mol Ecol 12:1361–1373

Chapter 7
Facilitation and Antagonism in Mycorrhizal Networks

Cameron Wagg, Rita Veiga and Marcel G.A. van der Heijden

Abstract Arbuscular mycorrhizal (AM) fungi are a group of soil and root inhabiting fungi that represent an ancient plant-fungi symbiosis. These fungi interconnect multiple plant individuals and species simultaneously generating a complex fungal network belowground that plays a significant role in shaping plant community composition and ecosystem productivity. However, the underlying mechanisms as to how AM fungal networks and their diversity influence plant performance and community structure are not always predictable and are frequently debated. Although all potential plant hosts may be able to associate with all AM fungi, plant-AM fungal associations can result in a range of AM fungal facilitative and antagonistic effects on plants. Although the facilitative effects of AM fungi have long been studied, the extent and mechanisms of AM fungal antagonistic effects are much less understood. Moreover, AM fungi are observed to vary in their functional properties and temporal patterns adding further complexity to the potential mechanisms by which AM fungi and the diversity of AM fungi determine plant community composition and productivity through their facilitative and antagonistic effects on plants. Here we review the potential mechanisms by which AM fungal communities facilitate greater diversity and productivity in plant communities, as well as the potential mechanisms by which AM fungi may be antagonistic to plant performance. Specifically we address how AM fungal communities might facilitate

C. Wagg · R. Veiga · M.G.A. van der Heijden
Plant Soil Interactions, Institute for Sustainability Sciences, Agroscope Reckenholz Tänikon, Research Station ART, Reckenholzstrasse 191, CH-8046 Zurich, Switzerland
e-mail: ritadaveiga@gmail.com

M.G.A. van der Heijden
e-mail: marcel.vanderheijden@agroscope.admin.ch

C. Wagg (✉) · M.G.A. van der Heijden
Institute of Evolutionary Biology and Environmental Studies, University of Zürich, Winterthurestrasse 190, CH-8057 Zurich, Switzerland
e-mail: cameron.wagg@ieu.uzh.ch

R. Veiga · M.G.A. van der Heijden
Plant-Microbe Interactions, Institute of Environmental Biology, Faculty of Science, Utrecht University, 3508 TC Utrecht, The Netherlands

T.R. Horton (ed.), *Mycorrhizal Networks*,
Ecological Studies 224, DOI 10.1007/978-94-017-7395-9_7

greater plant community performance through functional complementarity among AM fungi as a result of functional, spatial and temporal niche segregation. We also address facilitative and antagonistic aspects of AM fungi through their ability to allocate resources among plant community members that consequently facilitates plant recruitment and alters plant-plant competitive outcomes. By considering the multiple facets by which AM fungi may be facilitative or antagonistic to plants we identify potential knowledge gaps in mechanistically predicting how AM fungal communities shape plant community composition and maintain ecosystem productivity.

Keywords Facilitation • Antagonism • Mycorrhizal diversity • Functional complementarity • Community composition • Niche segregation • Phylogenetic dispersion • Competition • Biodiversity • Arbuscular mycorrhiza • Community ecology • Functional diversity

7.1 Introduction

It is thought that the first land plants formed arbuscular mycorrhizal (AM) fungal associations more than 400 million years ago and that this plant-fungal interaction played a significant role facilitating the establishment of the first terrestrial plants (Remy et al. 1994; Brundrett 2002). To date, the majority of land plants have maintained AM fungal associations (Harley and Harley 1987; Wang and Qui 2006; Smith and Read 2008). Only around 18–26 % of all vascular plants do not support symbiosis with AM fungi (Brundrett 2009). Some of these plants have developed alternative nutritional strategies (e.g. parasitism, carnivory, cluster roots), while others have lost the ability to become mycorrhizal by evolving under scenarios non-conducive for AM fungi and/or where AM fungal associations were no longer beneficial, but antagonistic (Brundrett 2009; Lambers et al. 2010; Lambers and Teste 2013). Several other mycorrhizal types developed over the past millennia, such as ericoid, ecto-, and orchid mycorrhizas (see Peterson et al. 2004 for an overview). Although these different types of mycorrhizal fungi hold key roles in ecosystems and perform specific functions, here we primarily focus on AM fungi.

AM fungi typically form direct symbiotic relationships with their host plants. This occurs via an intraradical nutrient exchange interface where photosynthetically derived carbons are allocated to the fungus, and in reciprocation, soil nutrients are provided to the plant (Smith and Read 2008). These soil nutrients are acquired by extensive AM fungal networks that interconnect roots of several different plants with potential impacts on the associating and non-associating plant communities (Reynolds et al. 2003; Leake et al. 2004; Selosse et al. 2006; van der Heijden and Horton 2009; Smith et al. 2009). The bidirectional relationship between AM fungi and plants is a key component of most terrestrial ecosystems as it shapes plant community composition, succession and productivity (Francis and Read 1994; Bever et al. 1997; van der Heijden et al. 1998; Bever et al. 2010).

However, the mechanisms by which AM fungal networks alter plant performance and plant community characteristics are more complex than a sole nutritional exchange between plants and fungi. It is now known that AM fungi can perform functions other than supplying limiting nutrients to plants (e.g. provide protection against pathogens, Newsham et al. 1995). A growing number of studies indicate that AM fungi can be associated with a decrease in plant productivity (Francis and Read 1995; Klironomos 2003; Rinaudo et al. 2010; Veiga et al. 2011, 2013). Such findings reveal the mycorrhizal symbiosis to be multi-functional and much more dynamic than previously thought. Consequently, a range of plant and fungal responses to the mycorrhizal association are potentially possible (Fig. 7.1; Francis and Read 1995; Johnson et al. 1997; Jakobsen and Hammer Chap. 4, this volume).

Plant—AM fungal associations are typically considered a mutualistic relationship implying both plant and fungal partners benefit from the association (Fig. 7.1). However, it is difficult to quantify the benefits and costs to AM fungi since these fungi are solely dependent on a plant host for carbon (Pfeffer et al. 1999). Thus, AM fungal responses to a plant host may be generally considered as positive (Smith and Read 2008; Whitfield 2007). For this reason we discuss facilitation and antagonism from a phytocentric viewpoint. We refer to facilitation as a positive plant response and antagonism as a negative response by a plant to interactions with AM fungi (Fig. 7.1). Plant responses to AM fungi are not easy to predict since the forces that govern facilitative and antagonistic effects of AM fungi are understood to be fairly complex and dynamic. For instance, the outcome of interactions between AM fungi and their host plants depend on a multitude of factors like plant species identity, the stage of development, the identity and diversity of the fungal partners, and the surrounding abiotic environment. In addition, both facilitative and antagonistic interactions between AM fungi and their plant hosts can occur via direct mechanisms, such as through an imbalance in the resource exchange between fungi and associated plants (Kiers et al. 2011; Walder et al. 2012; Merrild et al. 2013), as well as indirect mechanisms such as AM fungal mediated plant-plant competition (Fitter 1977; Zobel and Moora 1995; Wagg et al. 2011a; Merrild et al. 2013). Here we

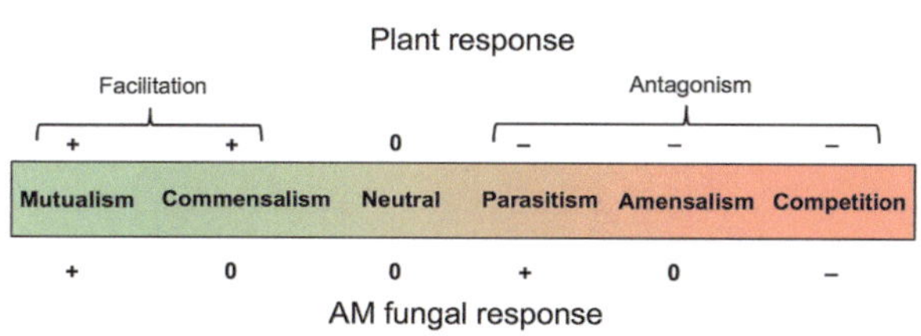

Fig. 7.1 The range of possible plant and fungal responses to mycorrhizal associations (modified from Francis and Read 1995). We refer to facilitation as a positive plant response and antagonism as a negative response by a plant to interactions with AM fungi. This effect can be mediated by many factors, such as the presence of neighboring plants or abiotic conditions

review the current knowledge and concepts regarding the mechanisms responsible for facilitative and antagonistic effects of AM fungi on associating plant communities. Finally we summarise the importance of furthering the current knowledge based on this dynamic plant-fungal relationship for the maintenance and productivity of both natural and managed ecosystems.

7.2 Facilitation

The facilitation of plant establishment and performance by AM fungi in ecosystems can occur through various mechanisms (see Table 2 in van der Heijden and Horton 2009 for an overview). Typically, plants associating with AM fungal partners benefit from a direct relationship where the fungus improves the ability of the host plant to acquire soil resources. Plant productivity may also be facilitated by the extensive hyphal networks in the soil that connects multiple plants; frequently referred to as a 'mycorrhizal network' (MN) in that a continuous fungal mycelium can connect to multiple plants of various species (e.g. Fig. 7.2; Selosse et al. 2006; Kiers et al. 2011; Walder et al. 2012). This MN can mediate the performance among plants (e.g. van der Heijden and Horton 2009). For instance, the support for AM hyphal proliferation and maintenance of the fungus by plants can increase the inoculation potential in the soil such that seedlings become more rapidly colonized, which can improve seedling establishment and may reduce the cost the establishing

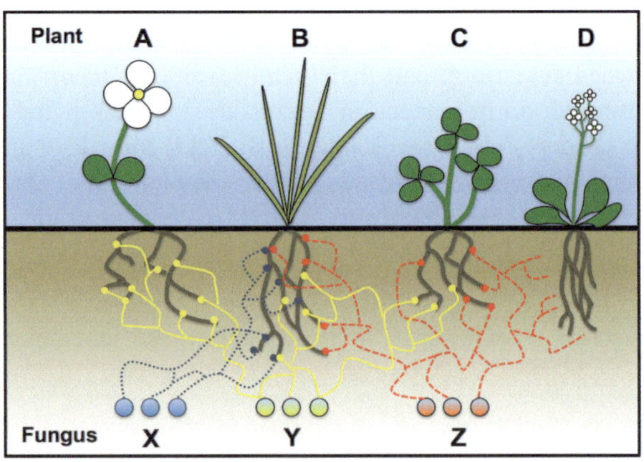

Fig. 7.2 An Illustration of associations between a plant community (species *A*, *B*, *C*, and *D*) and a fungal community (species *X*, *Y*, and *Z*). Hyphal networks are illustrated as different line types with colors corresponding to the originating fungus. *Points* where lines meet plant roots indicate the strength in association between the plant and fungus (i.e. fungus *Y* has 9 connections with plant *A*, while only 3 with plant *C*. Plant *D* does not associate with any AM fungus and is defined here as a non-host). *Note* the number of associations could represent abundance within the plant roots or a relative number of plant *A* individuals associating with fungus *Y*

plant needs to invest into the fungal association to improve its performance (e.g. Newman 1988; van der Heijden 2004; van der Heijden and Horton 2009). The production and maintenance of an extensive mycorrhizal network by plants can also have indirect effects on soil environmental characteristics that may be favourable to the establishment and performance of some plants, such as reduced nutrient leaching (van der Heijden and Horton 2009), improved stability in the soil structure (Rillig and Mummey 2006) and the liberation of nutrients from senescing roots and plant litter (Lindahl et al. 2007; Mikkelsen et al. 2008). It is important to note, however, that not all fungi function similarly or benefit plant hosts equally (Nara 2006). AM fungi can vary in their functional niche, from resource acquisition strategies to pathogen protection, and vary in their interaction with plants depending on the species identity of both the fungus and host plant (Klironomos 2003; Powell et al. 2009; Sikes et al. 2009; Hoeksema et al. 2010). As a result there has been a growing interest in whether the differing functions performed by various AM fungi complement each other such that a greater diversity of AM fungi might facilitate greater plant community performance as it is often observed that greater AM fungal richness can be associated with greater plant performance (van der Heijden et al. 1998; Maherali and Klironomos 2007; Jansa et al. 2008; Wagg et al. 2011b).

This positive AM fungal biodiversity effect on plant performance could occur by three mutually non-exclusive mechanisms; functional complementarity, spatial niche segregation and temporal niche segregation. These differences among AM fungi are the foundation by which they can facilitate plant performance and plant community composition. In this chapter we review (1) how the varying functions performed by different AM fungal taxa might be combined to improve the plant facilitative effects of an AM fungal community, (2) the potential for spatial and temporal niche segregation as mechanisms of complementarity among AM fungi, (3) how AM fungi relax plant-plant competitive interactions to promote plant species coexistence, and (4) how AM fungi facilitate plant communities through a common mycorrhizal network.

7.2.1 *Functional Complementarity in AM Fungal Communities*

The AM fungal association may benefit plants through the various different functions and characteristics associated with different AM fungal taxa (Powell et al. 2009). For example, plants have been shown to benefit from various AM fungal taxa that increase plant community biomass (van der Heijden et al. 1998), increase uptake of phosphorus and other soil resources (Joner and Jackobsen 1994; Smith et al. 2009; Marschner and Dell 1994; Hodge et al. 2001), and improve plant pathogen and pest protection (Gange and West 1994; Newsham et al. 1995; Azcón-Aguilar and Barea 1996). Additionally, AM fungi can vary in the strategies by which they acquire soil resources (Smith et al. 2000; Jansa et al. 2005; Thonar et al. 2010). These functional differences among AM fungal taxa could potentially complement each other to improve the overall facilitative effects of an AM fungal community on plant performance (Koide 2000; Fig. 7.3). For example, fungi co-colonising roots of the same host have been demonstrated to differ in phosphorus acquisition strategies, where one fungus acquires phosphorus within close proximity to the roots, while another acquires phosphorus from colonizing soil patches at greater distances from the root (Smith et al. 2000; Jansa et al. 2005; Thonar et al. 2010; see Wallander and Ekblad Chap. 3, this volume). Therefore increasing the number of AM fungi within the soil that interact with the plant community could theoretically increase the facilitative effects of the AM fungal community.

Indeed it has been shown that increasing AM fungal richness can be related to improved plant productivity in a number of studies suggesting that, in some cases, plants may gain greater benefits from multiple AM fungal associations (van der Heijden et al. 1998; Vogelsang et al. 2006; Maherali and Klironomos 2007; Wagg et al. 2011b). However, it is also often unclear if the functioning of the AM fungal community as a whole is truly greater than the sum it its parts. For example, some studies illustrate that a richer community of AM fungi provides similar benefits as the single most beneficial AM fungus in the more species rich community (Jansa et al. 2008; Wagg et al. 2011a, b). Subsequently, this has been associated with the

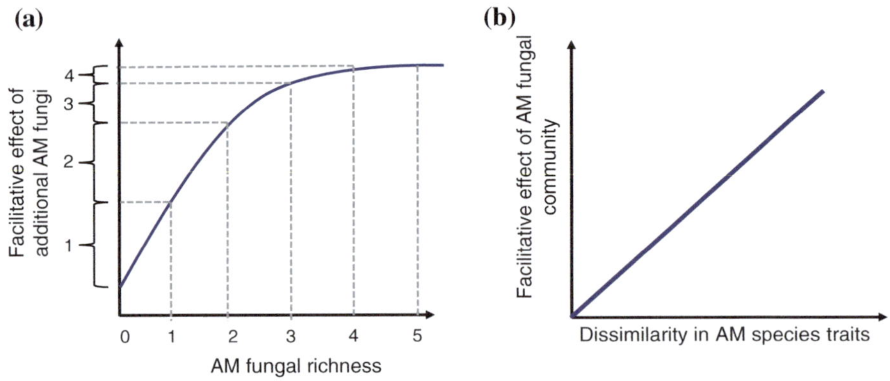

Fig. 7.3 The facilitative effects of additional AM fungi as AM fungal taxa richness increases are shown in panel (**a**). Each additional fungus is able to provide an additional service to plants and it thus, provides an additional improvement to plant performance (functional complementarity). Plant productivity and AM functions are not infinite. Therefore, facilitative effects plateau where all services AM fungi can provide to the plant host are saturated and additional fungi do not improve plant productivity (additional fungal taxa are functionally redundant in improving plant productivity). The facilitative effect of increasing AM fungal dissimilarity among traits (e.g. functional, spatial, or temporal variation among AM fungi) is shown in panel (**b**). Dissimilarity among AM fungi (i.e. plant host or soil substrate patch preference-specificity, as well as seasonal differences) could be represented within the dissimilarity along the *x*-axis. The facultative effect of the overall AM fungal community on the overall plant community is indicated on the *y*-axis

occurrence of a sampling probability or selection effect (e.g. Vogelsang et al. 2006). In general, this implies that a particularly effective fungus is supporting the overall facilitative effects of the AM fungal community, as opposed to a communal contribution of AM fungi to plant facilitation. Therefore additional AM fungal species increases the probability of including such a particularly effective fungus (Wardle 1999). Caution should hence be warranted when interpreting biodiversity effects of increasing AM fungal richness based on plant performance alone and the effect of each individual AM fungus should be considered to elucidate whether the greater facilitative effect of a more rich AM fungal community is driven by all or only a few of the fungal taxa present (see Wagg et al. 2011b).

Whether the facilitative effects of a more diverse AM fungal community are driven by the communal contribution of the fungi, or the probability of a particularly effective AM fungus, is likely context dependant. For instance, both scenarios have been shown to occur depending on the fertility and structure of the soil substrate (Wagg et al. 2011b). By assessing the community composition of AM fungi colonizing plants and the facilitative effects of all AM fungi independently in monoculture, Wagg et al. (2011b) were able to illustrate the mechanisms behind the facilitative effects of a more diverse AM fungal community. In a relatively nutrient poor sandy soil the fungus best fit for the environment was able to dominate the facilitative effects of more diverse AM fungal community—referred to as a "selection effect" (Loreau and Hector 2001). Conversely, in a relatively more fertile environment, the same fungal communities provided similar facilitative effects but were driven by a more even contribution of AM fungi—referred to as a "complementarity effect" (Loreau and Hector 2001). This indicates that the resources available in the environment can mediate AM fungal coexistence and their communal functioning.

It is also important to consider that the differences in services AM fungi provide for plants can be associated with phylogeny (Maherali and Klironomos 2007; Powell et al. 2009). In a keystone paper on the function of AM fungal diversity Maherali and Klironomos (2007) demonstrate the potential for using phylogenetic dispersion in AM fungal communities as a predictor of facilitative effects. They show that AM fungal communities consisting of a mixture of AM fungal taxa from the Gigasoporaceae, Acaulosporaceae, and Glomeraceae families could improve plant productivity more than an AM fungal community comprised of taxa from a single family. Perhaps even more intriguing in the study by Maherali and Klironomos (2007) is that the realized richness in the AM fungal community was greatest when the AM fungal community, initially consisting of 8 taxa, consisted of fungi from all three AM fungal families. This indicates greater phylogenetic dispersion as a mediator of greater coexistence among AM fungal taxa and their facilitative effects.

7.2.2 Spatial and Temporal Niche Segregation as a Mode of Complementarity

Results, such as those previously discussed by Maherali and Klironomos (2007) and Wagg et al. (2011b) provide evidence for a link between AM fungal coexistence and facilitation in AM fungal communities. Trade-offs between competition and performance are fairly common and an important aspect in understanding the functioning of communities (Herms and Mattson 1992; Mouquet et al. 2002). Bennett and Bever (2009) illustrate that the ability of an AM fungus to compete with another AM fungus has a trade-off with the ability to facilitate the performance of the host plant. Considering this, facilitation by an AM fungal community can be maintained at a higher level if competition among AM fungi is avoided by not only functional differences, but also by spatial segregation among AM fungi (Bever et al. 2009). This could arise from the colonization of different soil resource patches, such as mentioned previously concerning phosphorus acquisition strategies (e.g. Jansa et al. 2005), but also through host preference and functional compatibility between specific plant and fungal taxa (Ravnskov and Jakobsen 1995; see Molina and Horton Chap. 1, this volume). Although no true host specificity is known to commonly occur between AM fungi and plant taxa, certain plant-fungal species combinations have been known to be more effective than others (Klironomos 2003). Additionally, preferences toward particular host plants have been shown to occur in natural environments (Sanders 2003; Vandenkoornhuyse et al. 2003; Croll et al. 2008). This avoidance of competition among AM fungi through differences in plant host preferences may allow for the potential that different fungi differentially benefit the various potential host plants such that increasing AM fungal richness increases the potential of the AM fungal community to improve the overall plant community productivity.

The competition among AM fungi may also be avoided by their temporal life strategies. Different taxa of AM fungi are known to be active during different seasons (Gemma et al. 1989; Merryweather and Fitter 1998; Dumbrell et al. 2011) and show successional patterns across years (Oehl et al. 2009). Such functional, spatial, and temporal segregation could aid in the avoidance of competition among AM fungi and result in greater facilitative effects on host plants. This may be an explanation why certain combinations of AM fungi have been shown to be more beneficial than others (e.g. Wagg et al. 2011b). It is thought that temporal variation among plant species life strategies and performance is an underlying mechanism by which a greater diversity of plant species can coexist and contribute to the performance of the community (Loreau and de Mazancourt 2013). In a similar manner, the temporal variation among AM fungal taxa in their life strategies (e.g. Gemma et al. 1989; Merryweather and Fitter 1998; Oehl et al. 2009) may also reduce AM fungus-fungus competition and, as a result, consistently maintain the benefits the plant community gains from the AM fungal community over seasonal changes. However, it is not fully known whether such temporal asynchrony in the activity among AM fungi is a mechanism for reducing niche overlap and competition among AM fungi. Moreover, it has yet to be tested whether temporal variation among AM taxa is a potential mechanism by which an AM fungal community can function complementarily to help maintain ecosystem productivity.

Much research is still required to fully unravel the mechanisms by which the facilitative effects of AM fungal communities can be improved. However, current knowledge suggests that community dissimilarities—spatially, temporally and functionally—among individuals has much to offer to the understanding the mechanisms behind the performance of diverse AM fungal communities (see Fig. 7.3b). Specifically, (a) improving dissimilarity through reducing niche overlap spatially, temporally and functionally (b) improving the phylogenetic dispersion within an AM fungal community and (c) improving the resource heterogeneity available for AM fungi, may be particularly promising avenues for the future.

7.2.3 Facilitation Through Mediating Plant—Plant Interactions

An increase in the number of plant species in grasslands has been shown to result in increasing net plant productivity (e.g. Tilman 1996; Hector et al. 1999; Tilman et al. 2001). If this species richness productivity relationship is driven by the contribution of each additional plant species present, it requires that the additional plant species is able to capture the resource margin, such that it is able to improve the productivity of the community (Loreau and Hector 2001, also see Fig. 7.3a for an example). Differences among plant species in their ability to utilize different resource pools has previously been observed to be one mechanism by which sympatric species coexist and contribute to the overyielding in plant species mixtures (McKane et al. 2002; Harrison et al. 2007; Ashton et al. 2010). Considering AM fungal associations play a pivotal role in the ability of plants to acquire soil resources, there is a large potential for mycorrhizal fungal associations to alter plant coexistence, promote productivity in plant communities, and drive aboveground biodiversity ecosystem functioning relationships.

Indeed past studies have shown AM fungal associations are able to improve coexistence, productivity, and overyielding in plant communities (van der Heijden et al. 1998; Maherali and Klironomos 2007; Wagg et al. 2011a, b). Often the increased access to phosphorus and other soil resources through the direct association is attributed to the beneficial effects of AM fungal associations in plant communities, particularly when these resources are limiting for plant growth (Hoeksema et al. 2010; Johnson 2010). This ability of AM fungi to acquire and mobilize soil recourses otherwise unavailable to plants can therefore ease competitive interactions between plants by increasing the overall resource pool available to the plant community. Klironomos et al. (2000) nicely illustrate how the addition of AM fungi alters the resource space utilized by plant communities. They found the presence of AM fungi resulted in productivity reaching saturated levels at lower plant species richness. This demonstrates that fewer plant species were required to utilize the resources available and achieve maximum productivity when AM fungi were present. In comparison, the absence of AM fungi resulted in a linear relationship between plant richness and productivity showing that additional plant species were needed to capture marginal soil resources to improve the productivity of the community. One outcome of this may result from plants dependent on AM fungal associations performing poorly in plant mixtures in the absence of AM fungi as they are unable to access soil nutrients efficiently and are outcompeted by neighboring plants. It is generally established that in the presence of AM fungi, the fungi mediate the access to resources between competing plants by sequestering soil resources for the less competitive AM dependent plant (Fitter 1977; Hartnett et al. 1993; Zobel and Moora 1995; Urcelay and Diaz 2003; Scheublin et al. 2007; Collins and Foster 2009; Wagg et al. 2011a; Veiga et al. 2011). However, this can depend on the AM fungal dependency of the dominant species within a plant community (Urcelay and Diaz 2003). For instance, in a tall grass prairie Hartnett and Wilson (1999) found the suppression of AM fungi by the use of fungicide reduced the performance of the dominant C4 grasses resulting in a greater performance of the subdominant C3 grasses. This overall competitive shift resulting from reduced AM fungal abundance improved the overall diversity of the plant community.

It is important to also consider that not all coexisting plant species benefit equally from AM fungal associations and some plants depend more upon AM fungal associations for maintaining productivity than others (van der Heijden 2002; Klironomos 2003). For example, some plant species show little response to AM fungal associations, such as C3 grasses (Hetrick et al. 1990). Plants appearing not to benefit directly from a mycorrhizal association may indirectly benefit from AM fungal associations. For instance, plants supporting an AM fungal network may indirectly benefit by facilitating neighboring plants that improve the environment, such as by improving defense against a common predator or N fixation by legumes (see Fig. 7.4 for an example). Many studies have illustrated that plants can facilitate the establishment and performance of other plant species through supporting a mycorrhizal network demonstrating that plant-plant facilitation through a common mycorrhizal network to be an important ecological determinant of plant communities (van der Heijden and Horton 2009).

7.2.4 AM Fungi as a Support Network for Plants?

An AM hyphal network can co-colonize numerous intra- and interspecific plants through a common mycorrhizal network (Fig. 7.2). A number of studies have illustrated that seedlings benefit through establishing an association with an AM fungal network supported by a pre-established plant (see van der Heijden and Horton 2009 and citations therein; Nara Chap. 6, this volume). This may be a

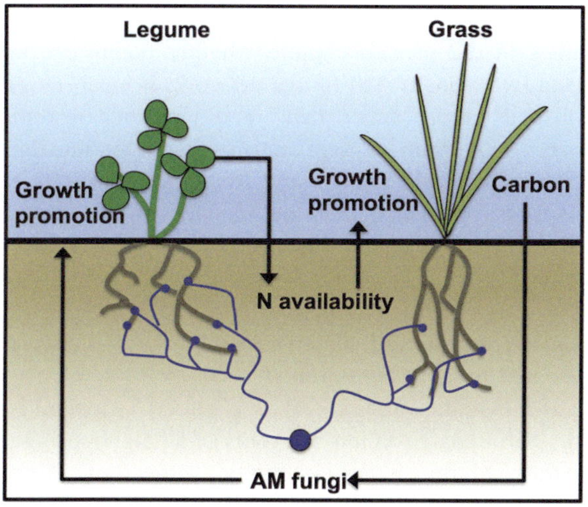

Fig. 7.4 An example of a hypothetical indirect mechanism by which AM fungi facilitate plant community productivity. For instance, facultative mycorrhizal plants, such as C3 grasses, do not typically seem to benefit directly from supporting an AM fungal association. However, considering grass-legume plant mixtures exhibit overyielding effects in agricultural and natural environments, the nitrogen demands of the grass are subsidized by a neighbouring nitrogen fixing legume that depend heavily on AM fungal associations for productivity and to support their nitrogen fixing bacteria. Therefore by supporting the development of an AM fungal community the grass may indirectly benefit from increased nitrogen availability through the improved growth of its neighbouring legume

consequence of pre-established plants investing carbon into the development and proliferation of a mycorrhizal network that other plants may benefit from (Newman 1988). This facilitative influence of an AM fungal network shared between plants can occur through an imbalance in investment to support AM fungi that co-connects plants. For instance, one plant may invest more carbon into the AM fungal association and receive little nutrients in exchange, while another plant invests little carbon while receiving greater amounts of resources (Walder et al. 2012). This is thought to be one of the potential mechanisms behind the facilitation in establishment of seedlings unable to invest large amounts of carbon to an AM fungal association (van der Heijden 2004). This facilitative effect may be of particular importance for seedling recruitment in environments where fungal networks are limiting to plant establishment, such as during succession and post disturbance (Gange et al. 1990, 1993; Korb et al. 2004; Simard and Durall 2004; Selosse et al. 2006; Nara 2006).

It has been proposed that plants associating through a MN are able to translocate carbon and nutrients along fungal networks such that the fungus is able to redistribute resources throughout the plant community, from nutrient sufficient to nutrient deficient plants (Newman 1988; Simard et al. Chap. 5, this volume). This has inspired studies in the past to explore whether plants can aid neighbouring plants co-colonized by a shared AM fungal network. It has been proposed that the previously established larger "nurse plant" could provide resources, such as nitrogen, phosphorus and carbon, to neighbouring plants via the shared mycorrhizal network (Chiarello et al. 1982; Francis and Read 1984; Francis et al. 1986). For instance, there is some evidence that the transfer of phosphorus and nitrogen between plants through a shared mycorrhizal network can occur (He et al. 2003; Wilson et al. 2006). However, although resources may be transferred between plants via an AM fungal network, there is little supporting evidence that these resources are actually transferred directly from the AM fungi to the host and incorporated into plant organs aboveground.

It has been known that atmospheric carbon captured by a plant is allocated to the AM fungal partner and transferred into the roots of a neighbouring plant through its incorporation in the AM fungal tissue colonizing the roots (Graves et al. 1997; Fitter et al. 1998; Zabinski et al. 2002). However, these studies conclude that the carbon is retained in the fungal tissue and not transferred to the host plant, therefore providing no direct facilitative benefit to the plant. Overall, however, it is unclear whether shared AM fungal mycorrhizal networks provide means for a direct reallocation of resources between plants (Robinson and Fitter 1999; Selosse et al. 2006). The translocation of resources between plants by AM fungi may result indirectly through hyphal turnover and microbial mediated diffusion through the soil (Robinson and Fitter 1999). In general the consensus regarding the direct translocation of carbon among plants via AM fungal networks for the benefit of a neighbouring connected plant does not seem to hold up as an ecologically meaningful mechanism by which AM fungi facilitate plant community composition and productivity (Robinson and Fitter 1999; Bever et al. 2010; van der Heijden and Horton 2009). However, the translocation of nitrogen and phosphorus between plants via AM fungal networks and the ecological consequences of resource redistribution throughout the soil via the extensive AM hyphal networks remain largely unresolved (but see Lekberg et al. 2010 and Weremijewicz and Janos 2013).

It is important to consider that carbon allocation to the fungus does not necessarily reflect a cost to the plant. In many ecosystems nutrients are limiting plant growth and genes responsible for photosynthesis are down-regulated as carbon (e.g. as starch) accumulates in the plant. Hence, if this is the case, carbon can be considered a luxury good for the plant and investment into mycorrhizal networks may not produce a cost for the plant (e.g. Kiers and van der Heijden 2006). As such fungi merely stimulate the sink strength and do not reduce the amount of carbon available for growth. This scenario is most likely to be relevant for strongly nutrient limited ecosystems where photosynthesis does not run at full capacity (e.g. Qui and Israel 1992, Poorter and de Jong 1999). Thus, the facilitative effects of mycorrhizal networks are likely to be related to soil nutrient availability.

7.3 Antagonism

Although facilitative effects of AM fungi are commonly observed, AM fungi have also been known to be associated with growth depressions and other negative effects on affiliated plants (Jakobsen and Hammer Chap. 4, this volume). The fact that such effects have not only been reported in pot experiments, but also in field conditions, suggests that AM fungal antagonism is not a mere artifact (Smith and Smith 2011). Nonetheless, as in AM fungal facilitation, AM fungal antagonism is very context dependent and may change through time (Johnson 2010). Moreover, most studies usually focus on one or few indicators of plant performance, usually growth responses or nutrient uptake (typically phosphorus; Johnson and Graham 2013). Hence, an apparently plant growth antagonistic AM fungal association might may be regarded as beneficial if other AM fungal functions are considered throughout the life cycle of the plant (e.g. seedling establishment, protection against pathogens, drought tolerance, fecundity, etc.).

Recently, there has been growing interest in the antagonistic effects of AM fungi on host plants and on the drivers that may cause this typically beneficial association to become antagonistic. Therefore, several mechanisms by which AM fungal associations incur growth depressions in plants have been proposed. Many of these are similar to those by which AM fungi facilitate plant productivity, but with the opposite effect. However, generally, very little is known about the mechanisms responsible for antagonistic effects of AM fungal associations relative to facilitative effects. In this section we review mechanisms by which AM fungi may function as plant antagonists: (1) nutrient exchange imbalance between plants and fungi, (2) AM fungal mediated enhanced performance of competitors, (3) AM fungal allelopathy and activation of defence responses.

7.3.1 Plants at the Losing End of Resource Exchange

Perhaps the most well known driver of AM fungal antagonism is a consequence of the resources available to the plant for direct uptake from the soil (Johnson 2010). For instance, soils with high levels of plant available nutrients, particularly nitrogen and phosphorus, can result in a negative effect of AM fungi on plant productivity (Hoeksema et al. 2010; Johnson 2010). This effect may be particularly evident if plants do not require an AM fungal partner for acquiring resources to achieve optimal growth. Thus, AM fungal associations may provide no benefit, or incur a cost, to the plant for it to maintain the fungal association under certain abiotic conditions that limit the mycorrhizal association (Johnson et al. 1997; Graham and Eissenstat 1998; Johnson 2010). However, when conditions are not optimal for the host plant to maintain the AM fungal association, such as under high phosphorus relative to nitrogen nutrient conditions, a reduced AM fungal colonization of roots is commonly observed (Jasper et al. 1979; Thomas et al. 1986; Johnson 2010).

This would reflect a more complex resource supply and demand scenario between plants and fungi based on the plant available soil resources (Johnson 2010; Kiers et al. 2011).

Negative AM fungal effects on plant growth are hence often explained by carbon demands of the fungus that exceed any reciprocal benefit, primarily through enhanced phosphorus uptake by the plant (Tinker 1975; Graham and Abbott 2000; Smith et al. 2009). Consequently, AM fungi that induce growth depressions are usually regarded as "cheaters" or "parasites" (e.g. Johnson et al. 1997). In fact, fungal carbon demands on a plant host have been estimated to be as high as 15–20 % of its total carbon budget (Jakobsen and Rosendahl 1990; Wright et al. 1998). Additionally, some studies suggest that mycorrhizal C costs are linked to lower plant allocation to growth and defense (Buwalda and Goh 1982; Peng et al. 1993; Graham and Abbott 2000; Jifon et al. 2002; Vannette and Hunter 2011).

Recent experiments using radioactive phosphorus (^{32}P or ^{33}P) have shown that in some unresponsive or negatively-responsive plants, a great percentage of plant phosphorus is derived from the AM fungal partner (Smith et al. 2003; Li et al. 2006; Grace et al. 2009). In such situations, it seems that the AM fungus is not acting like a "parasite" by delivering little or no phosphorus to the host. It may be that the growth of such plants is nitrogen limited and additional phosphorus supply by the fungus does not lead to enhanced plant growth. Alternatively, it has been proposed that growth depressions may arise from phosphorus deficiency as a result of a decrease in direct phosphorus uptake (via plant roots) in mycorrhizal plants, that is not compensated for by the phosphorus delivered via the AM fungus (Grace et al. 2009; Smith et al. 2009; Smith and Smith 2011, 2012). Furthermore, it has been observed that, in some plants, photosynthetic rates can be stimulated by the carbon sink exerted by AM fungal colonization, counteracting mycorrhizal carbon costs (Kaschuk et al. 2009; Lendenmann et al. 2012). Together, these observations have led to an ongoing debate on how imbalanced carbon-for-phosphorus trade might be an over simplified model that cannot universally explain negative effects of AM fungi on plants and on whether the fungus is truly antagonistic (Smith and Smith 2012; Johnson and Graham 2013; Smith and Smith 2013). Regardless, nutritional exchanges between a plant and its fungal partner(s) are still a key factor determining plant responses (Johnson 2010), and might be especially important in mycorrhizal plant communities where plants are linked by a mycorrhizal network (MN).

Previously we discussed how a MN could have facilitative effects through the translocation of resources among plants. However, the MN interconnecting two (or more) plants may also have antagonistic effects. A number of studies have shown that competition between plants for soil resources can be altered due to the presence of a MN, where two neighboring plants unequally invest and benefit from a MN (Janouskova et al. 2011; Walder et al. 2012; Weremijewicz and Janos 2013; Merrild et al. 2013). Typically, a plant host will invest carbon into the AM fungal association, but the nutrient rewards from the AM fungus may be preferentially allocated to a neighbouring plant connected to the AM fungal network (Lekberg et al. 2010; Kiers et al. 2011). This could have potential negative effects to the carbon-investing plant. In these cases, the plants receiving the unwarranted benefits are referred to as "cheaters". Walder et al. (2012) nicely illustrate how carbon and phosphorus exchange rates between plant and fungus can be altered by the presence of other plants tapping into the mycorrhizal fungal network. They found that when sorghum (*Sorghum bicolor*) and flax (*Linum usitatissimum*) were interconnected by a MN, sorghum provided large amounts of carbon to the fungus and received little nutrients in return, performing marginally poorer as a result. Conversely, flax invested little carbon but gained much more in nutritional benefits from the fungus and its performance was enhanced.

Interestingly, negative effects of AM fungi on plant growth have also been observed in non-mycorrhizal plants (Allen et al. 1989; Sanders and Koide 1994; Francis and Read 1995; Rinaudo et al. 2010; Veiga et al. 2012, 2013) and in mycorrhiza-defective mutants (Neumann and George 2005; Facelli et al. 2010). These and other studies usually report AM fungal structures colonizing roots of typically non-AM fungal host plants (Ocampo et al. 1980; Horton et al. 1998; Smith et al. 1998; Wagg et al. 2008, 2011c). In nearly all cases arbuscules, the primary site for nutrient exchange between plant and fungus (Parniske 2008; Bonfante and Genre 2010), are reported as being absent. Although it is generally unknown what functional role this atypical colonization has on the plant, the absence of arbuscules suggests the lack of a trophic interaction and root infection may be more associated with the life strategy of the fungus (Wagg et al. 2011c). Therefore, imbalanced nutrient exchanges between the plant and AM fungus likely do not explain antagonistic effects of AM fungi on non-mycorrhizal plants and it is plausible that other mechanisms are involved.

7.3.2 AM Mediated Plant—Plant and Plant—AM Fungal Competition

Competition for soil resources between plants and between plants and AM fungi can result in indirect antagonistic effects of AM fungi. As discussed earlier some plants depend more upon AM fungal associations to acquire soil resources and improve their competitive ability than others. The improved resource uptake by the more AM fungal dependent plant can come at a cost to the neighbouring plants as resources are depleted. For instance, both Scheublin et al. (2007) and Wagg et al. (2011b) showed that the presence of AM fungi changed the competitive relationship between a grass and a legume species, favoring the more AM-dependent species, the legume, to the detriment of the grass. AM mediated plant-plant competition might also explain why negative effects of AM fungi on the growth of non-mycorrhizal plants (or mycorrhiza-defective mutants) are frequently observed when they co-occur with a mycorrhizal species (e.g. Sanders and Koide 1994; Facelli et al. 2010; Veiga et al. 2012, 2013, see Fig. 7.5 for an example). This mechanism is probably one of the

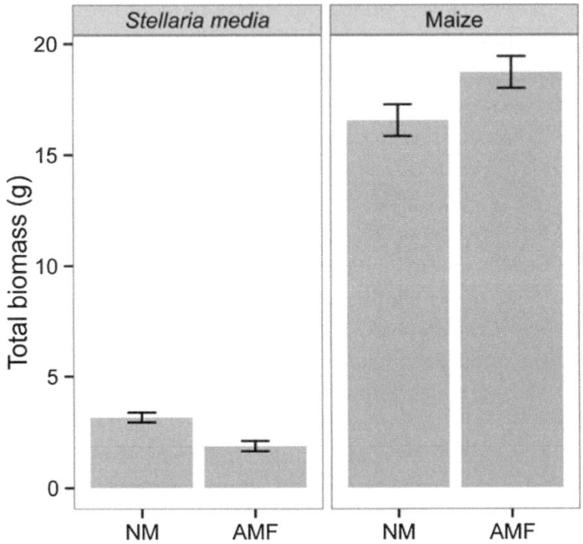

Fig. 7.5 Example of a growth depression of a non-host plant (*Stellaria media*) in the presence of an AM fungal (*AMF*) network supported by a co-existing host species (*Zea maize*). Roots of the two plant species were separated by a hyphal mesh (30 μm in size) to restrict direct root competition (see Veiga et al. 2011 for similar Methods). *Bars* represent the means ± SEM ($n = 7$) *of S. media* or maize in the presence (*AMF*) or absence (NM) of AM fungi. The total biomass of *S. media* in the presence of an *AMF* network supported by maize was lower ($P < 0.05$) than the control *S. media* plants grown in the absence of AM fungi. By contrast, the host plant maize grew better ($P = 0.05$) when AM fungi were present. This suggests that the host, maize, benefited from AM fungi through increased access to soil resources that, consequently, became unavailable to the neighbour non-host, *S. media*. (R. Veiga Unpublished data.)

primary mechanisms for plant growth depressions in a mixed plant community in the presence of AM fungi (Fitter 1977; Hartnett et al. 1993; Zobel and Moora 1995; Urcelay and Diaz 2003; Scheublin et al. 2007; Collins and Foster 2009; Wagg et al. 2011a; Veiga et al. 2011).

It is known that soil microbes can compete with plants for soil resources such as nitrogen (Kaye and Hart 1997; Schimel and Bennett 2004; Dunn et al. 2006; Harrison et al. 2007). In a similar manner, AM fungi compete with plants by taking up soil nutrients for their own growth and development (Treseder and Allen 2002; Hodge and Fitter 2010). For instance, AM fungi may compete with plants for N as considerable amounts of N have been found in mycorrhizal fungal networks (Hodge and Fitter 2010). Therefore AM fungal nutrient immobilization may also have a deleterious effect on plant performance.

7.3.3 *AM Fungal Allelopathy and Activation of Defence Responses*

Less explored avenues by which AM fungi may be antagonistic is through the release of toxic compounds and/or by activating plant defence responses. For instance, Francis and Read (1994, 1995) observed that aqueous extracts of soil containing AM mycelium had a direct inhibitory effect on the development of non-mycorrhizal seedlings, suggesting that AM fungi may produce allelopathic compounds. However, recently, by using two different methods Veiga et al. (2012) found no evidence that AM fungi produce compounds that suppress the growth of AM non-hosts.

In addition to growth depressions and seedling mortality, some studies have reported abnormal root structures and root development of non-mycorrhizal species in the presence of AM fungi. Allen et al. (1989) observed that inoculation of the non-host *Salsola kali* with AM fungi resulted in browning and death of infected root segments while Francis and Read (1995) reported swellings and distortion of the meristems of several non-host roots in the presence of AM mycelium. Such effects resemble a hypersensitive-like plant response (García-Garrido and Ocampo 2002), probably to limit colonization by the fungus. Such a response could directly impair the plants' capacity to take up soil resources. Furthermore, it is known that defence responses can entail costs derived from trade-offs between investment in defence and allocation of resources to plant growth and development (Walters and Heil 2007). Therefore, activation of defence responses could be another mechanism explaining growth depression and abnormalities in the non-host plants in the presence of AM fungi. This hypothesis has not been further investigated in any great detail. Interestingly however, there are indications that similar antagonistic effects of AM fungi can be reproduced on the non-host *Arabidopsis thaliana*, which is the model organism for plant molecular biology and genetics (Veiga et al. 2013). *Arabidopsis thaliana* may thereby in the future serve as a valuable tool in unravelling the molecular basis of incompatible plant—AM fungi interactions.

7.4 Concluding Remarks and Future Considerations

The facilitative and antagonistic interactions between plant communities aboveground and AM fungal communities and networks belowground are more dynamic and context dependent than previously thought. Unravelling the complexities of such interactions are pertinent for the remediation of grassland biodiversity and agricultural management. This entails understanding the functional compatibility between plants and AM fungi. The way forward in attaining a predictive understanding of AM fungal antagonistic and facilitative effects for ecological application may be in the assessment of phylogenetic and functional changes in the AM fungal community after anthropogenic disturbance. For instance, tillage damages the richness of AM fungi and potentially results in the loss of some of the services AM fungi provide to agriculturally desirable plant hosts. The phylogenetic under dispersion in an AM fungal community, indicated by the presence of few AM fungal genera or families, could be used as an indicator of where AM fungal communities may be limited in functional diversity, if phylogeny can be linked with functional complementarity and AM fungal coexistence (e.g. Maherali and Klironomos 2007). Finally, the application of niche theory and temporal community dynamics to AM fungal communities would greatly improve the knowledge base as to how environmental heterogeneity may support AM fungal diversity and thus the overall facilitative effects of an AM fungal community. Nonetheless, although the facilitative, and to a lesser extent antagonistic effects of AM fungi have been long studied, the mechanisms that control them in a predictable fashion for future ecological application have yet to be fully realized.

References

Allen MF, Allen EB, Friese CF (1989) Responses of the non-mycotrophic plant *Salsola kali* to invasion by vesicular-arbuscular mycorrhizal fungi. New Phytol 111:45–49

Ashton IW, Miller AE, Bowman WD, Suding KN (2010) Niche complementarity due to plasticity in resource use: plant portioning of chemical N forms. Ecology 91:3252–3260

Azcón-Aguilar C, Barea JM (1996) Arbuscular mycorrhizas and biological control of soil-borne plant pathogens—an overview of the mechanisms involved. Mycorrhiza 6:457–464

Bennett AE, Bever JD (2009) Trade-offs between arbuscular mycorrhizal fungal competitive ability and host growth promotion in *Plantago lanceolata*. Oecoloigia 160:807–816

Bever JD, Westover KM, Antonvics J (1997) Incorporating the soil community into plant population dynamics: the utility of the feedback approach. J Ecol 85:561–573

Bever JD, Richardson SC, Lawrence BM, Holmes J, Watson M (2009) Preferential allocation to beneficial symbiont with spatial structure maintains mycorrhizal mutualism. Ecol Lett 12:13–21

Bever JD, Dickie IA, Facelli E, Facelli JM, Klironomos J, Moora M, Rillig MC, Stock WD, Tibbett M, Zobel M (2010) Rooting theories of plant community ecology in microbial interactions. Trends Ecol Evol 25:468–478

Bonfante P, Genre A (2010) Mechanisms underlying beneficial plant-fungus interactions in mycorrhizal symbiosis. Nat Commun 1:48

Brundrett MC (2002) Coevolution of roots and mycorrhizas of land plants. New Phytol 154:275–304

Brundrett MC (2009) Mycorrhizal associations and other means of nutrition of vascular plants: understanding the global diversity of host plants by resolving conflicting information and developing reliable means of diagnosis. Plant Soil 320:37–77

Buwalda JG, Goh KM (1982) Host-fungus competition for carbon as a cause of growth depressions in vesicular-arbuscular mycorrhizal ryegrass. Soil Biol Biochem 14:103–106

Chiarello N, Hickman JC, Mooney HA (1982) Endomycorrhizal role for interspecific transfer of phosphorus in a community of annual plants. Science 217:941–943

Collins CD, Foster BL (2009) Community-level consequences of mycorrhizae depend on phosphorus availability. Ecology 90:2567–2576

Croll D, Wille L, Gamper GA, Mathimaran N, Lammers PJ, Corradi N, Sanders IR (2008) Genetic diversity and host plant preferences revealed by simple sequence repeat and mitochondrial markers in a population of the arbuscular mycorrhizal fungus *Glomus intraradices*. New Phytol 178:672–687

Dumbrell AJ, Ashton PD, Aziz N, Feng G, Nelson M, Dytham C, Fitter AH (2011) Distinct seasonal assemblages of arbuscular mycorrhizal fungi revealed by massively parallel pyrosequencing. New Phytol 190:794–804
Dunn R, Mikola J, Bol R, Bardgett RD (2006) Influence of microbial activity on plant-microbial competition for organic and inorganic nitrogen. Plant Soil 289:321–334
Facelli E, Smith SE, Facelli JM, Christophersen HM, Smith FA (2010) Underground friends or enemies: model plants help to unravel direct and indirect effects of arbuscular mycorrhizal fungi on plant competition. New Phytol 185:1050–1061
Fitter AH (1977) Influence of mycorrhizal infection on competition for phosphorus and potassium by two grasses. New Phytol 79:119–125
Fitter AH, Graves JD, Watkins NK, Robinson D, Scrimgeour CM (1998) Carbon transfer between plants and its control in networks of arbuscular mycorrhizas. Funct Ecol 12:406–412
Francis R, Read DJ (1984) Direct transfer of carbon between plants connected by vesicular-arbuscular mycorrhizal mycelium. Nature 307:53–56
Francis R, Finlay RD, Read DJ (1986) Vesicular-arbusuclar mycorrhizal in natural vegetation systems. IV. Transfer of nutrients in inter- and intra-specific combinations of host plants. New Phytol 102:103–111
Francis R, Read DJ (1994) The contribution of mycorrhizal fungi to the determination of plant community structure. Plant Soil 159:11–25
Francis R, Read DJ (1995) Mutualism and antagonism in the mycorrhizal symbiosis, with special reference to impacts on plant community structure. Can J Bot 73:1301–1309
Gange AC, Brown VK, Farmer LM (1990) At test of mycorrhizal benefit in an early successional plant community. New Phytol 115:85–91
Gange AC, Brown VK, Sinclair A (1993) Vesicular-arbuscular mycorrhizal fungi: a determinant of plant community structure in early succession. Funct Ecol 7:616–622
Gange AC, West HM (1994) Interactions between arbuscular mycorrhizal fungi and foliar- feeding insects in *Plantago lanceolata* L. New Phytol 128:79–87
García-Garrido JM, Ocampo JA (2002) Regulation of the plant defence response in arbuscular mycorrhizal symbiosis. J Exp Bot 53:1377–1386
Gemma JN, Koske RE, Carreiro M (1989) Seasonal dynamics of selected species of VA mycorrhizal fungi in a sand dune. Mycol Res 92:317–321
Grace EJ, Cotsaftis O, Tester M, Smith FA, Smith SE (2009) Arbuscular mycorrhizal inhibition of growth in barley cannot be attributed to extent of colonization, fungal phosphorus uptake or effects on expression of plant phosphate transporter genes. New Phytol 181:938–949
Graham JH, Abbott LK (2000) Wheat responses to aggressive and non-aggressive arbuscular mycorrhizal fungi. Plant Soil 220:207–218
Graham JH, Eissenstat DM (1998) Field evidence for the carbon cost of citrus mycorrhizas. New Phytol 140:103–110
Graves JD, Watkins NK, Fitter AH, Robinson D, Scrimgeour C (1997) Interspecific transfer of carbon between plants linked by a common mycorrhizal network. Plant Soil 192:153–159
Harley JL, Harley EL (1987) A check-list of mycorrhiza in the British flora. New Phytol 105:1–102
Harrison KA, Bol R, Bardgett RD (2007) Preferences for different nitrogen forms by coexisting plant species and soil microbes. Ecology 88:989–999
Hartnett DC, Hetrick BAD, Wilson GWT, Gibson DJ (1993) Mycorrhizal influence on intra and interspecific neighbor interactions among co-occurring prairie grasses. J Ecol 81:787–795
Hartnett DC, Wilson GWT (1999) Mycorrhizae influence plant community structure and diversity in tallgrass prarie. Ecol 80:1187–1195
He XH, Critchley C, Bledsoe C (2003) Nitrogen transfer within and between plants through common mycorrhizal networks (CMNs). Crit Rev Plant Sci 22:531–567
Hector A, Schmid B, Beierkuhnlein C, Caldeira MC, Diemer M, Dimitrakopoulos PG, Finn JA, Freitas H, Giller PS, Good J, Harris R, Hogberg P, Huss-Danell K, Joshi J, Jumpponen A, Korner C, Leadley PW, Loreau M, Minns A, Mulder CPH, O'Donovan G, Otway SJ, Pereira JS, Prinz A, Read DJ, Scherer-Lorenzen M, Schulze ED, Siamantziouras ASD,

Spehn EM, Terry AC, Troumbis AY, Woodward FI, Yachi S, Lawton JH (1999) Plant diversity and productivity experiments in European grasslands. Science 286:1123–1127
Herms DE, Mattson WJ (1992) The dilemma of plants—to grow or defend. Quart Rev Biol 67:283–335
Hetrick BAD, Wilson GWT, Todd TC (1990) Differential response of C3 and C4 grasses to mycorrhizal symbiosis, P fertilization and soil microorganisms. Can J Bot 68:461–467
Hodge A, Campbell CD, Fitter AH (2001) An arbuscular mycorrhizal fungus accelerates decomposition and acquires nitrogen directly from organic material. Nature 413:297–299
Hodge A, Fitter AH (2010) Substantial nitrogen acquisition by arbuscular mycorrhizal fungi from organic material has implications for N cycling. PNAS 107:13754–13759
Hoeksema JD, Chaudhary VB, Gehring CA, Johnson NC, Karst J, Koide RT, Pringle A, Zabinski C, Bever JD, Moore JC, Wilson GWT, Klironomos JN, Umbanhowar J (2010) A meta-analysis of context-dependency in plant response to inoculation with mycorrhizal fungi. Ecol Lett 13:394–407
Horton TR, Cázares E, Bruns TD (1998) Ectomycorrhizal, vesicular—arbuscular and dark septate fungal colonization of bishop pine (*Pinus muricata*) seedlings in the first 5 months of growth after wildfire. Mycorrhiza 8:11–18
Jakobsen I, Rosendahl L (1990) Carbon flow into soil and external hyphae from roots of mycorrhizal cucumber plants. New Phytol 115:77–83
Janouskova M, Rydlova J, Puschel D, Szakova J, Vosatka M (2011) Extraradical mycelium of arbuscular mycorrhizal fungi radiating from large plants depresses the growth of nearby seedlings in a nutrient deficient substrate. Mycorrhiza 21:641–650
Jansa J, Mozafar A, Frossard E (2005) Phosphorus acquisition strategies within arbuscular mycorrhizal fungal community of a single field site. Plant Soil 276:163–176
Jansa J, Smith FA, Smith SA (2008) Are there benefits of simultaneous root colonization by different arbuscular mycorrhizal fungi? New Phytol 177:779–789
Jasper DA, Robson AD, Abbott LK (1979) Phosphorus and the formation of the vesicular-arbuscular mycorrhizas. Soil Biol Biogeochem 11:501–505
Jifon JL, Graham JH, Drouillard DL, Syvertsen JP (2002) Growth depression of mycorrhizal citrus seedlings grown at high phosphorus supply is mitigated by elevated CO_2. New Phytol 153:133–142
Johnson NC (2010) Resource stoichiometry eludicates the structure and function of arbuscular mycorrhizas arcoss scales. New Phytol 185:631–647
Johnson NC, Graham JH (2013) The continuum concept remains a useful framework for studying mycorrhizal functioning. Plant Soil 363:411–419
Johnson NC, Graham JH, Smith FA (1997) Functioning of mycorrhizal associations along the mutualism-parasitism continuum. New Phytol 135:575–586
Joner EJ, Jackobsen I (1994) Contribution by two arbuscular mycorrhizal fungi to P uptake by cucumber (*Circumis sativus* L.) from ^{32}P labeled organic matter during mineralization in soil. Plant Soil 163:203–209
Kaschuk G, Kuyper TW, Leffelaar PA, Hungria M, Giller KE (2009) Are the rates of photosynthesis stimulated by the carbon sink strength of rhizobial and arbuscular mycorrhizal symbioses? Soil Biol Biochem 41:1233–1244
Kaye JP, Hart C (1997) Competition for nitrogen between plants and soil microorganisms. Trends Ecol Evol 12:139–143
Kiers ET, Duhamel M, Beesetty Y, Mensah JA, Franken O, Verbruggen E, Fellbaum CR, Kowalchuk G, Hart MM, Bago A, Palmer T, West SA, Vandenkoornhuyse P, Jansa J, Bücking H (2011) Reciprocal rewards stabilize cooperation in the mycorrhizal symbiosis. Science 333:880–882
Kiers TR, van der Heijden MGA (2006) Mutualistic stability in the arbuscular mycorrhizal symbiosis: exploring hypothesis of evolutionary cooperation. Ecology 87:1627–1636
Klironomos JN (2003) Variation in plant response to native and exotic arbuscular mycorrhizal fungi. Ecology 84:2292–2301

Klironomos JN, McCune J, Hart M, Neville J (2000) The influence of arbuscular mycorrhizae on the relationship between plant diversity and productivity. Ecol Lett 3:137–141

Koide RT (2000) Functional complementarity in the arbuscular mycorrhizal symbiosis. New Phytol 147:233–235

Korb JE, Johnson NC, Covington WW (2004) Slash pile burning effects on soil biotic and chemical properties and plant establishment: recommendations for amelioration. Restor Ecol 12:52–63

Lambers H, Brundrett MC, Raven JA, Hopper SD (2010) Plant mineral nutrition in ancient landscapes: high plant species diversity on infertile soils is linked to functional diversity for nutritional strategies. Plant Soil 334:11–31

Lambers H, Teste FP (2013) Interactions between arbuscular mycorrhizal and non-mycorrhizal plants: do non-mycorrhizal species at both extremes of nutrient availability play the same game? Plant Cell Environ 36:1911–1915

Leake JR, Johnson D, Donnelly D, Muckle GE, Boddy L, Read DJ (2004) Networks of power and influence: the role of mycorrhizal mycelium in controlling plant communities and agro-ecosystem functioning. Can J Bot 82:1016–1045

Lekberg Y, Hammer EC, Olsson PA (2010) Plants as resource islands and storage units—adopting the mycocentric view of arbuscular mycorrhizal networks. FEMS Microbiol Ecol 74:336–345

Lendenmann M, Thonar C, Barnard RL, Salmon Y, Werner RA, Frossard E, Jansa J (2012) Symbiont identity matters: carbon and phosphorus fluxes between *Medicago truncatula* and different arbuscular mycorrhizal fungi. Mycorrhiza 21:689–702

Li HY, Smith SE, Holloway RE, Zhu YG, Smith FA (2006) Arbuscular mycorrhizal fungi contribute to phosphorus uptake by wheat grown in a phosphorus-fixing soil even in the absence of positive growth responses. New Phytol 172:536–543

Lindahl BD, Ihrmark K, Boberg J, Trumbore SE, Hogberg P, Stenlid J, Finlay RD (2007) Spatial separation of litter decomposition and mycorrhizal nitrogen uptake in a boreal forest. New Phytol 173:611–620

Loreau M, Hector A (2001) Partitioning selection and complementarity in biodiversity experiments. Nature 412:72–76

Loreau M, de Mazancourt C (2013) Biodiversity and ecosystem stability: a synthesis of underlying mechanisms. Ecol Lett 16:106–115

Maherali H, Klironomos JN (2007) Influence of phylogeny on fungal community assembly and ecosystem functioning. Science 316:1746–1748

Marschner H, Dell B (1994) Nutrient uptake in mycorrhizal symbiosis. Plant Soil 159:89–201

Merrild MP, Ambus P, Rosendahl S, Jakobsen I (2013) Common arbuscular mycorrhizal networks amplify competition for phosphorus between seedlings and established plants. New Phytol 200:229–240

Merryweather J, Fitter A (1998) The arbuscular mycorrhizal fungi of Hyacinthoides non-scripta II. Seasonal and spatial patterns of fungal populations. New Phytol 138:131–142

McKane RB, Johnson LC, Shaver GR, Nadelhoffer KJ, Rastetter EB, Fry B, Giblin AR, Kielland K, Kwiatkowski BL, Laundre JA, Murray G (2002) Resource-based niches provide a basis for plant species diversity and dominance in arctic tundra. Nature 425:68–71

Mikkelsen Rosendahl S, Jakobsen I (2008) Underground resource allocation between individual networks of mycorrhizal fungi. New Phytol 180:890–898

Mouquet N, Moore JL, Loreau M (2002) Plant species richness and community productivity: why the mechanisms that promotes coexistence matters. Ecol Lett 5:55–65

Nara K (2006) Ectomycorrhizal networks and seedling establishment during early primary succession. New Phytol 169:169–178

Neumann E, George E (2005) Does the presence of arbuscular mycorrhizal fungi influence growth and nutrient uptake of a wild-type tomato cultivar and a mycorrhiza-defective mutant, cultivated with roots sharing the same soil volume? New Phytol 166:601–609

Newman E (1988) Mycorrhizal links between plants: their functioning and ecological significance. Adv Ecol Res 18:243–271

Newsham KK, Fitter AH, Watkinson AK (1995) Arbuscular mycorrhiza protect an annual grass from root pathogenic fungi in the field. J Ecol 83:991–1000
Ocampo JA, Martin J, Hayman DS (1980) Influence of plant interactions on vesicular-arbuscular mycorrhizal infections. I. Host and non-host plants grown together. New Phytol 84:27–35
Oehl F et al (2009) Distinct sporulation dynamics of arbuscular mycorrhizal fungal communities from different agroecosystems in long-term microcosms. Agric Ecosyst Environ 134:257–268
Parniske M (2008) Arbuscular mycorrhiza: the mother of plant root endosymbiosis. Nature Rev Microbiol 6:763–775
Peng S, Eissenstat DM, Graham JH, Williams K, Hodge NC (1993) Growth depression in mycorrhizal citrus at high-phosphorus supply. Plant Phys 101:1063–1071
Peterson RL, Massicotte HB, Melville LH (2004) Mycorrhizas: anatomy and cell biology. NRC Research Press, Ottawa
Pfeffer PE, Douds DD Jr, Bécard G, Shachar-Hill Y (1999) Carbon uptake and the metabolism and transport of lipids in an arbuscular mycorrhiza. Plant Phys 120:587–598
Poorter H, de Jong R (1999) A comparison of specific leaf area, chemical composition and leaf construction costs of field plants from 15 habitats differing in productivity. New Phytol 143:163–176
Powell JR, Parrent JL, Hart MM, Klironomos JN, Rillig MC, Maherali H (2009) Phylogenetic trait conservatism and the evolution of functional trade-offs in arbuscular mycorrhizal fungi. Proc R Soc B 276:4237–4245. doi:10.1098/rspb.2009.1015
Qui J, Israel DW (1992) Diurnal starch accumulation and utilization in phosphorus-deficient soybean plants. Plant Phys 98:316–323
Ravnskov S, Jakobsen I (1995) Functional compatibility in arbuscular mycorrhizas measured as hyphal P transport to the plant. New Phytol 129:611–618
Remy W, Taylor TN, Hass H, Kerp H (1994) Four hundred-million-year-old vesicular arbuscular mycorrhizae. PNAS 91:11841–11843
Reynolds HL, Packer A, Bever JD, Clay K (2003) Grassroots ecology: plant-microbe-soil interactions as drivers of plant community structure and dynamics. Ecology 84:2281–2291
Rillig MC, Mummey DL (2006) Mycorrhizas and soils structure. New Phytol 171:41–53
Rinaudo V, Bàrberi P, Giovanneti M, van der Heijden MGA (2010) Mycorrhizal fungi suppress aggressive agricultural weeds. Plant Soil 333:7–20
Robinson D, Fitter A (1999) The magnitude and control of carbon transfer between plants linked by a common mycorrhizal network. J Exp Bot 50:9–13
Sanders IR (2003) Preference, specificity and cheating in the arbuscular mycorrhizal symbiosis. Trends Plant Sci 8:143–145
Sanders IR, Koide RT (1994) Nutrient acquisition and community structure in co-occuring mycotrophic and non-mycotrophic old-field annuals. Funct Ecol 8:77–84
Scheublin TR, van Logtestijn RSP, van der Heijden MGA (2007) Presence and identity of arbuscular mycorrhizal fungi influence competitive interactions between plant species. J Ecol 95:631–638
Schimel JP, Bennett J (2004) Nitrogen mineralization: challenges of a changing paradigm. Ecology 85:591–602
Selosse M-A, Richard F, He X, Simmard SW (2006) Mycorrhizal networks: des liaisons dangereuses? Trends Ecol Evol 21:621–628
Sikes BA, Cottenie K, Klironomos JN (2009) Plant fungal identity determines pathogen protection of plant roots by arbuscular mycorrhizas. J Ecol 97:1274–1280
Simard SM, Durall DM (2004) Mycorrhizal networks: a review of their extent, function, and importance. Can J Bot 82:1140–1165
Smith FA, Jackobsen I, Smith SE (2000) Spatial differences in acquisition of soil phosphate between two arbuscular mycorrhizal fungi in symbiosis with *Medicago truncatula*. New Phytol 147:357–366
Smith FA, Grace EJ, Smith SE (2009) More than a carbon economy: nutrient trade and ecological sustainability in facultative arbuscular mycorrhizal symbioses. New Phytol 182:347–358

Smith FA, Smith SE (2013) How useful is the mutualism-parasitism continuum of arbuscular mycorrhizal functioning? Plant Soil 363:7–18

Smith JE, Johnson KA, Cázares E (1998) Vesicular mycorrhizal colonization of seedlings of Pinaceae and Betulaceae after spore inoculation with *Glomus intraradices*. Mycorrhiza 7:279–285

Smith SE, Read DJ (2008) Mycorrhizal symbiosis. Academic Press, Amsterdam

Smith SE, Smith FA, Jakobsen I (2003) Mycorrhizal fungi can dominate phosphate supply to plants irrespective of growth responses. Plant Phys 133:16–20

Smith SE, Smith FA (2011) Roles of arbuscular mycorrhizas in plant nutrition and growth: new paradigms from cellular to ecosystem scales. Ann Rev Plant Biol 62:227–250

Smith SE, Smith FA (2012) Fresh perspectives on the roles of arbuscular mycorrhizal fungi in plant nutrition and growth. Mycologia 104:1–13

Thonar C, Schnepf A, Frossard E, Roose T, Jansa J (2010) Traits related to differences in function among three arbuscular mycorrhizal fungi. Plant Soil 312:85–99

Tilman D (1996) Biodiversity: population versus ecosystem stability. Ecology 77:350–363

Tilman D, Reich PB, Knops J, Wedin D, Mielke T, Lehman C (2001) Diversity and productivity in a long-term grassland experiment. Science 294:835–843

Tinker PB (1975) Soil chemistry of phosphorus and mycorrhizal effects on plant growth. In: Sanders FE, Mosse B, Tinker PB (eds) Endomycorrhizas. Academic Press, London, pp 353–371

Thomas BD, Robson AD, Abbott LK (1986) Effects of phosphorus on the formation of mycorrhizas by *Gigaspora calospora* and *Glomus fasciculatum* in relation to root charbohydrates. New Phytol 103:751–765

Treseder KK, Allen MF (2002) Direct nitrogen and phosphorus limitation of arbuscular mycorrhizal fungi: a model and field test. New Phytol 155:507–515

Urcelay C, Díaz S (2003) The mycorrhizal dependence of subordinates determines the effect of arbuscular mycorrhizal fungi on plant diversity. Ecol Lett 6:388–391

van der Heijden MGA (2002) Arbuscular mycorrhizal fungi as a determinant of plant diversity: in search for underlying mechanisms and general principles. In: van der Heijden MGA, Sanders IR (eds) Mycorrhizal ecology (ecological studies), vol 157. Springer, Heidelberg, pp 243–265

van der Heiden MGA, Horton TR (2009) Socialism in soil? The importance of mycorrhizal fungal networks for facilitation in natural ecosystems. J Ecol 97:1139–1150

van der Heijden MGA, Klironomos JN, Ursic M, Moutoglis P, Streitwolf-Engel R, Boller T, Wiemken A, Sanders IR (1998) Mycorrhizal fungal diversity determines plant biodiversity, ecosystem variability and productivity. Nature 396:69–72

van der Heijden MGA (2004) Arbuscular mycorrhizal fungi as support systems for seedling establishment in grassland. Ecol Lett 7:293–303

Vandenkoornhuyse P, Ridgeway KP, Watson IJ, Fitter AH, Young JPW (2003) Co-existing grass species have distinctive arbuscular mycorrhizal communities. Mol Ecol 12:3085–3095

Vannette RL, Hunter MD (2011) Plant defence theory re-examined: nonlinear expectations based on the costs and benefits of resource mutualisms. J Ecol 99:66–76

Veiga RSL, Jansa J, Frossard E, van der Heijden MGA (2011) Can arbuscular mycorrhizal fungi reduce the growth of agricultural weeds? PLoS One 6:e27825. doi:10.1371/journal.pone.0027825

Veiga RSL, Howard K, van der Heijden MGA (2012) No evidence for allelopathic effects of arbuscular mycorrhizal fungi on the non-host Stellaria media. Plant Soil 360:319–331

Veiga RSL, Faccio A, Genre A, Pieterse CM, Bonfante P, van der Heijden MG (2013) Arbuscular mycorrhizal fungi reduce growth and infects roots of the non-host plant *Arabidopsis thaliana*. Plant Cell Environ 36:1926–1937

Vogelsang K, Reynolds HL, Bever JD (2006) Mycorrhizal fungal identity and richness determine the diversity and productivity of a tallgrass prairie system. New Phytol 172:554–562

Wagg C, Pautler M, Massicotte HB, Peterson RL (2008) The co-occurrence of ectomycorrhizal, arbuscular mycorrhizal, and dark septate fungi in seedlings of four members of the Pinaceae. Mycorrhiza 18:103–110

Wagg C, Jansa J, Stadler M, Schmid B, van der Heijden MGA (2011a) Mycorrhzal fungal identity and diversity relaxes plant-plant competition. Ecology 92:1303–1313

Wagg C, Jansa J, Schmid B, van der Heijden MGA (2011b) Belowground biodiversity effects of plant symbionts support aboveground productivity. Ecol Lett 14:1001–1009

Wagg C, Antunes PM, Peterson RL (2011c) Arbuscular mycorrhizal fungal phylogeny related interactions with a non-host. Symbiosis 53:41–46

Walder F, Niemann H, Natarajan M, Lehmann MF, Boller T, Wiemken A (2012) Mycorrhizal networks: common goods of plants shared under unequal terms of trade. Plant Physiol 59:789–797

Walters D, Heil M (2007) Costs and trade-offs associated with induced resistance. Physiol Mol Plant Pathol 71:3–17

Wang B, Qiu Y-L (2006) Phylogenetic distribution and evolution of mycorrhizas in land plants. Mycorrhiza 16:299–363

Wardle DA (1999) Is "sampling effect" a problem for experiments investigating biodiversity-ecosystem function relationships? Oikos 87:403–407

Weremijewicz JJ, Janos DPD (2013) Common mycorrhizal networks amplify size inequality in *Andropogon gerardii* monocultures. New Phytol 198:203–213

Whitfield J (2007) Fungal roles in soil ecology: underground networking. Nature 449:136–138

Wilson GWT, Hartnett DC, Rice CW (2006) Mycorrhizal-mediated phosphorus transfer between the tallgrass prairie plants *Sorghastrum nutans* and *Artemisia ludovicana*. Funct Ecol 20:427–435

Wright DP, Read DJ, Scholes JD (1998) Mycorrhizal sink strength influences whole plant carbon balance of *Trifolium repens* L. Plant Cell Environ 21:881–891

Zabinski C, Quinn L, Callaway R (2002) Phosphorus uptake, not carbon transfer, explains arbuscular mycorrhizal enhancement of *Centaurea maculosa* in the presence of native grassland species. Funct Ecol 16:758–765

Zobel M, Moora M (1995) Interspecific competition and arbuscular mycorrhiza: importance for the coexistence of calcareous grassland species. Folia Geobotanica Phytotaxonomica 30:223–230

Chapter 8
Interspecific Mycorrhizal Networks and Non-networking Hosts: Exploring the Ecology of the Host Genus *Alnus*

Peter G. Kennedy, Jennifer K.M. Walker and Laura M. Bogar

Abstract While the dominant ectomycorrhizal (ECM) fungi in most temperate and tropical forests have low host specificity, a commonly cited exception to this pattern is the ECM fungal community associated with the host genus *Alnus*. In this chapter, we discuss multiple hypotheses that have been put forth to explain the specificity of the *Alnus* ECM symbiosis and consider their strengths and weaknesses in light of current research on the topic. In addition to reviewing the range of suggested explanations, we also propose and discuss a new alternative explanation of *Alnus* ECM specificity involving three-way interactions among *Alnus* plants, ECM fungi, and *Frankia* bacteria. With specific regard to common mycorrhizal networks (CMNs), we believe they may play an important role in the specificity observed in the *Alnus* ECM system. To understand that role in the larger context of research on *Alnus* ECM fungal communities, we begin our chapter with a synopsis of the studies documenting the unique specificity pattern. From there, we discuss why it appears to be advantageous for *Alnus* plants not to participate in interspecific CMNs. Finally, we elaborate on how specificity may be established and maintained in the *Alnus* ECM system and suggest what we consider to be promising future research directions.

Keywords Mycorrhizal specificity · Mutualisms · Partner choice · Common mycorrhizal networks · Tri-partite interactions

---

P.G. Kennedy (✉)
Department of Plant Biology, University of Minnesota, 1445 Gortner Ave, St. Paul, MN 55108, USA
e-mail: kennedyp@umn.edu

P.G. Kennedy
Department of Ecology, Evolution, and Behavior, University of Minnesota, St. Paul, MN, USA

J.K.M. Walker
Hawkesbury Institute for the Environment, Western Sydney University, Richmond 2753, NSW, Australia

L.M. Bogar
Department of Biology, Stanford University, Palo Alto, CA, USA

T.R. Horton (ed.), *Mycorrhizal Networks*,
Ecological Studies 224, DOI 10.1007/978-94-017-7395-9_8

8.1 Introduction

A fundamental prerequisite for the formation of interspecific common mycorrhizal networks (CMNs) is the ability of mycorrhizal fungi to associate with multiple species of co-occurring host plants. Many ectomycorrhizal (ECM) fungi have broad host ranges (Molina et al. 1992; Molina and Horton, Chap. 1, this volume) and form compatible mycorrhizal relationships with many distantly related plant genera and species. At the same time, ECM fungi specific to a particular host plant genus are common, even on ECM host plants with broad fungal compatibility (Molina et al. 1992). ECM hosts also have variable degrees of receptivity to fungal associates. For example, some ECM hosts, such as *Pseudotsuga menziesii*, are reported to be receptive to colonization by thousands of fungal species (Trappe and Fogel 1977); others, such as *Pisonia grandis*, appear to associate with only a very limited number of ECM fungi (Suvi et al. 2010).

Based on a range of recent studies, it appears that the most frequent and/or abundant ECM fungi in most temperate and tropical forests have low host specificity (Horton and Bruns 1998; Horton et al. 1999; Cullings et al. 2000; Kennedy et al. 2003; Nara and Hogetsu 2004; Ishida et al. 2007; Twieg et al. 2007; Tedersoo et al. 2008; Richard et al. 2009; Smith et al. 2011; but see Smith et al. 2009). A commonly cited exception to this pattern, however, is ECM fungi associated with the host genus *Alnus*. Unlike other ECM fungi-plant host systems, *Alnus* ECM fungal communities have been consistently characterized by low species richness and a high proportion of genus-specific species (Molina 1979; Tedersoo et al. 2009; Walker et al. 2014). While other ECM hosts do associate with ECM fungi that are also genus-specific (e.g. *Rhizopogon* and ECM hosts genera within the Pinaceae), they are rarely the dominant fungi present in mature forests.

The factors contributing to the reciprocal specificity of the *Alnus* ECM system have been the subject of considerable speculation. In this chapter, we highlight a number of hypotheses that have been put forth to explain this specificity and consider their strengths and weaknesses in light of current research on the topic. We begin with a synopsis of the studies documenting the unique specificity pattern. From there, we discuss why it appears to be advantageous for *Alnus* plants not to participate in interspecific CMNs. Finally, we elaborate on how specificity may be established and maintained in the *Alnus* ECM system and suggest what we consider to be promising future research directions.

8.2 Documenting the *Alnus* ECM Specificity Pattern

Frank (1888) was the first to determine that fungal colonization of the roots of *Alnus* trees was ectomycorrhizal in nature (Table 8.1). It took many more years, however, before distinct morphologies were identified (Masui 1926) and fungal species identities were reported (Favre 1948, Singer 1950). An early review by Trappe

(1962) cited 14 ECM fungal species associated with *Alnus* hosts based on morphological characterization, including members of the genera *Alnicola*, *Russula*, *Lactarius*, *Gyrodon*, and *Cenococcum*. Additional ECM fungal morphotypes were later observed on field-collected *Alnus* roots (Horak 1963; Neal et al. 1968; Mejstrik and Benecke 1969), some of which were initially identified as *Cortinarius*, *Paxillus*, and *Alpova*. Molina (1979, 1981) and Godbout and Fortin (1983) found that fungal species consistently observed with *Alnus* trees as sporocarps (Neal et al. 1968; Trappe 1975) formed ectomycorrhizas in pure culture synthesis assays, while others not observed to be associated with *Alnus* did not typically form ectomycorrhizas, or formed ones that were anatomically anomalous. Intriguingly, the presence of *Paxillus involutus* (now recognized as a species complex; Jargeat et al. 2014) as an *Alnus* associate in field settings remained unclear, but additional pure culture work indicated that this species could form functional ectomycorrhizas with *Alnus* species in lab settings (Chatarpaul et al. 1989; Arnebrant et al. 1993; Massicotte et al. 1999). Despite further detailed morphotyping analyses of ECM root tips (Miller et al. 1991), the global total of ECM fungal species thought to associate with *Alnus* trees by the mid-1990s was fewer than 50 (Molina et al. 1994).

DNA-based analyses of the *Alnus* ECM system have largely confirmed previous work based on other methods. Pritsch et al. (1997) were the first to use these methods by matching RFLP patterns of *Alnus* ECM morphotypes with sporocarps present in *Alnus* forests. Although those authors did not detect any new *Alnus*-associated ECM fungal genera, they did increase the number of species present on *Alnus* roots. In subsequent studies where DNA extraction was followed by sequencing of the fungal ITS and/or LSU region, some new genera and lineages were identified (Tedersoo et al. 2009; Kennedy and Hill 2010; Kennedy et al. 2011a; Bogar and Kennedy 2013). Those studies also increased the number of ECM fungal species associated with *Alnus* hosts, but not in a way that significantly altered the general pattern of low richness and high specificity. Interestingly, Tedersoo et al. (2009) found that the majority of ascomycete species (4 of 6) associated with *A. glutinosa* and *A. incana* in Estonia were also found in association with other ECM hosts, although those species made a minor component of the communities identified in that study. Rochet et al. (2011) summarized much of this molecular work by noting that there appear to be six dominant Basidiomycete genera (*Tomentella*, *Alnicola*, *Lactarius*, *Cortinarius*, *Alpova*, and *Russula*), a few other Basidiomycete genera not consistently found as ectomycorrhizas (e.g. *Paxillus*, *Hebeloma*, *Inocybe*, and *Pseudotomentella*), and a number of unknown members of the Helotiales associated with *Alnus* hosts. At present, the best estimate of the number of *Alnus*-associated ECM fungal species comes from the global-scale study by Polme et al. (2013), which suggested total richness to be around 200 species.

Table 8.1 Chronological summary of observations and experiments on ectomycorrhizal (ECM) fungi associated with *Alnus* species, with a focus on those reporting new species or genera

Publication	Alnus species	Number of taxa[a]	New genera[b]	Reported indentity of fungal associate[c]	Identification method	Study type
Frank (1888)	*A. viridis*	~	~	the presence of fungal taxa described	ECM morphology	Field
Masui (1926)	*Alnus* spp.	6	~	6 mycorrhizal types distiguished	ECM morphology	Field
Favre (1948)	*A. viridis, A. incana*	~	1	several ***Alnicola*** spp. observed	Fruitbody	Field
Singer (1950)	*A. acuminata*	~	2	survey included ***Russula*** spp.	Fruitbody	Field
Trappe (1962)	*A. incana, A. nigra, A. viridis, A. jorullensis, A. crispa, Alnus* spp.	14	5	*Alnicola diplocystis, A. melinoides, A. umbrina,* ***Cenococcum*** *graniforme,* ***Lactarius*** *controversus, glyciosmus, lilacinus, obscuratus,* ***Gyrodon*** *lividus, montana, monticola, Russula emetica, montivaga, decolorans*	Reported associations	Review
Horak (1963)	*A. viridis*	8	8	*Alnicola melinoides, submelinoides, suavis,* ***Hydrocybe*** (sic) *atropusillus, Lactarius tabidus (theiogalus),* ***Paxillus*** *involutus, Phlegmacium moseri* = ***Cortinarius****, Russula alnetorum*	ECM morphology	Field
Trappe (1964)	*A. rubra*	1	8	*Cenococcum graniforme = geophilum*	ECM morphology	field
Neal et al. (1968)	*A. rubra*	2	9	Dark brown clavate = ***Alpova*** *diplophloeus*, Pale brown glabrous = *Lactarius obscuratus*	ECM morphology	Field
Mejstrik and Benecke (1969)	*A. viridis*	3	9	Subtype B, F, K	ECM morphology	Field/bioassay
Froidevaux (1973)	*A. rubra*	1	9	*Lactarius obscuratus*	ECM morphology	Field

(continued)

Table 8.1 (continued)

Publication	Alnus species	Number of taxa[a]	New genera[b]	Reported indentity of fungal associate[c]	Identification method	Study type
Molina (1979, 1981)	*A. rubra, A. glutinosa, A. incana, A. rhombifolia, A. sinuata*	6	13	*Alpova diplophloeus*, ***Astraeus*** *pteridis, (**Laccaria** laccata), (Paxillus involutus), (**Pisolithus** tinctorius),* ***Scleroderma*** *hypogaeum*	ECM formation	Pure culture assay
Godbout and Fortin (1983)	*A. crispa, A. rugosa*	10	15	*Alpova diplophloeus, Cenococcum geophilum, Cortinarius* cf. *subporphyropus,* ***Hebeloma*** cf. *crustuliniforme, Laccaria laccata,* ***Leccinum*** *subleucopheum, holopus, Paxillus involutus, Pisolithus tintorius, Schleroderma* (sic) *citrinum*	ECM formation	Pure culture assay
Miller et al. (1991)	*A. rubra*	11	16	*Alpova diplophloeus, Cortinarius bibulus, Hebeloma* cf. *crustuliniforme, Laccaria laccata, Lactarius obscuratus, Paxillus involutus, Thelephora terrestris = Alnicola* spp, ***Tomentella*** *sublilacina*	ECM morphology	Field/bioassay pure culture assay
Pritsch et al. (1997)	*A. glutinosa*	16	16	*Alnirhiza cystidiobrunnea = (Tomentella* aff. *sublilacina, lilacina, violaceae, suffusa, cana, texta, cremicolor, atroverrucosa, Cortinarius* cf. *helvelloides, alneus, Lactarius obscuratus, omphaliformes, lilacinus, Naucoria escharoides = Alnicola melinoides, subconspersa, Russula pumila*	ECM formation/RFLP	Field
Becerra et al. (2002, 2005)	*A. acuminata*	10	16	*Alnirhiza silkacea = Gyrodon monticola, Cortinarius helodes, tucumanensis, Lactarius omphaliformis,* sp., *Naucoria escharoides, Russula alnijorullensis, Tomentella* spp.	ECM formation/RFLP	Field

(continued)

Table 8.1 (continued)

Publication	Alnus species	Number of taxa[a]	New genera[b]	Reported indentity of fungal associate[c]	Identification method	Study type
Tedersoo et al. (2009)	*A. glutinosa, A. incana*	40	21	*Alnicola* spp., *Cortinarius alnetorum,* sp., *(Geopyxis* cf. *carbonaria), Gyrodon lividus, Hebeloma* aff. *helodes, Helvella* aff. *elastica,* sp., *unknown Humaria* sp., *Inocybe* sp., *Lactarius cyathuliformis, lilacinus,* sp., *unknown Pachyphloeus* spp., *Paxillus filamentosus, rubicudulus, Peziza michelii, Pseudotomentella* spp., *Russula pumila, Tomentella* aff. *sublilacina, stuposa, coerulea, ellisii, lateritia, terrestris,* sp.	DNA sequencing of ITS and LSU	Field
Kennedy and Hill (2010)	*A. rubra*	14	22	*Alnicola escharoides,* sp., *Alpova diplophloeus, Cortinarius* spp., *Inocybe* sp., *Lactarius* cf. *obscuratus, Pseudotomentella* sp., *Tomentella* spp, ***Xerocomus* sp.**	DNA sequencing of ITS	Field
Kennedy et al. (2011)	*A. jorullensis, A. acuminata*	23	24	*Alnicola* sp., *Alpova sp.,* ***Clavulina*** sp., *Cortinarius* spp., *Hymenogastraceae* sp., *Inocybe* spp., *Lactarius* spp., ***Sebacinacea*** spp., *Tomentella* spp.	DNA sequencing of ITS and LSU	Field
Bogar and Kennedy (2013)	*A. rhombifolia*	13	25	*Alnicola* spp., *Alpova diplophloeus* spp., *Boletaceae* sp., *Clavulina* sp, ***Laccaria laccata***, *Russula* sp., *Tomentella ellisii,* spp., *Thelephorales* sp., *Tuber* sp.	DNA sequencing of ITS	Field/bioassay

[a]Number of unique groups, morphotypes, or species reported
[b]Running total of new genera is an estimate, and has been adjusted based on corrected reports
[c]First reports of new genera are in bold type; species in brackets formed incomplete mycorrhizas

8.3 Could the *Alnus* ECM Specificity Pattern Be an Artifact?

The atypical specificity of the *Alnus* ECM system has been observed in many different studies and experimental settings. The consistency of those results presented a striking pattern, but some aspects of previous work leave open the possibility that the currently accepted specificity paradigm could be artifactual. For example, the study of *Alnus* ECM fungal communities has focused largely on temperate geographic regions, but *Alnus* species also occur at tropical latitudes in Central and South America. Studies from many groups of organisms have shown that species richness tends to be higher in tropical regions and decreases as one moves towards the poles (i.e. the latitudinal gradient of species diversity—LGD) (Townsend et al. 2008). As such, part of the current perception that *Alnus* ECM fungal communities are species poor may be related to the temperate bias of past *Alnus* studies. Similarly, nearly all previous studies involving comparisons between *Alnus* and other ECM fungal communities have involved distantly related ECM hosts, such as *Pseudotsuga menziesii* (Miller et al. 1992), multiple Pinaceae species (Massicotte et al. 1994), and *Pinus montezumae* (Kennedy et al. 2011a). Because ECM fungal community similarity has been shown to be lower when comparing more distantly related hosts (Ishida et al. 2007), the observed specificity of the *Alnus* system could also be an artifact of the types of host comparisons made thus far. Below we discuss three of our own studies that recently examined these issues to test the robustness of the *Alnus* ECM specificity pattern (Kennedy et al. 2011a; Bogar and Kennedy 2013).

Two *Alnus* species, *A. jorullensis* and *A. acuminata*, grow in montane tropical forests in central Mexico, either alone or with other ECM host species such as *Pinus montezumae*. Kennedy et al. (2011a) sampled the ECM fungal communities present at multiple sites for each *Alnus* species. We found that, like their temperate counterparts, the *Alnus* ECM fungal communities in Mexico had relatively low species richness. Interestingly, many of the ECM fungi present in the Mexican *Alnus* forests were strikingly similar to those present in *Alnus* forests in other parts of the world. For example, in the genus *Tomentella*, the five most abundant species in Mexico had sequences that matched much better to sequences of *Tomentella* species sampled in forests in the United States, Europe, and Argentina than to other *Tomentella* species sampled in Mexico (Fig. 8.1). The sequence matches were very high (>97 %), suggesting that *Alnus* species may associate with many of the same ECM fungi globally. A similar pattern was also evident in the ECM fungal genera *Cortinarius*, *Lactarius*, and *Inocybe*. In addition, we identified notably higher species richness on *P. montezumae* in ten-fold fewer ECM fungal root tips; 24 ECM fungal species were identified from 42 *Pinus montezumae* ECM root tips, compared to only 21 ECM fungal species detected on over 400 concurrently sampled *Alnus* ECM root tips. This result reinforces the depauperate nature of *Alnus* ECM fungal communities compared with other ECM hosts. More importantly, despite a clear intermingling of root systems at two of the study sites, there were no species shared

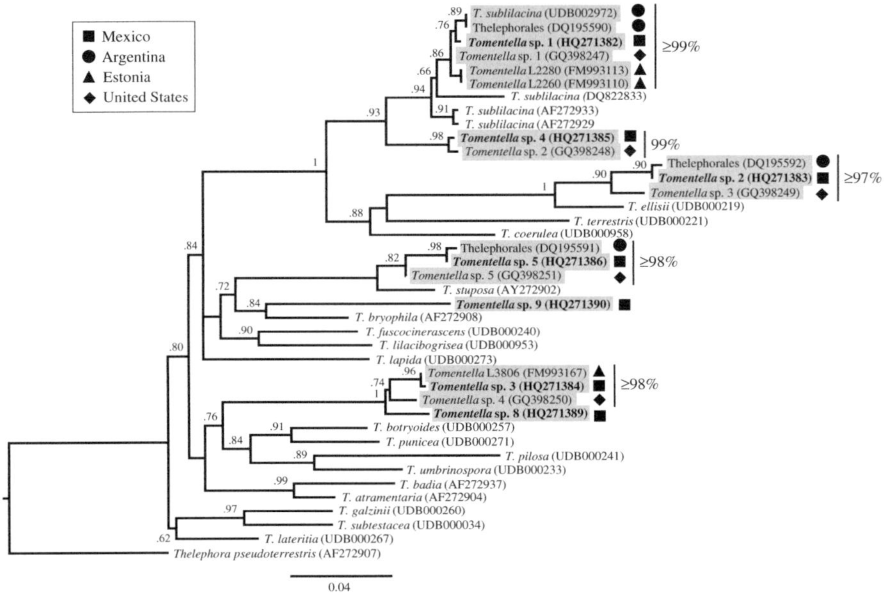

Fig. 8.1 Phylogenetic reconstruction of taxa in the ECM genus *Tomentella* based on rDNA ITS sequences. Nodes are labeled with aLRT scores from the maximum likelihood analysis above 0.60. Species are labeled with species names or unique identifier and GenBank or UNITE number in parentheses. *Alnus*-associated species are designated in *gray boxes*, with the Mexican *Alnus*-associated species in *bold*. Symbols next to selected *Alnus*-associated species indicate the geographic area from which they were obtained. The percentage values for the selected groups represent pair-wise comparisons between all group members. *Thelephora pseudoterrestris* was designated as the outgroup for rooting

between the ECM fungal communities on *A. jorullensis* and *P. montezumae*. The low species richness and reciprocal specificity observed on *Alnus* species in this tropical-based study, suggests that the unique pattern present in the *Alnus* ECM system is consistent regardless of geographic location of study. Polme et al. (2013) confirmed this conclusion with a comprehensive spatial sampling of 22 *Alnus* species over 96 geographic locations covering a wide range of latitudes.

We also examined whether host evolutionary relationships might explain *Alnus* ECM specificity by comparing *Alnus* and *Betula* ECM fungal communities. We predicted that if *Alnus* ECM fungi are specific to the family Betulaceae, rather than only to *Alnus*, they would be expected to associate with both genera. To test this hypothesis, we examined the ECM fungal communities on *Alnus rhombifolia* and *Betula occidentalis* in western Idaho, USA (Bogar and Kennedy 2013). We characterized the communities in a riparian habitat where the hosts co-occur, comparing the ECM fungal communities present when *Alnus* and *Betula* roots overlapped to those present on each host alone. In order to accomplish this, we collected soil cores at the base of trees of each host species in pure stands (i.e. from solitary host trees >2 m from the other host type) and soil cores from between the two host species (i.e. we dug directly in between alternate hosts that were ≤30 cm from each other) in mixed stands. We found that the ECM fungal community associated with *A. rhombifolia* was relatively species poor and host-specific as compared to the ECM fungal community on *B. occidentalis* (i.e. overall, there were fewer ECM fungal species on *A. rhombifolia*, and the same ECM fungal species were not found on *B. occidentalis* in this study, Fig. 8.2). This was true even when overlapping root systems of the two hosts were sampled (Fig. 8.2). In comparison to the results of Kennedy et al. (2011a), the specificity of the *Alnus*-associated ECM fungal community was not as strong: the two host species associated with six of the same ECM fungal species across the site. However, overlapping root systems (i.e. those found in the same core) of *A. rhombifolia* and *B. occidentalis* almost never associated with the same fungal species. Thus, even though host specificity may not be absolute for some *Alnus*-associated ECM fungi, it appears that association with *Alnus* may preclude simultaneous association with *Betula*. In short, this study suggests that the unique specificity observed in the *Alnus* ECM system is not a byproduct of previous comparisons involving more distantly related hosts.

More recently, we also examined the ECM fungal communities associated with *Alnus glutinosa* in New Zealand, which is a non-native tree invader on the North and South Islands (Bogar et al. 2015). We speculated that by sampling outside the native range of *Alnus* trees, the *A. glutinosa* individuals present in New Zealand might be 'forced' into associating with a broader suite of ECM fungi, particularly those present on native ECM hosts (e.g. *Nothofagus* spp.). We found, however, that the ECM fungal communities present on *A. glutinosa* in New Zealand were notably species poor (only 9 species present across over 300 root tips sampled) and completely dominated by European *Alnus*-associated ECM species (which is the native range of *A. glutinosa*). In fact, we found no ECM fungal species present on *A. glutinosa* that appeared to be associated with native New Zealand ECM hosts. This result further reiterates the globally anomalous nature of the *Alnus*-associated ECM system and suggests that even well outside their native ranges, the specificity of the plants and fungi involved in this symbiosis remains intact.

Taken together, we believe there is abundant evidence that *Alnus* ECM fungal communities are both species poor (Masui 1926; Horak 1963; Neal et al. 1968; Mejstrik and Benecke 1969; Brunner et al. 1990; Miller et al. 1992) and highly host specific (Molina 1979, 1981; Molina et al. 1992; Godbout and Fortin 1983; Pritsch et al. 1997; Tedersoo et al. 2009; Kennedy and Hill 2010; Bent et al. 2011) and that this unique pattern is not based on sampling artifact (Kennedy et al. 2011a; Bogar and Kennedy 2013; Bogar et al. 2015).

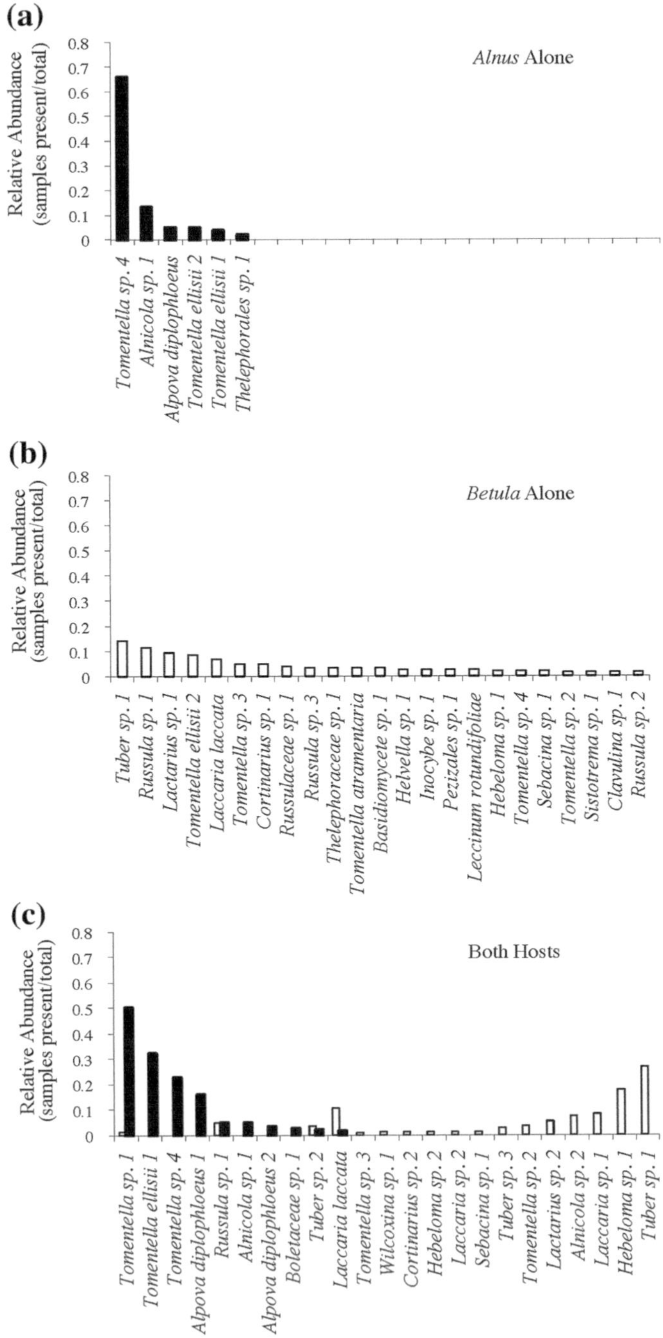

Fig. 8.2 Ranked relative abundance of ECM species richness on *Betula occidentalis* (*open bars*) and *Alnus rhombifolia* (*filled bars*) at a field site in western Idaho, USA. Sampling found six fungal taxa that were present on both *Betula* and *Alnus* (names in *bold*)

8.4 Why or Why not Participate in CMNs?

To better understand why reciprocal specificity among *Alnus* and their associated ECM fungi makes them unlikely to participate in interspecific CMNs, it is helpful to briefly revisit the ecological benefits provided by CMNs. One of the most widely cited benefits of CMNs for both plants and fungi is access to a larger resource pool. By joining an extensive established mycelial network, connected plants have the ability to draw from a much larger soil volume than unconnected plants (Newman 1988). This benefit appears to be particularly important for seedlings, which lack well-developed root systems (van der Heijden and Horton 2009). With connections to a variety of host species (Booth 2004), or to the same host species at a range of growth stages (Teste et al. 2010; Beiler et al. 2010; Booth and Hoeksema 2010), ECM fungi also receive carbon from multiple sources. This redundancy may provide an important buffer against spatially or temporally variable host inputs (e.g. deciduous versus evergreen hosts, canopy versus understory individuals).

A related proposed benefit of CMNs is inter- or intraspecific plant facilitation (Molina and Horton, Chap. 1, this volume; Nara Chap. 6, this volume). Molina and Trappe (1982) hypothesized that the resprouting ability of certain plant species in forests of the Pacific Northwest, USA allows the ECM fungal community to be maintained directly after fire or clear-cutting. The presence of compatible fungi benefits subsequent colonization of later seral plants by providing those individuals with access to established mycelial networks. Seedlings of the resprouting plants then reciprocally establish in the understory of those later seral forests and therefore benefit from CMNs in the same way. Evidence supporting CMN-mediated interspecific plant facilitation has been documented in California (Horton et al. 1999), Japan (Nara and Hogetsu 2004) and Corsica (Richard et al. 2009), and recent work in the dry forests of western Canada indicates that CMNs can also facilitate the establishment of conspecific *Pseudotsuga menziesii* seedlings (Teste et al. 2010). Fungal benefit in the above scenarios comes from the ability to maintain a constant carbon source during disturbance-associated host species regeneration.

CMNs may also benefit plants by mediating nutrient transfer among connected individuals (Simard et al. Chap. 5, this volume). This benefit has been most clearly documented among mycoheterotrophic plants, which received all of their carbon from CMNs connected to adjacent autotrophic plants (Bidartondo 2005). The transfer of carbon has also been documented among autotrophic plants, although the levels of movement among autotrophic individuals appear to be much lower than to both mycoheterotrophic or mixotrophic plants (Simard et al. 2012). In addition to carbon, other resources can also move among CMN-linked plants, including nitrogen (Arnebrant et al. 1993; He et al. 2004, 2005), phosphorus (Finlay and Read 1986), water, and defense compounds (Song et al. 2010; Johnson and Gilbert 2015). Although we are unaware of studies demonstrating beneficial movement of resources among fungal individuals through linked plants, that pathway may exist, especially for a resource that would be lost or not transferable through soil. Finally, we believe it is important to stress that the three aforementioned benefits are not mutually exclusive; plants and fungi may benefit in multiple simultaneous ways from CMNs (Simard et al. 2012).

Notwithstanding potential intraspecific networks (i.e. connections among *Alnus* individuals) and given the aforementioned benefits provided by CMNs, why do *Alnus* plants and their associated ECM fungi remain unconnected to co-occurring non-*Alnus* ECM hosts? One reason is likely related to the general life history of this host genus. As typically pioneer successional species, *Alnus* individuals establish in habitats where, in many cases, other ECM hosts are not already present. Doing so reduces or eliminates the opportunity for *Alnus* plants to join established mycelial networks or to benefit from CMN-mediated facilitation. (Because *Alnus* seedlings are shade intolerant, and do not occur under an established *Alnus* overstory, they also do not have immediate access to intraspecific CMNs.) While *Alnus* forests tend to be mono-dominant initially, there is establishment by other ECM hosts (e.g. those in the Pinaceae) over time (Miller et al. 1992). While the presence of other ECM host species provides the potential for CMNs and the transfer of resources between connected individuals, the dynamics of CMN nutrient transfer appears to be unfavorable for *Alnus* plants. Simard et al. (1997) showed that interspecies CMN-mediated resource transfer follows a source-sink pattern, with net carbon movement towards shaded individuals. Connecting to CMNs with understory species would therefore represent a carbon loss for *Alnus* individuals, as it would for other pioneer species. In addition, since *Alnus* seedlings tend not to establish under canopies, there are also no reciprocal opportunities for this genus to regain carbon benefits from CMNs (unlike the scenario discussed by Molina and Trappe (1982) above).

The forests of central Mexico provide an interesting exception to this pattern. At some locations, *Alnus jorullensis* persists under a *Pinus montezumae* canopy, resulting in *Alnus* as the potential carbon sink (i.e. a favorable situation for *Alnus*; Kennedy et al. 2011a). As noted earlier, however, *Alnus* and *Pinus* individuals at those sites do not appear to share any common ECM fungi, therefore no CMNs between *Alnus* and *Pinus* are possible. This finding suggests the absence of CMNs among *Alnus* trees and other ECM hosts is not solely driven by unfavorable carbon-based source-sink dynamics.

We believe that a second key factor discouraging the formation of CMNs for *Alnus* plants is their co-association with nitrogen-fixing *Frankia* bacteria. These bacteria provide *Alnus* with a unique source of nitrogen relative to co-occurring ECM host plants. Although interspecific CMNs involving *Alnus* plants appear to be functionally non-existent in natural settings, in a laboratory study, Arnebrant et al. (1993) showed that substantial amounts ($\sim$20 %) of fixed nitrogen could move through CMNs from *Alnus glutinosa* to *Pinus contorta*. Similar results were obtained by Ekblad and Huss-Danell (1995), who observed that up to 9.5 % of the nitrogen in CMN-linked *Pinus sylvestris* seedlings was derived from *Frankia*-based nitrogen fixation. Given the substantial carbon allocation by *Alnus* plants towards *Frankia* bacteria (see below), and the value of nitrogen as a resource, the absence of CMNs between *Alnus* and non-*Alnus* individuals would prevent co-occurring plants from directly accessing this commodity. Intriguingly, He et al. (2004, 2005) used labeled isotopes to show a net movement of nitrogen through CMNs from non-*Frankia*-associated *Eucalyptus maculata* to *Frankia*-associated *Casuarina cunninghamiana* individuals. Since these latter results conflict with the findings of Arnebrant et al. (1993) and Ekblad and Huss-Danell (1995), additional studies, particularly in field settings, are needed to further define the patterns and drivers of nitrogen transfer dynamics.

8.5 Establishment and Maintenance of the *Alnus*-ECM Fungus Specificity Pattern

Although the absence of CMNs may be selectively advantageous for *Alnus* plants based on their life history and relationship with *Frankia* bacteria, questions remain about how specificity in the *Alnus*-ECM fungus system is established and maintained. Many authors have discussed this system from a co-evolutionary standpoint (Molina et al. 1994; Moreau et al. 2006; Kennedy and Hill 2010) and there is evidence to support its role in driving patterns of co-speciation (Rochet et al. 2011). Our interests, however, lie in the more proximate causes of the observed specificity. As such, we focus the remainder of the chapter on a number of hypotheses that may explain how current interactions among *Alnus* trees and their associated ECM fungi reinforce their unique specificity pattern.

8.5.1 Alnus-*ECM Fungus Specificity: Signaling and Sanctioning Hypothesis*

Before ECM host plants and fungi begin to interact with one another, each symbiont is confronted with incomplete information about the other partner. For example, how do *Alnus* plants identify which of the fungi in the ECM community pool have the right characteristics to meet their needs? Similarly, how do ECM fungi differentiate *Alnus* roots from those of other co-occurring hosts? The latter issue is partially resolved by the fact that *Alnus* trees often establish in mono-dominant stands, but there are a number of situations in which *Alnus* individuals do co-occur with other host species (Tedersoo et al. 2009; Kennedy et al. 2011a; Bogar and Kennedy 2013). This problem of asymmetric information can be resolved in two ways (Archetti et al. 2011). The first is to choose partners before the interaction is established. This mechanism, known as partner choice (Bull and Rice 1991), can be accomplished by signaling. Under this scenario, *Alnus* plants would broadcast information about their own attributes, and the ECM fungi would respond by associating or not based on that signal. The experimental study of Massicotte et al. (1994) showed strong indirect support for chemical signaling between *Alnus* plants and ECM fungi. Those authors observed that *Alpova diplophloeus*, an *Alnus*-specific ECM fungus, germinated readily in the presence of *Alnus* roots (as determined by subsequent root tip colonization) but never in the sole presence of roots of a number of other ECM host species. Conversely, no non-*Alnus* associated ECM fungus germinated in the presence of *Alnus* roots alone, but most did germinate in the presence of their preferred host, infrequently colonizing a secondary host. Collectively, these data suggest that *Alnus* roots release a unique chemical cue that induces spore germination of only the fungi having attributes beneficial to *Alnus* (and perhaps only eliciting a response from those fungi that may also benefit from resources associated with *Alnus*). Analogous signaling that induces partner germination has been observed among mycoheterotrophic plants and their associated ECM fungi (Bruns and Read 2000; Bidartondo and Bruns 2005) as well as with the conifer-induced germination of other host-specific ECM fungi in the genus *Suillus* (Fries et al. 1987).

A second way that the asymmetric information problem can be resolved is by monitoring the interaction after it has been established. This kind of monitoring is commonly referred to as host sanctioning and typically involves some form of punishment of "misbehaving" symbionts (Kiers and Denison 2008). One example of host sanctioning comes from the soybean-rhizobia symbiosis, where the soybean host is able to selectively decrease oxygen availability to nodules that are not fixing nitrogen (Kiers et al. 2003). While nitrogen fixation is a tightly controlled anaerobic process mediated by plant leghemoglobin, oxygen is still required by these bacteria as a terminal electron acceptor, therefore reduced oxygen impedes rhizobial performance (Kiers et al. 2003). In the *Alnus* ECM system, some of the results of Molina (1979) are consistent with host sanctioning. He found that two ECM fungi not typically associated with *Alnus rubra*, *Paxillus involutus* and *Astraeus pteridis*, were able to establish mycorrhizas with this host species in pure culture synthesis assays. Interestingly, Molina (1979) found that cross-sections of the *A. rubra*-*P. involutus* mycorrhizas had high concentrations of phenolics in root cortical cells that were not present in the comparable mycorrhizas of *Alpova diplophloeus*. This was interpreted as the result of *Alnus* recognizing *P. involutus* as the "wrong" symbiont and attempting to decrease subsequent colonization. Similar results were reported by Malajczuk et al. (1982) involving interactions between multiple *Eucalyptus* host species and *Pinus*-specific ECM fungi. However, if this mechanism of sanctioning was the primary way that *Alnus* plants avoid significant colonization by the "wrong" fungi, a similar pattern should have also been observed in the mycorrhizas of *Astraeus pteridis*. Instead, phenolic concentrations in *A. rubra*-*A. pteridis* mycorrhizas were low, suggesting that this "wrong" symbiont (1) was able to meet host needs and prevent sanctioning, (2) was subject to sanctioning at some other time or under some environmental condition not captured in that experiment, or (3) had some way of remaining undetected despite being the "wrong" symbiont.

These two mechanisms, partner choice and host sanctioning, could also work in concert to create the unique specificity observed in the *Alnus* ECM system. The collective results of the two aforementioned studies suggest that partner choice likely plays a significant role in preferentially inducing the germination of ECM fungi recognized by *Alnus* as beneficial, while host sanctioning might be an important mechanism available to minimize or eliminate any "incorrect" *Alnus* ECM interactions. Considering the relative strengths of the two mechanisms, colonization of *Alnus* plants by the "wrong" ECM fungi seems either non-existent or to occur only very rarely in field settings (Kennedy et al. 2011a; Bogar and Kennedy 2013; Polme, personal communication). If host sanctioning were the dominant mechanism driving specificity, one would expect to find more ECM fungi forming mycorrhizal associations with *Alnus* trees, at least initially. Since this is not normally the case, it seems that pre-interaction partner recognition is most likely the dominant mechanism affecting *Alnus* ECM specificity, with interaction-based host sanctioning playing a limited secondary role.

8.5.2 **Alnus**-*ECM Fungus Specificity: Interspecific Competition Hypothesis*

The specificity of the *Alnus*-ECM fungus system may also be mediated by competition between *Alnus* and co-occurring ECM host plants either directly or via ECM fungi. Both hosts and fungi could escape a certain amount of competitive pressure by restricting the set of symbionts with which they associate. This applies particularly to situations in which hosts or fungi are adapted to colonize soon after disturbance events. To fully appreciate why a set of symbionts might not participate in local CMNs, it is important to consider selection acting on both the hosts and the fungi individually since it occurs at distinctly different spatial and temporal scales.

As noted previously, CMN connectivity could be helpful to ECM host species that establish under the canopy of other trees. By maintaining broad receptivity to many different ECM fungi, a later successional ECM host has a greater chance of joining an established mycelial network early in development. This scenario was discussed by Kropp and Trappe (1982) with respect to *Tsuga heterophylla*, a late-successional, broadly receptive ECM host in northwest North America. In contrast, by denying later successional seedlings access to CMNs by associating with host-specific ECM fungal communities, early successional hosts would suppress the establishment of competing hosts and maintain their own dominance in a stand. In the case of *Alnus*, whose dominance is limited by its short-lived nature, not participating in CMNs would also prevent any 'facultative epiparasitism' (sensu Bruns et al. 2002) of fixed nitrogen by co-occurring ECM host plants. Intraspecific competition, of course, would be unaffected by this specificity, and would remain an important ecological force in these situations (also noted by Bruns et al. 2002). Kropp and Trappe (1982) and Molina et al. (1992) both noted that pioneer tree species often do associate with communities of host-specific ECM fungi (e.g. *Pseudotsuga*, *Alnus*), supporting the hypothesis that these early successional settings encourage specialization.

Selection on the fungi must also influence whether or not a set of symbionts will participate in local CMNs. As discussed above, in most situations, ECM fungi would benefit by connecting to multiple host species. This would expand the effective resource pool available to a given fungal genet, and provide the fungus with insurance in the event that resources were no longer provided by a primary host. Competitive dynamics, however, have led some ECM fungi to specialize on particular hosts. Bruns et al. (2002) discuss the case of the genus *Rhizopogon*, species of which dominate both the "spore bank" and the below-ground communities of their ECM hosts (*Pinus* and *Pseudotsuga*) early in forest succession at Point Reyes, CA, USA. Over time, this group of fungi becomes less common on their hosts, suggesting that they are weaker competitors relative to the other fungi with which the hosts associate (Bruns et al. 2002). It seems possible, then, that these fungi have specialized on early successional hosts as a consequence of competition: a combination of long-lived propagules (Bruns et al. 2009) and well-timed, host-specific germination (Massicotte et al. 1994) could allow these fungi to guarantee themselves a host with relatively little competition from other fungi, at least early in succession (see further discussion of this dynamic in Kennedy 2010; Kennedy et al. 2011b). While the competitive dynamics of *Alnus*- and non-*Alnus*-associated ECM fungi have not been examined, a similar spore longevity pattern to *Rhizopogon* has been noted for the *Alnus*-specific species *Alpova diplophloeus* (Miller et al. 1994).

On the whole, both ECM hosts and fungi may experience competitive pressure to specialize—and thus, evade CMN participation—under a number of circumstances, but particularly early-successional situations and settings in which a symbiont has enhanced access to a particular set of resources.

8.5.3 **Alnus**-*ECM Fungus Specificity: Soil Chemistry Hypothesis*

One of the ways that fidelity (see Molina and Horton, Chap. 1, this volume) could be reinforced is by some form of environmental filtering. A widely noted environmental parameter with respect to *Alnus* forests is their soil chemistry (Hibbs et al. 1994; Becerra et al. 2005; Tedersoo et al. 2009; Yarwood et al. 2010). *Alnus* soils are typically characterized by low pH, which is a byproduct of the hydrogen production associated with nitrification (Bormann et al. 1994). Both high acidity and high nitrate levels may represent a formidable combination of environmental filters, as both have been shown to the limit the growth of a variety of ECM fungi (Hung and Trappe 1983; Lilleskov et al. 2002; Avis et al. 2003; Trudell and Edmonds 2004; Cox et al. 2010). To experimentally test their effects in the *Alnus* ECM system, Huggins et al. (2014) manipulated the pH and nitrate concentrations present in the liquid media of a suite of *Alnus*- and non-*Alnus* ECM fungal species. They found that the growth of *Alnus* ECM fungi were not, on average, affected by high acidity, while non-*Alnus* ECM fungi had a significantly negative growth response under the same conditions. Similarly, when grown at high nitrate, non-

Alnus ECM fungi also generally performed more poorly. Taken together, the results of Huggins et al. (2014) are consistent with soil pH and nitrate concentrations being important environmental filters that may underlie the specificity in the *Alnus* ECM system. At the same time, multiple lines of other evidence do not clearly support this mechanism. Sites initially dominated by *Alnus* trees are readily replaced by other ECM host species over time and if high soil acidity and nitrate levels are strongly inhibitory to non-*Alnus* ECM fungi, one would expect that ECM fungal colonization of other hosts to be low in *Alnus*-influenced soils. Miller et al. (1992), however, observed that *Pseudotsuga menziesii* seedlings grown in soils from both young and older *Alnus* forests were similarly well colonized with a diverse range of ECM fungi as *P. menziesii* seedlings grown in young and older *P. menziesii* forest soils. Data from the recent field study of *Alnus* and *Betula* ECM fungal communities also indicates that non-*Alnus* ECM fungi can survive on their preferred hosts even when occupying the same soil as *Alnus* roots (Bogar and Kennedy 2013). If the specificity of *Alnus* ECM fungal communities is strongly driven by soil chemistry alone, the ECM fungal community on *Betula* roots should have been substantially changed when overlapping with *Alnus* roots relative to the community on *Betula* roots in the absence of *Alnus*. It was not, however, suggesting that either *Betula* ECM fungal associates are tolerant of similar soil conditions as *Alnus* ECM fungal associates or conditions were not changed enough in mixed settings to shift community composition significantly. It also appears that at least some *Alnus*-associated ECM fungi are negatively affected by high nitrate concentrations. For example, Koo et al. (1995) found that colonization of *Alpova diplophloeus* on *Alnus rubra* seedlings was significantly decreased in highly mineral nitrogen-amended soils and Huggins et al. (2014) also showed that some *Alnus* ECM fungi performed poorly at high nitrate levels. Taken together, these studies suggest that *Frankia*-induced changes in pH and soil nitrogen concentrations can affect ECM fungal colonization and community structure, but do not appear to be solely responsible for the atypical composition of *Alnus* ECM fungal communities.

Along with pH and nitrogen, *Alnus* trees are known to influence other aspects of soil chemistry as well. Specifically, soils in *Alnus* forests can be low in inorganic phosphorus (Giardina et al. 1995; Compton and Cole 1998, 2001), and enriched in organic phosphorus (Zou et al. 1995). Tedersoo et al. (2009) hypothesized that ECM fungal communities associated with *Alnus* species may be strongly associated with soil phosphorus concentrations due to the phosphorus demands of co-occurring *Frankia* (see below). Their community analyses, however, indicated that soil phosphorus levels had no statistically significant effects on ECM fungal community structure. Koo et al. (1996) also found that mineral phosphorus fertilization did not decrease mycorrhizal colonization in a greenhouse study. Collectively, these studies suggest that soil phosphorus concentration does not strongly influence *Alnus* ECM fungal community composition or colonization, however, our more recent work suggests that *Alnus* ECM fungal communities may have a unique physiological response to soil phosphorus availability (see below).

8.5.4 Alnus-*ECM Fungus Specificity: Host Metabolic Hypothesis*

A different way that the presence of *Frankia* bacteria may affect *Alnus* ECM fungal communities is by shifting host nutritional needs in a way that favors fungi adept at acquiring nutrients aside from nitrogen. In particular, nitrogen-fixing plants are often limited by phosphorus (Benson and Clawson 2002), so *Alnus* individuals may selectively associated with ECM fungi that have enhanced enzymatic abilities towards phosphorus acquisition. Indirect support for this hypothesis was shown by Ekblad et al. (1995), who found that in microcosms containing *Alnus incana* and *Pinus sylvestris* seedlings colonized by *Paxillus involutus*, fungal biomass peaked in low phosphorus soils. Another study found that *Alnus* seedlings colonized by both *Frankia* and ECM fungi could have higher phosphorus tissue concentrations when grown in certain types of soils than seedlings colonized by just *Frankia* alone (Yamanaka et al. 2003). However, because the presence of ECM fungal colonization tends to raise seedling phosphorus levels on other hosts (Smith and Read 2008), it is unclear whether the response seen in those *Alnus*-based studies is due simply to ECM fungal colonization or colonization by ECM fungi specialized on greater phosphorus acquisition.

To test whether *Alnus* ECM fungi have different enzymatic capabilities relative to ECM fungi associated with other host trees, direct assays of enzyme production from ECM root tips are necessary. The logistics and throughput capacities of ECM root tip enzyme assays have improved significantly in recent years (Courty et al. 2005; Pritsch et al. 2011) and a growing body of literature is developing around these techniques (Courty et al. 2010; Pritsch and Garbaye 2011; Jones et al. 2012). With regard to the *Alnus*-ECM fungus system, we recently compared the enzyme activity of ECM fungal root tips sampled from pure stands of *Alnus rubra* and *Pseudotsuga menziesii* at the Cascade Head and H.J. Andrews Experimental Forests in Oregon, USA (Walker et al. 2014). Excised ECM fungal root tips were tested for acid phosphatase (phosphorus) and leucine aminopeptidase (nitrogen) activity and DNA was extracted for molecular identification based on the rRNA ITS gene region.

From those samples, we were able to molecularly identify 62 and 75 % of the ECM fungal root tips sampled from *A. rubra* and *P. menziesii* plots, respectively. The ITS sequences of 18 different ECM fungal species were recovered from *A. rubra* root tips, while 76 ECM fungal species were detected on *P. menziesii* root tips, and an additional four species were shared. These levels of species richness correspond well with previous studies of both *A. rubra* and *P. menziesii* ECM fungal communities (Kennedy et al. 2003; Cline et al. 2005; Horton et al. 2005; Kennedy and Hill 2010). In support of the aforementioned hypothesis, the *A. rubra*-associated ECM fungal community had significantly higher acid phosphatase activity than the ECM fungi associated with *P. menziesii*, while the leucine aminopeptidase of *A. rubra*-ECM fungal root tips was significantly lower at the nitrogen-rich site (Fig. 8.3). Collectively, these results indicate that *A. rubra*-associated ECM fungi appear to

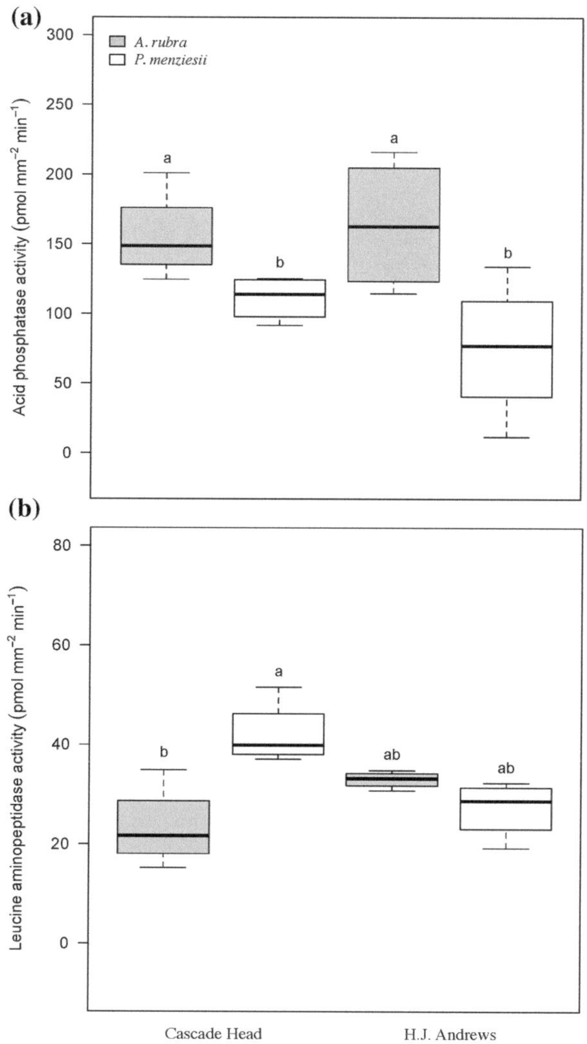

Fig. 8.3 Differences in potential (*a*) acid phosphatase and (*b*) leucine aminopeptidase activity between the ECM fungal community on *Alnus rubra* (grey) as compared to *Pseudotsuga menziesii* (white) at Cascade Head and H.J. Andrews, Oregon, USA. Raw data is presented in the figure, but all data were cube root transformed in order to meet assumptions of normality for statistical analyses. Lower case letters designate significant differences at $P \leq 0.05$ detected by univariate ANOVAs and subsequent Tukey's HSD tests. Boxes surrounding median values represent the first and third quartiles, while whiskers show the smaller (and larger) of either the maximum (and minimum) values or 1.5× the interquartile range (approximately ±2 SD); N = 2

have enhanced phosphorus acquisition abilities, and that host nitrogen status may mediate ECM fungal physiological response as demonstrated by the elevated levels of organic nitrogen acquisition by the *P. menziesii* ECM fungal community in the absence of *Frankia*-derived nitrogen.

8.5.5 Alnus-*ECM Fungus Specificity: A Host-Fungus Reward System Based on Nitrogen?*

We formalize an additional hypothesis regarding *Alnus* ECM specificity: that *Alnus* plants may also provide a reward to ECM fungi to help maintain specificity in this system. A reward system may be particularly important for *Alnus* individuals because they may provide less carbon to ECM fungi than other hosts, due to their simultaneous interaction with *Frankia* bacteria. While this speculation about carbon allocation has yet to be tested, the photosynthetic rates of *Alnus* species are similar to non-*Frankia*-associated broad-leaved species (Agren and Ingestad 1987; Koike 1990). As such, *Alnus* individuals do not appear to have a larger carbon pool from which to allocate to their dual symbionts. Since carbon allocation to *Frankia* and ECM fungi has been estimated at ~ 15 % per symbiosis (Tjepkema et al. 1986; Smith and Read 2008), it seems likely that, relative to other hosts, *Alnus* plants may provide less carbon to ECM fungi. In light of this carbon dynamic, what might *Alnus* plants offer to prevent defection to more carbon generous hosts?

We suggest that *Alnus* may provide its chosen ECM fungi with direct access to the nitrogen fixed by the *Frankia* bacteria. While this would represent a reversal of the way nitrogen is typically traded between plants and ECM fungi (nitrogen is usually provided to the plant by the fungus), the unique ecology of this tri-partite symbiosis may favor this change in partner trading dynamics. From the fungal perspective, getting nitrogen from the host would decrease the need to scavenge nitrogen from the soil. Although *Alnus* individuals may provide less carbon to the fungi, the fungi may not need to invest as much carbon in nitrogen-scavenging enzymes as they would when colonizing a non-nitrogen-fixing host. Furthermore, since organic matter may be limited in early successional settings, it would be easier for the fungi to get nitrogen from the host instead of relying on organic sources in soil. From the host perspective, it may be advantageous to provide *Frankia*–derived nitrogen to ECM fungi unlikely to participate in networks so that no nitrogen is lost through CMNs to other host species. While providing nitrogen to their fungi would represent a cost to *Alnus* plants (because of their carbon investment in the *Frankia* bacteria) all else being equal, multiple studies have shown that *Alnus* individuals colonized by both *Frankia* bacteria and ECM fungi can be larger than plants with only a single symbiont (Chatarpaul et al. 1989; Koo et al. 1995). This suggests the putative benefits of host-ECM nitrogen provisioning outweigh its costs. It has also been noted that plants can have higher phosphorus concentrations in their tissues when colonized by *Frankia* bacteria and ECM fungi (Yamanaka et al. 2003). Since phosphorus has been demonstrated to be a limiting resource for nitrogen fixation (Jha et al. 1993; Uliassi and Ruess 2002), the presence of ECM fungi would be beneficial to the *Frankia* bacteria as well, if the plant is able to allocate greater phosphorus to bacterial nodules.

In support of this nitrogen reward hypothesis, Arnebrant et al. (1993) found that many of the amino acids in the ECM fungus used in their study, *P. involutus*, contained nitrogen originally fixed by *Frankia*. Given the short-time scale of their experiment (ten weeks, seven day labeling period) and the fact that live *Frankia* nodules are not known to excrete nitrogen into their external environment, it seems very likely that the nitrogen was passed from *Frankia* to the plant and then onto the ECM fungus. Ekblad and HussDanell (1995) obtained comparable results, although amino acids were not directly assayed in that study. The results of two additional studies are also consistent with a potential transfer of nitrogen from host to fungus in *Alnus* ECM interactions. Koo et al. (1995) found that *Alpova diplophloeus* colonized only 10 % of the root systems of *Alnus rubra* seedlings when they were non-nodulated, but 65 % when *Frankia* nodules were present. Similarly, Yamanaka et al. (2003) observed no colonization by *Alpova diplophloeus* on non-nodulated *Alnus tenuifolia* seedlings, but between 75 and 100 % colonization when *Frankia* nodules were present. Although neither of these studies directly indicates that nitrogen is the resource responsible for higher ECM fungal colonization, the results are consistent with a significant benefit provided to the fungus by co-colonization with *Frankia* bacteria. It should also be noted that in experimental settings, *Frankia* are the first of the two microbial symbionts to colonize *Alnus* seedlings (Miller et al. 1992; Koo et al. 1995), and that all *Alnus* individuals in field conditions are colonized by *Frankia* bacteria (Benson and Dawson 2007). These latter findings indicate that the ability of *Alnus* individuals to readily access nitrogen for a reward system for ECM fungi appears to be the default state in nature.

If nitrogen is provided from the plant to the fungus, monitoring its consumption could also be a way in which *Alnus* individuals control which fungi are sanctioned. Presumably, the nitrogen demands of ECM fungi engaged in CMNs with other hosts would be higher than for non-networked species [due to demand from the other hosts and the source-sink dynamics of CMN resource transfer (Simard et al. 1997)]. By limiting colonization of CMN-forming fungi, *Alnus* individuals may be able to prevent any facultative epiparasitism via CMNs. It should be noted, however, that this method of sanctioning would not prevent host generalist ECM fungi from colonizing *Alnus* plants. In fact, if a host generalist ECM fungus were only associating with an *Alnus* individual, its nitrogen demands should be similar to that of typically *Alnus*-associated species and therefore it would likely avoid sanctions. Data from our study of *Alnus rhombifolia* and *Betula occidentalis* ECM fungal communities show some support for this scenario (Bogar and Kennedy 2013). We found that there were six ECM species associated with both *Alnus rhombifolia* and *Betula occidentalis*. Five of the six ECM species were, however, never found on the roots of both hosts within the same soil core. This suggests while some fungal species could associate with both hosts, different individuals of those fungal species were present on each host. We did, however, find one species, *Laccaria laccata*, that was present on both *Alnus rhombifolia* and *Betula occidentalis* ECM root tips in the same soil core. If those tips were colonized by the same fungal individual and our logic about sanctioning based on nitrogen allocation is accurate, we would expect that *Alnus* ECM roots in that core would begin to reject colonization by *L. laccata* over time due to excess nitrogen consumption. A key untested assumption of this logic is the spatial scale over which sanctioning is occurring. If it occurs at the individual tip scale, it seems unlikely that *L. laccata* would be able to establish since excess nitrogen consumption should begin immediately if source-sink dynamics drive CMN resource transfer (Simard et al. 1997). In contrast, if it occurs at the multi-tip scale, it seems possible that *L. laccata* could establish on *Alnus*, but once sensed as a significant nitrogen drain, would be rejected as a preferred symbiont.

8.6 Future Research Directions

As shown in this chapter, the reciprocal specificity of the *Alnus*-ECM fungus system is well established. The mechanisms responsible for creating and maintaining this specificity and how it may reinforce non-participation in CMNs, however, still require further study. Based on the hypotheses discussed above, we believe research in the following areas will be particularly important: (1) determining the full carbon budget for *Alnus* plants colonized with both *Frankia* and ECM fungi (to test the primary assumption of the rewards system hypothesis), (2) further examining the enzymatic capacities of *Alnus*-associated and non-*Alnus*-associated ECM fungal root tips (to reinforce our recent findings supporting the host metabolic hypothesis), (3) defining the signal used by *Alnus* plants to induce specific spore germination (to validate the role of partner recognition), (4) exploring the growth of more *Alnus*-associated and non-*Alnus*-associated fungi under a range of acidity and nitrogen concentrations in the same experimental setting (to better test the role of environmental filtering), and (5) assessing the competitive dynamics between *Alnus*- and non-*Alnus*-associated fungi (to assess the influence of interspecific competition). In addition, if *Frankia* bacteria play a central role in the specificity patterns observed in the *Alnus* ECM system, a similar pattern should be seen in other systems where all three symbionts are present. Members of the plant genera *Allocasuarina, Casuarina Cercopcarpus and Dryas* are known ECM fungal hosts that also associate with *Frankia* bacteria. Therefore, examining patterns of ECM fungal richness and host specificity in these host systems would be helpful in generalizing about the putatively distinctive nature of the *Alnus-Frankia*-ECM fungus tri-partite symbiosis.

Acknowledgments The authors thank T. Horton, S. Trudell, and two anonymous reviewers for a number of constructive suggestions on previous drafts of this manuscript. The authors also gratefully acknowledge the intellectual encouragement of R. Molina, who pioneered much of the work on host specificity in the *Alnus* ECM system. Financial support was provided by the National Science Foundation (DEB #: 1020735 to PGK).

References

Agren GI, Ingestad T (1987) Root: shoot ratios as a balance between nitrogen productivity and photosynthesis. Plant Cell Environ 10:579–586. doi:10.1111/1365-3040.ep11604105

Archetti M, Scheuring I, Hoffman M, Frederickson ME, Pierce NE, Yu DW (2011) Economic game theory for mutualism and cooperation. Ecol Lett 14:1300–1312. doi:10.1111/j.1461-0248.2011.01697.x

Arnebrant K, Ek H, Finlay RD, Söderström B (1993) Nitrogen translocation between *Alnus glutinosa* (L.) Gaertn. seedlings inoculated with *Frankia* sp. and *Pinus contorta* Doug. ex Loud seedlings connected by a common ectomycorrhizal mycelium. New Phytol 124:231–242. doi:10.1111/j.1469-8137.1993.tb03812.x

Avis PG, McLaughlin D, Dentinger B, Reich PB (2003) Long-term increase in nitrogen supply alters above- and below-ground ectomycorrhizal communities and increases the dominance of *Russula* spp. in a temperate oak savanna. New Phytol 160:239–253. doi:10.1046/j.1469-8137.2003.00865.x

Becerra A, Zak M, Horton TR, Micolini J (2005) Ectomycorrhizal and arbuscular mycorrhizal colonization of *Alnus acuminata* from Calilegua National Park (Argentina). Mycorrhiza 15:525–531. doi:10.1007/s00572-005-0360-7

Beiler KJ, Durall DM, Simard SW, Maxwell SA, Kretzer AM (2010) Architecture of the wood-wide web: *Rhizopogon* spp. genets link multiple Douglas-fir cohorts. New Phytol 185:543–553. doi:10.1111/j.1469-8137.2009.03069.x

Benson DR, Clawson ML (2002) Evolution of the actinorhizal plant symbioses. In: Triplett EW (ed) Prokaryotic nitrogen fixation: a model system for analysis of biological process. Horizon Scientific Press, Wymondham, pp 207–224

Benson DR, Dawson JO (2007) Recent advances in the biogeography and genecology of symbiotic *Frankia* and its host plants. Physiol Plantarum 130:318–330. doi:10.1111/j.1399-3054.2007.00934.x

Bent E, Kiekel P, Brenton R, Taylor DL (2011) Root-associated ectomycorrhizal fungi shared by various boreal forest seedlings naturally regenerating after a fire in interior Alaska and correlation of different fungi with host growth responses. Appl Environ Microbiol 77:3351–3359. doi:10.1128/AEM.02575-10

Bidartondo MI (2005) The evolutionary ecology of myco-heterotrophy. New Phytol 167:335–352. doi:10.1111/j.1469-8137.2005.01429.x

Bidartondo M, Bruns T (2005) On the origins of extreme mycorrhizal specificity in the Monotropoideae (Ericaceae): performance trade-offs during seed germination and seedling development. Mol Ecol 14:1549–1560. doi:10.1111/j.1365-294X.2005.02503.x

Bogar LM, Kennedy PG (2013) New wrinkles in an old paradigm: neighborhood effects modify the structure and specificity of *Alnus*-associated ectomycorrhizal fungal communities. FEMS Microbiol Ecol 83:767–777. doi:10.1111/1574-6941.12032

Bogar LM, Dickie IA, Kennedy PG (2015) Testing the co-invasion hypothesis: ectomycorrhizal fungal communities on *Alnus glutinosa* and *Salix fragilis* in New Zealand. Divers Distrib. 21:268–278. doi:10.1111/ddi.12304

Booth MG (2004) Mycorrhizal networks mediate overstorey-understorey competition in a temperate forest. Ecol Lett 7:538–546. doi:10.1111/j.1461-0248.2004.00605.x

Booth MG, Hoeksema JD (2010) Mycorrhizal networks counteract competitive effects of canopy trees on seedling survival. Ecology 91:2294–2302. doi:10.1890/09-1139.1

Bormann B, Cromack K, Russell W (1994) Influences of red alder on soils and long-term ecosystem productivity. In: Hibbs D, DeBell D, Tarrant R (eds) The biology and management of red alder. Oregon State University Press, Corvallis, pp 47–56

Brunner IL, Brunner F, Miller OK (1990) Ectomycorrhizal synthesis with Alaskan *Alnus tenuifolia*. Can J Bot 68:761–767. doi:10.1139/b90-101

Bruns T, Read D (2000) In vitro germination of nonphotosynthetic, myco-heterotrophic plants stimulated by fungi isolated from the adult plants. New Phytol 148:335–342. doi:10.1046/j.1469-8137.2000.00766.x
Bruns TD, Bidartondo MI, Taylor DL (2002) Host specificity in ectomycorrhizal communities: what do the exceptions tell us? Integr Comp Biol 42:352–359. doi:10.1093/icb/42.2.352
Bruns TD, Peay KG, Boynton PJ, Grubisha LC, Hynson NA, Nguyen NH, Rosenstock NP (2009) Inoculum potential of *Rhizopogon* spores increases with time over the first 4 yr of a 99-yr spore burial experiment. New Phytol 181:463–470. doi:10.1111/j.1469-8137.2008.02652.x
Bull JJ, Rice WR (1991) Distinguishing mechanisms for the evolution of co-operation. J Theor Biol 149:63–74. doi:10.1016/S0022-5193(05)80072-4
Chatarpaul L, Chakaravarty P, Subramaniam P (1989) Studies in tetrapartite symbioses. 1. Role of ectomycorrhizal and endomycorrhizal fungi and *Frankia* on the growth-performance of *Alnus incana*. Plant Soil 118:145–150. doi:10.1007/BF02232800
Cline E, Ammirati J, Edmonds R (2005) Does proximity to mature trees influence ectomycorrhizal fungus communities of Douglas-fir seedlings? New Phytol 166:993–1009. doi:10.1111/j.1469-8137.2005.01387.x
Compton J, Cole D (1998) Phosphorus cycling and soil P fractions in Douglas-fir and red alder stands. For Ecol Manage 110:101–112. doi:10.1016/S0378-1127(98)00278-3
Compton J, Cole D (2001) Fate and effects of phosphorus additions in soils under N-2-fixing red alder. Biogeochemistry 53:225–247. doi:10.1023/A:1010646709944
Courty PE, Pritsch K, Schloter M, Hartmann A, Garbaye J (2005) Activity profiling of ectomycorrhiza communities in two forest soils using multiple enzymatic tests. New Phytol 167:309–319. doi:10.1111/j.1469-8137.2005.01401.x
Courty PE, Buee M, Diedhiou AG, Frey-Klett P, Le Tacon F, Rineau F, Turpault M, Uroz S, Garbaye J (2010) The role of ectomycorrhizal communities in forest ecosystem processes: new perspectives and emerging concepts. Soil Biol Biochem 42:679–698. doi:10.1016/j.soilbio.2009.12.006
Cox F, Barsoum N, Bidartondo MI, Borja I, Lilleskov EA, Nilsson LO, Rautio P, Tubby K, Vesterdal L (2010) A leap forward in geographic scale for forest ectomycorrhizal fungi. Ann For Sci 67:200. doi:10.1051/forest/2009107
Cullings KW, Vogler D, Parker V, Finley S (2000) Ectomycorrhizal specificity patterns in a mixed *Pinus contorta* and *Picea engelmannii* forest in Yellowstone National Park. Appl Environ Microbiol 66:4988–4991. doi:10.1128/AEM.66.11.4988-4991.2000
Ekblad A, HussDanell K (1995) Nitrogen fixation by *Alnus incana* and nitrogen transfer from *A. incana* to *Pinus sylvestris* influenced by macronutrients and ectomycorrhiza. New Phytol 131:453–459. doi:10.1111/j.1469-8137.1995.tb03082.x
Ekblad A, Wallander H, Carlsson R, HussDanell K (1995) Fungal biomass in roots and extramatrical mycelium in relation to macronutrients and plant biomass of ectomycorrhizal *Pinus sylvestris* and *Alnus incana*. New Phytol 131:443–451. doi:10.1111/j.1469-8137.1995.tb03081.x
Favre J (1948) Les associations fongiques des haut-marais jurassiens et de quelques regions voisines. Mat. Flor. Crypt Suisse. 10
Finlay R, Read D (1986) The structure and function of the vegetative mycelium of ectomycorrhizal plants 2: the uptake and distribution of phosphorus by mycelial strands interconnecting host plants. New Phytol 103:157–165. doi:10.1111/j.1469-8137.1986.tb00604.x
Frank B (1888) Über die physiologische Bedeutung der Mykorriza. Ber. dt. bot. Ges 6:248
Fries N, Serck-Hanssen K, Dimberg L, Theander O (1987) Abietic acid, an activator of basidiospore germination in ectomycorrhizal species of the genus *Suillus* (Boletaceae). Exp Mycol 11:360–363
Giardina CP, Huffman S, Binkley D, Caldwell BA (1995) Alders increase soil phosphorus availability in a Douglas-fir plantation. Can J For Res 25:1652–1657. doi:10.1139/x95-179
Godbout C, Fortin JA (1983) Morphological features of synthesized ectomycorrhizae of *Alnus crispa* and *A. rugosa*. New Phytol 94:249–262. doi:10.1111/j.1469-8137.1983.tb04498.x

He X, Critchley C, Ng H, Bledsoe C (2004) Reciprocal N ((NH4+)-N-15 or (NO3-)-N-15) transfer between nonN(2)-fixing *Eucalyptus maculata* and N-2-fixing *Casuarina cunninghamiana* linked by the ectomycorrhizal fungus *Pisolithus* sp. New Phytol 163:629–640. doi:10.1111/j.1469-8137.2004.01137.x

He X, Critchley C, Ng H, Bledsoe C (2005) Nodulated N(2)-fixing *Casuarina cunninghamiana* is the sink for net N transfer from non-N(2)-fixing Eucalyptus maculata via an ectomycorrhizal fungus *Pisolithus* sp using (15)NH(4)(+) or (15)NO(3)(-) supplied as ammonium nitrate. New Phytol 167:897–912. doi:10.1111/j.1469-8137.2005.01437.x

Hibbs D, DeBell D, Tarrant R (1994) The biology and management of red alder. Oregon State University Press, Corvallis

Horak E (1963) Pilzökologische Untersuchungen in der subalpinen stufe (*Piceetum subalpinum* und *Rhodoreto-Vaccinietum*) der Rhatischen Alpen. Mitt. Schweiz. Anst. Forstl. Versuch. 39:1–112

Horton TR, Bruns TD (1998) Multiple-host fungi are the most frequent and abundant ectomycorrhizal types in a mixed stand of Douglas fir (*Pseudotsuga menziesii*) and bishop pine (*Pinus muricata*). New Phytol 139:331–339. doi:10.1046/j.1469-8137.1998.00185.x

Horton TR, Bruns TD, Parker V (1999) Ectomycorrhizal fungi associated with *Arctostaphylos* contribute to *Pseudotsuga menziesii* establishment. Can J Bot 77:93–102. doi:10.1139/b98-208

Horton TR, Molina RJ, Hood K (2005) Douglas-fir ectomycorrhizae in 40-and 400-year-old stands: mycobiont availability to late successional western hemlock. Mycorrhiza 15:393–403. doi:10.1007/s00572-004-0339-9

Huggins JA, Talbot JM, Gardes M, Kennedy PG (2014) Unlocking environmental keys to specificity: different tolerance of acidity and nitrate of *Alnus*-associated ectomycorrhizal fungi. Fungal Ecology 12(12):51–62

Hung L, Trappe J (1983) Growth variation between and within species of ectomycorrhizal fungi in response to pH in vitro growth. Mycologia 75:234–241

Ishida TA, Nara K, Hogetsu T (2007) Host effects on ectomycorrhizal fungal communities: insight from eight host species in mixed conifer-broadleaf forests. New Phytol 174:430–440. doi:10.1111/j.1469-8137.2007.02016.x

Jargeat P, Chaumeton J-P, Navaud O, Vizzini A, Gryta H (2014) The *Paxillus involutus* (Boletales, Paxillaceae) complex in Europe: Genetic diversity and morphological description of the new species *Paxillus cuprinus*, typification of *P. involutus* s.s., and synthesis of species boundaries. Fungal Biol 118:12–31

Jha D, Sharma G, Mishra RR (1993) Mineral nutrition in the tripartite interaction between Frankia, Glomus and Alnus at different soil phosphorus regimes. New Phytol 123:307–311. doi:10.1111/j.1469-8137.1993.tb03740.x

Johnson D, Gilbert L (2015) Interplant signaling through hyphal networks. New Phytol 205:1444–1453

Jones MD, Phillips LA, Treu R, Ward V, Berch SM (2012) Functional responses of ectomycorrhizal fungal communities to long-term fertilization of lodgepole pine (*Pinus contorta* Dougl. Ex Loud. Var. latifola Engelm.) stands in central British Columbia. Appl Soil Ecol 60:29–40. doi:10.1016/j.apsoil.2012.01.010

Kennedy PG, Hill LT (2010) A molecular and phylogenetic analysis of the structure and specificity of *Alnus rubra* ectomycorrhizal assemblages. Fungal Ecol 3:195–204. doi:10.1016/j.funeco.2009.08.005

Kennedy PG, Izzo AD, Bruns TD (2003) There is high potential for the formation of common mycorrhizal networks between understorey and canopy trees in a mixed evergreen forest. J Ecol 91:1071–1080. doi:10.1046/j.1365-2745.2003.00829.x

Kennedy PG, Garibay-Orijel R, Higgins LM, Angeles-Arguiz R (2011a) Ectomycorrhizal fungi in Mexican *Alnus* forests support the host co-migration hypothesis and continental-scale patterns in phylogeography. Mycorrhiza 21:559–568. doi:10.1007/s00572-011-0366-2

Kennedy PG, Higgins LM, Rogers RH, Weber MG (2011b) Colonization-competition tradeoffs as a mechanism driving successional dynamics in ectomycorrhizal fungal communities. PLoS ONE 6:e25126. doi:10.1371/journal.pone.0025126

Kiers ET, Denison RF (2008) Sanctions, cooperation, and the stability of plant-rhizosphere mutualisms. Ann Rev Ecol Evol S 39:215–236. doi:10.1146/annurev.ecolsys.39.110707.173423

Kiers E, Rousseau R, West S, Denison R (2003) Host sanctions and the legume-rhizobium mutualism. Nature 425:78–81. doi:10.1038/nature01931

Koike T (1990) Autumn coloring, photosynthetic performance and leaf development of deciduous broad-leaved trees in relation to forest succession. Tree Physiol 7:21–32

Koo CD, Molina R, Miller S (1995) Effects of light and inoculation of *Frankia* and *Alpova diplophoeus* on the tripartite symbioses development in *Alnus rubra* Bong. seedlings. J Korean Forest Soc 84:306–318

Koo CD, Molina R, Miller S, Li CY (1996) Effects of Nitrogen and Phosphorus fertilization on ectomycorrhiza development, N-fixation and growth of red alder seedlings. J Korean Forest Soc 85:96–106

Kropp BR, Trappe JM (1982) Ectomycorrhizal fungi of *Tsuga heterophylla*. Mycologia 74:479–488

Lilleskov EA, Fahey T, Horton TR, Lovett G (2002) Belowground ectomycorrhizal fungal community change over a nitrogen deposition gradient in Alaska. Ecology 83:104–115. doi:10.2307/2680124

Malajczuk N, Molina R, Trappe J (1982) Ectomycorrhiza formation in Eucalyptus. I. Pure culture synthesis, host specificity and mycorrhizal compatibility with *Pinus radiata*. New Phytol 91:467–482

Massicotte HB, Molina RJ, Luoma DL, Smith JE (1994) Biology of the ectomycorrhizal genus, *Rhizopogon* II. Patterns of host-fungus specificity following spore inoculation of diverse hosts grown in monoculture and dual culture. New Phytol 126:677–690. doi:10.1111/j.1469-8137.1994.tb02962.x

Massicotte HB, Molina RJ, Tackaberry L, Smith JE, Amaranthus MP (1999) Diversity and host specificity of ectomycorrhizal fungi retrieved from three adjacent forest sites by five host species. Can J Bot 77:1053–1076. doi:10.1139/b99-115

Masui K (1926) A study of the ectotrophic mycorrhiza of Alnus. Mem. Coll Sci Kyoto Imp Univ B 2:189–209

Mejstrik V, Benecke U (1969) The ectotrophic mycorrhizas of *Alnus viridis* (Chaix.) D.C. and their significance in respect to phosphorus uptake. New Phytol 68:141–149. doi:10.1111/j.1469-8137.1969.tb06427.x

Miller S, Koo CD, Molina R (1991) Characterization of red alder Ectomycorrhizae—a preface to monitoring belowground ecological responses. Can J Bot 69:516–531. doi:10.1139/b91-071

Miller S, Koo CD, Molina R (1992) Early colonization of red alder and Douglas fir by ectomycorrhizal fungi and *Frankia* in soils from the Oregon coast range. Mycorrhiza 2:53–61. doi:10.1007/BF00203250

Miller S, Torres P, McClean TM (1994) Persistence of basidiospores and sclerotia of ectomycorrhizal fungi and *Morchella* in soil. Mycologia 86:89–95. doi:10.2307/3760722

Molina R (1979) Pure culture synthesis and host specificity of red alder mycorrhizae. Can J Bot 57:1223–1228

Molina R (1981) Ectomycorrhizal specificity in the genus *Alnus*. Can J Bot 59:325–334

Molina RJ, Trappe JM (1982) Lack of mycorrhizal specificity by the ericaceous hosts *Arbutus menziesii* and *Arctostaphylos uva-ursi*. New Phytol 90:495–509. doi:10.1111/j.1469-8137.1982.tb04482.x

Molina RJ, Massicotte HB, Trappe JM (1992) Specificity phenomena in mycorrhizal symbioses: community-ecological consequences and practical applications. In: Allen MF (ed) Mycorrhizal functioning, an integrative plant-fungal process. Chapman and Hall, New York, pp 357–423

Molina R, Myrold D, Li CY (1994) Root symbioses of red alder: technological opportunities for enhanced regeneration and soil improvement. In: Hibbs D, DeBell D, Tarrant R (eds) The biology and management of red alder. Oregon State University Press, Corvallis, pp 23–46

Moreau P, Peintner U, Gardes M (2006) Phylogeny of the ectomycorrhizal mushroom genus *Alnicola* (Basidiomycota, Cortinariaceae) based on rDNA sequences with special emphasis on host specificity and morphological characters. Mol Phylogenet Evol 38:794–807. doi:10.1016/j.ympev.2005.10.008

Nara K, Hogetsu T (2004) Ectomycorrhizal fungi on established shrubs facilitate subsequent seedling establishment of successional plant species. Ecology 85:1700–1707. doi:10.1890/03-0373

Neal J, Trappe J, Lu K, Bollen W (1968) Some ectotrophic mycorrhizae of *Alnus rubra*. In: Trappe J, Franklin J, Tarrant R, Hansen G (eds) Biology of Alder. USDA Forest Service, Portland, pp 179–184

Newman EI (1988) Mycorrhizal links between plants—their functioning and ecological significance. Adv Ecol Res 18:243–270. doi:10.1016/S0065-2504(08)60182-8

Pritsch K, Garbaye J (2011) Enzyme secretion by ECM fungi and exploitation of mineral nutrients from soil organic matter. Ann For Sci 68:25–32. doi:10.1007/s13595-010-0004-8

Pritsch K, Munch J, Buscot F (1997) Identification and differentiation of mycorrhizal isolates of black alder by sequence analysis of the ITS region. Mycorrhiza 10:87–93. doi:10.1007/s005720000063

Pritsch K, Courty PE, Churin J, Cloutier-Hurteau B, Ali MA, Damon C, Duchemin M, Egli S, Ernst J, Fraissinet-Tachet L, Kuhar F, Legname E, Marmeisse R, Mueller A, Nikolova P, Peter M, Plassard C, Richard F, Schloter M, Selosse M, Franc A, Garbaye J (2011) Optimized assay and storage conditions for enzyme activity profiling of ectomycorrhizae. Mycorrhiza 21:589–600. doi:10.1007/s00572-011-0364-4

Richard F, Selosse M, Gardes M (2009) Facilitated establishment of *Quercus ilex* in shrub-dominated communities within a Mediterranean ecosystem: do mycorrhizal partners matter? FEMS Microbiol Ecol 68:14–24. doi:10.1111/j.1574-6941.2009.00646.x

Rochet J, Moreau P, Manzi S, Gardes M (2011) Comparative phylogenies and host specialization in the alder ectomycorrhizal fungi *Alnicola*, *Alpova* and *Lactarius* (Basidiomycota) in Europe. BMC Evol Biol 11:40. doi:10.1186/1471-2148-11-40

Simard SW, Perry DA, Jones MD, Myrold DD, Durall DM, Molina RJ (1997) Net transfer of carbon between ectomycorrhizal tree species in the field. Nature 388:579–582. doi:10.1038/41557

Simard SW, Beiler KJ, Bingham MA, Deslippe JR, Philip LJ, Teste FP (2012) Mycorrhizal networks: mechanisms, ecology, and modelling. Fungal Biol Rev 26:39–60

Singer R (1950) New and interesting Basidiomycetes VIII. Sydowia 4:130–157

Smith SE, Read DJ (2008) Mycorrhizal symbiosis, 3rd edn. Elsevier, New York

Smith ME, Douhan GW, Fremier AK, Rizzo DM (2009) Are true multihost fungi the exception or the rule? Dominant ectomycorrhizal fungi on *Pinus sabiniana* differ from those on co-occurring *Quercus* species. New Phytol 182:295–299. doi:10.1111/j.1469-8137.2009.02801.x

Smith ME, Henkel TW, Aime MC, Fremier AK, Vilgalys R (2011) Ectomycorrhizal fungal diversity and community structure on three co-occurring leguminous canopy tree species in a Neotropical rainforest. New Phytol 192:699–712. doi:10.1111/j.1469-8137.2011.03844.x

Song YY, Zeng RS, Xu JF, Li J, Shen X, Yihdego WG (2010) Interplant communication of tomato plants through underground common mycorrhizal networks. PLoS ONE 5:e13324

Suvi T, Tedersoo L, Abarenkov K, Beaver K, Gerlach J, Koljalg U (2010) Mycorrhizal symbionts of *Pisonia grandis* and *P. sechellarum* in Seychelles: identification of mycorrhizal fungi and description of new Tomentella species. Mycologia 102:522–533. doi:10.3852/09-147

Tedersoo L, Jairus T, Horton BM, Abarenkov K, Suvi T, Saar I, Koljalg U (2008) Strong host preference of ectomycorrhizal fungi in a Tasmanian wet sclerophyll forest as revealed by DNA barcoding and taxon-specific primers. New Phytol 180:479–490. doi:10.1111/j.1469-8137.2008.02561.x

Tedersoo L, Suvi T, Jairus T, Ostonen I, Polme S (2009) Revisiting ectomycorrhizal fungi of the genus *Alnus*: differential host specificity, diversity and determinants of the fungal community. New Phytol 182:727–735. doi:10.1111/j.1469-8137.2009.02792.x

Teste FP, Simard SW, Durall DM, Guy RD, Berch SM (2010) Net carbon transfer between *Pseudotsuga menziesii var. glauca* seedlings in the field is influenced by soil disturbance. J Ecol 98:429–439. doi:10.1111/j.1365-2745.2009.01624.x

Tjepkema JD, Schwintzer CR, Benson DR (1986) Physiology of actinorhizal nodules. In: Briggs WR (ed) Ann Rev Plant Physiol 37:209–232. doi:10.1146/annurev.pp.37.060186.001233

Townsend CR, Begon M, Harper JL (2008) Essentials of ecology. Blackwell, Oxford

Trappe JM (1964) Mycorrhizal hosts and distribution of *Cenococcum graniforme*. Lloydia 27:100–106

Trappe JM (1975) A revision of the genus *Alpova* with notes on *Rhizopogon* and Melanogastraceae. Nova Hedgwigia 51:279–309

Trappe JM, Fogel R (1977) Ecosystematic functions of mycorrhizae. Colorado State Univ Range Sci Dept Sci Ser 26:205–214

Trudell SA, Edmonds RL (2004) Macrofungus communities correlate with moisture and nitrogen abundance in two old-growth forests, Olympic National Park, Washington. Can J Bot 82:781–800. doi:10.1139/B04-057

Twieg BD, Durall DM, Simard SW (2007) Ectomycorrhizal fungal succession in mixed temperate forests. New Phytol 176:437–447. doi:10.1111/j.1469-8137.2007.02173.x

Uliassi DD, Ruess RW (2002) Limitations to symbiotic nitrogen fixation in primary succession on the Tanana River floodplain. Ecology 83:88–103. doi:10.2307/2680123

van der Heijden MGA, Horton TR (2009) Socialism in soil? The importance of mycorrhizal fungal networks for facilitation in natural ecosystems. J Ecol 97:1139–1150. doi:10.1111/j.1365-2745.2009.01570.x

Walker JKM, Cohen H, Higgins L, Kennedy PG (2014) Testing the link between community structure and function for ectomycorrhizal fungi involved in a global tri-partite symbiosis. New Phytol 102:287–296

Yamanaka T, Li CY, Bormann BT, Okabe H (2003) Tripartite associations in an alder: effects of *Frankia* and *Alpova diplophloeus* on the growth, nitrogen fixation and mineral acquisition of *Alnus tenuifolia*. Plant Soil 254:179–186. doi:10.1023/A:1024938712822

Yarwood SA, Bottomley PJ, Myrold DD (2010) Soil microbial communities associated with Douglas-fir and red alder stands at high- and low-productivity forest sites in Oregon, USA. Microb Ecol 60:606–617. doi:10.1007/s00248-010-9675-9

Zou X, Binkley D, Caldwell BA (1995) Effects of dinitrogen-fixing trees on phosphorus biogeochemical cycling in contrasting forests. Soil Sci Soc Am J 59:1452–1458

Chapter 9
Experimentally Testing Effects of Mycorrhizal Networks on Plant-Plant Interactions and Distinguishing Among Mechanisms

Jason D. Hoeksema

Abstract Plants of the same and different species are often linked by common mycorrhizal networks (CMNs), and there is substantial disagreement in the literature about whether these linkages have important effects on plant-plant interactions, beyond simply providing mycorrhizal inoculum. Here, I attempt to reconcile opposing viewpoints by reviewing available evidence for three distinct mechanisms by which CMNs can affect plant-plant interactions. I also analyze the details of manipulative field experiments that have been conducted to test CMN effects on plant-plant interactions, and make recommendations for the kinds of future studies that will be most useful in moving forward. I argue that few experiments have unequivocally tested whether CMNs have unique effects on plant-plant interactions, and that these experiments have largely been ignored in favor of debates about the magnitude of resource flows (especially carbon) from plant to plant through CMNs. I suggest that progress on the debate will only be made through more thorough testing of alternative mechanisms besides plant-to-plant carbon flow, especially coupled with experimental manipulations of CMNs to test for consequences on specific aspects of plant community ecological processes.

Keywords Experimental design · Mycorrhizal fungi · Mycorrhizal network · Plant ecology · Plant competition · Plant community

---

J.D. Hoeksema (✉)
Department of Biology, University of Mississippi, PO Box 1848, University MS 38677, USA
e-mail: hoeksema@olemiss.edu

T.R. Horton (ed.), *Mycorrhizal Networks*,
Ecological Studies 224, DOI 10.1007/978-94-017-7395-9_9

9.1 Introduction

Many mycorrhizal ecologists agree that common mycorrhizal networks (CMNs), i.e., physical linkages among plant individuals via the mycelia of mycorrhizal fungi, are likely common in nature (Newman 1988; Molina et al. 1992; Francis and Read 1994; Leake et al. 2004; Simard and Durall 2004; Simard et al. 2012; Molina and Horton Chap. 1, this volume; Giovannetti et al. Chap. 2, this volume). The observation that species of arbuscular mycorrhizal (AM) and ectomycorrhizal (EM) fungi are often compatible with multiple host plant species, coupled with the ability of genetically compatible hyphae to anastomose (see Giovannetti et al. Chap. 2, this volume), suggest that CMNs are probably ubiquitous, although confirmation of this conjecture will require direct evidence for these linkages in the field, including careful population genetic studies (Leake et al. 2004; Simard and Durall 2004; Selosse et al. 2006). Hundreds of achlorophyllous plants likely depend on CMN connections with green plants for 100 % of their carbon supply (Leake et al. 2004; Bidartondo 2005), and such 'mycoheterotrophy' also exists to various degrees in some green orchid species, especially at the seedling stage (e.g., Julou et al. 2005; Zimmer et al. 2008). What is less clear is the degree to which these CMN linkages have unique consequences for the ecology of typical green plants, and especially whether mechanisms and consequences (such as community composition) of plant-plant interactions are commonly altered by these linkages. Are physical mycelial linkages between plants per se, beyond just the provisioning of mycorrhizal fungi, generally important for plant ecology?

A variety of approaches, ranging from isotope tracer studies to field manipulations of CMNs, have been used to directly or indirectly address this question, and results of these efforts have generated significant debate. In reviewing some aspects of the evidence, some authors have concluded that CMNs can have "profound effects on plant communities" (Selosse et al. 2006), while others have argued that there is nothing unique about these physical linkages that separates them from other kinds of mutualisms in which plants engage and thus the term "common mycorrhizal networks" is misleading (Bever et al. 2010). Here, I attempt to reconcile these opposing viewpoints and make recommendations for the kinds of future studies that will be most useful in moving forward. I argue that few experiments have unequivocally tested whether CMNs have unique effects on plant-plant interactions, and that these experiments have largely been ignored in favor of debates about the magnitude of resource flows (especially carbon) from plant to plant through CMNs. I suggest that progress on the debate will only be made through more thorough testing of alternative mechanisms besides plant-to-plant carbon flow, especially coupled with experimental manipulations of CMNs to test for consequences on specific aspects of plant community ecological processes. This review is not intended as a comprehensive survey of studies on CMNs; rather, I attempt a conceptual analysis that may be useful as a roadmap for design and interpretation of future studies.

9.1.1 Mechanisms Are Central to the Debate

At least three distinct mechanisms have been hypothesized by which CMNs may affect plant-plant interactions:

Mechanism 1: Flow of resources from one plant to another through the CMN. Under this mechanism, one plant may benefit if it receives a net flow through the CMN of limiting resources including carbon, nitrogen, phosphorus, or water.
Mechanism 2: Unequal contributions of carbon to the CMN by different plants. Under this mechanism, some plants (e.g., recruiting seedlings) may receive the benefits of association with a CMN while contributing a less than proportional share of carbon to the build the CMN.
Mechanism 3: Unequal distribution of a resource by the CMN to different plants. Under this mechanism, a common resource (regardless of where it was obtained by the fungus) may be distributed unequally to different plants.

Newman (1988), in a seminal review that stimulated a great deal of work on the function and significance of CMNs, highlighted five potential "profound implications" of CMNs for the functioning of ecosystems. The first of these implications was that seedlings might be able to join a pre-existing hyphal network and benefit from it at an early stage, which falls under Mechanism 2 above. His second and fourth implications were that organic and mineral nutrients, respectively, could flow from plant to plant and alter the performance of the receiving plant or the balance of plant-plant interactions. These phenomena fall under Mechanism 1 above. Newman's third implication was that plant competition could be altered if competing plants are receiving nutrients from a commonly shared fungal network, rather than taking up nutrients independently, which falls under Mechanism 3 above. Newman's fifth implication was that nutrients could flow from dying plants through CMNs directly to living plants, which has interesting implications for ecosystem cycling of nutrients but does not relate directly to our discussion here on the implications for plant ecology.

Much debate over the importance of CMNs for plant ecology has centered around one specific version of Mechanism 1: net flow of carbon from one plant to another through a CMN (see Simard et al. Chap. 5, this volume). After reviewing early studies that showed radio-labeled carbon could potentially flow both directions from plant to plant through a CMN, Newman (1988) argued that key next steps would be to test whether flow was more significant in one direction than another (i.e., net flow) and to quantify how large is the carbon gain by the receiver plant compared to its gain from photosynthesis. When net flow of carbon through a CMN was first demonstrated in an ectomycorrhizal system, the resulting paper appeared in a high profile international journal (Simard et al. 1997a) and was greeted with significant enthusiasm and discussion. Subsequently, however, there has been significant debate about whether the amounts of carbon flowing through a CMN are likely to be ecologically meaningful (i.e., to have consequences for individual plant growth, populations, or communities), whether the data distinguish between carbon flow through the CMN versus other pathways (such as respiration from CMN fungi and subsequent fixation by plants linked to the CMN), and why some other studies have failed to find net transfer of carbon (Fitter et al. 1999; Robinson and Fitter 1999; Pfeffer et al. 2004; Bever and Schultz 2005; Whitfield 2007; Bever et al. 2010 but see, e.g., Song et al. 2015). Indeed, it has been argued that since the evidence for net carbon flow from plant to plant is so mixed, CMNs may not have unique consequences for general plant ecology that require recognition of the physical linkages among green plants (Bever et al. 2010).

9.1.2 Density Effects and Plant-Soil Feedbacks: General Phenomena that Make CMNs Irrelevant?

Specifically, Bever et al. (2010) argued that the dynamics of plant interactions with mycorrhizal fungi (and other above- and belowground mutualists such as animal pollinators and n-fixing bacteria) can be effectively explained by a traditional population dynamics framework, regardless of whether CMNs form physical linkages among plants (see also Bever and Schultz 2005). Under this view, Mechanisms 2 and 3 may influence plant ecology, but the physical connection among plants through a CMN is not important conceptually. For example, one of two plant species may have a disproportionately positive effect on the density of a particular shared mycorrhizal fungal species in the soil (Mechanism 2), which then has a disproportionately positive effect on the second plant species (Mechanism 3). Although the physical linkage provides the means by which one plant may disproportionately contribute to the density of shared mycorrhizal fungi, or by which a shared fungus may contribute disproportionately to one particular host plant (e.g., Fellbaum et al. 2014), such disproportionate effects on shared interactors is not unique to plant-plant interactions mediated by CMNs. The sharing of mycorrhizal fungi between two plants is conceptually no different than, say, the sharing of the same pollinator by two plants. Species of plants and mycorrhizal fungi simply interact and affect each other's densities, and the strength of indirect interactions among plants and consequent community composition of both plants and fungi depend on the relative strengths of specific direct interactions between particular plant and fungal species (Fig. 9.1). This view has been synthesized more generally, taking into account other soil organisms besides mycorrhizal fungi, as the plant-soil feedbacks approach (Bever 2003).

There are significant advantages to this perspective. Perhaps most importantly, it allows effects of mycorrhizal fungi on plant ecology to be modeled, both conceptually and mathematically, in a very general existing theoretical framework that also applies to other kinds of interactions. It allows us to avoid the messy mixture of concepts (density/population effects and physical engineering/physiological effects) involved when we must consider that mycorrhizal fungi may act as physical conduits for resources among plants. Moreover, plant-soil feedback phenomena can be measured experimentally with straightforward protocols.

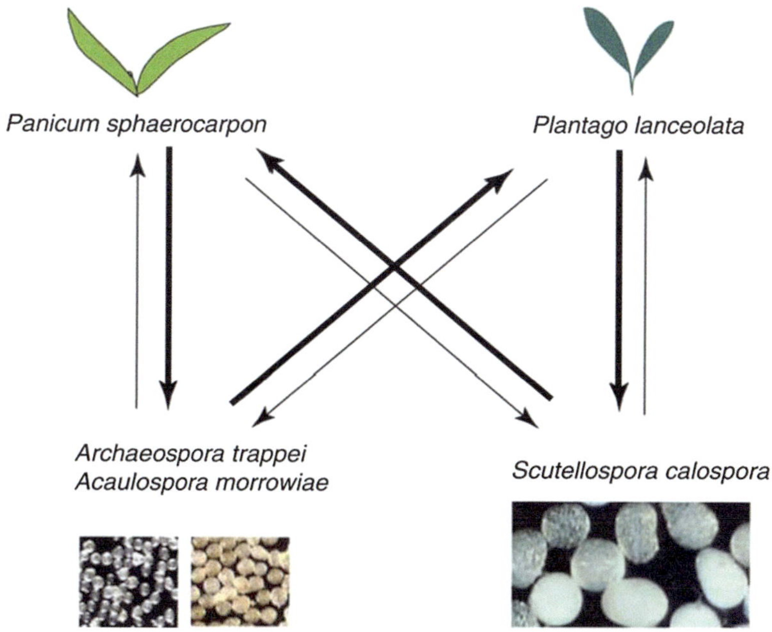

Fig. 9.1 Net pairwise negative feedback between *Panicum sphaerocarpon* and *Plantago lanceolata* generated by changes in composition of AM fungi (Bever 2002). The thickness of arrows represents the relative strengths of benefit between individual species of plants and AM fungi. *Scutellospora calospora* has high fitness with *Plantago*, but *Plantago* does not grow well with *Sc. calospora*. Rather, *Plantago* has highest growth rates in association with AM fungi, *Archaeospora trappei* and *Acaulospora morrowiae*, which themselves have high fitness in association with *Panicum*. The asymmetric fitness relationships generate negative feedback which can contribute to coexistence of these competing plant species. Modified from Fig. 9.2d in Bever et al. (2010), and used with permission from Elsevier

I suggest, however, that available evidence does not support abandoning the idea that CMNs and the physical connections among green plants have unique general effects on plant ecology. Although the plant-soil feedbacks approach can usually account for Mechanisms 2 and 3, it does not capture Mechanism 1. Even if we accept the argument that net carbon flows between green plants through CMNs are usually not significant enough to be ecologically important, and we set aside the examples of achlorophyllous plants and green orchids, compared to carbon flow, relatively little attention has been paid to plant-to-plant flow of other plant resources, such as nitrogen (N), phosphorus (P), and water (Simard et al. 2012; see Simard et al. Chap. 5, this volume); or of non-resource molecules such as defense or stress signaling compounds (e.g., Song et al. 2014, 2015) and allelopathic chemicals (Achatz and Rillig 2014; see Jakobsen and Hammer Chap. 4, this volume). If these flows are important, then CMNs may significantly affect plant ecology via Mechanism 1, and those effects cannot be captured in the population dynamics of mycorrhizal mutualists. Moreover, experiments manipulating CMNs and measuring plant ecological outcomes (such as the strength of plant-plant interactions) have found evidence that CMNs have unique effects on these outcomes in some contexts, and Mechanism 1 has found support or has not been ruled out in those cases. While such experiments are still relatively few, their importance, design, and strengths and weaknesses have been given short shrift relative to studies of resource flow in the debate about CMNs in plant ecology.

In the remainder of this chapter, I have two objectives. First, I will summarize the potential importance for plant ecology of nitrogen, phosphorus, and water flows from plant to plant through CMNs. Second, I will analyze the details and outcomes of experiments designed to test effects of CMNs on plant-plant interactions, focusing especially on manipulative field experiments, and including discussion of how such experiments can help to clarify the role of CMNs in plant ecology.

9.1.3 Nitrogen, Phosphorus, and Water Flow Through CMNs

Nitrogen. N flow from plant to plant through both AM and EM CMNs can sometimes be substantial (reviewed by He et al. 2009; see Simard et al. Chap. 5, this volume). For example, a pair of laboratory microcosm experiments (He et al. 2004, 2005) examined the potential for net N transfer through CMNs between two ectomycorrhizal plants, *Eucalyptus maculata* and *Casuarina cunninghamiana*. The authors grew these plants in compartments separated by 37 μm mesh, which allowed hyphal growth between plants in adjacent compartments, and a 5 mm air gap, which reduced diffusion of water and N between the compartments. Prior studies had used this approach to study one-way N transfer through CMNs (e.g., Bethlenfalvay et al. 1991), but He et al. moved the field forward by applying these methods to test for bi-directional and net flow between two plant species. They also included high water-holding capacity crystals in the soil medium to reduce water diffusion, and included non-mycorrhizal control treatments. As *Casuarina* engages in a nitrogen-fixing symbiosis with *Frankia* bacteria living in root nodules, non-nodulated treatments were also included to test for the effects of nodulation on N transfer. These experiments demonstrated the potential for substantial amounts of N transfer from plant to plant through CMNs, with up to 39 % of ^{15}N-labeled ammonium being transferred from mycorrhizal *Eucalyptus* to *Casuarina* and only 10 % in the reverse direction (He et al. 2005). In the same subset of plants, approximately 30 % of *Casuarina*'s N was derived from transfer through the CMN.

In a field study of an AM system, Moyer-Henry et al. (2006) used ^{15}N natural abundance studies to show that two different AM weed species received up to 80 % of their N via transfers from leguminous crops. Two different non-mycorrhizal weeds in the same experiment received negligible N transfer from the crops, suggesting that the main pathway of N transfer from the crops to the weeds was via AM hyphal networks. Not all studies have found such substantial amounts of N transferred through CMNs (e.g., Shen and Chu 2004), but these examples show plant-to-plant N transfer in amounts that certainly have the potential to be ecologically significant in N-limited environments, in support of Mechanism 1 above.

Phosphorus. Ecologically meaningful quantities of P have now also been demonstrated to move from plant to plant, with support for the hypothesis that CMNs provide a primary pathway for such flow (see Simard et al. Chap. 5, this volume). For example, Wilson et al. (2006) used a creative combination of CMN-manipulation treatments (No roots/+CMN vs. No roots/No CMN) and mycorrhizal suppression treatments (Benomyl fungicide or no fungicide), along with ^{32}P-labeling, to estimate P flows through multiple pathways between two AM plants, Indian Grass (*Sorghastrum nutans*) and the perennial forb Louisiana Sagewort (*Artemisia ludoviciana*). They found that Indian Grass received more than 50 % of its P via interplant transfers through AM fungal CMNs, whereas Louisiana Sagewort received less than 20 % of its P via CMN transfer, and that transfer through a soil pathway and diffusion through the mesh barrier were negligible. This experimental design and associated results build on earlier studies that used ^{32}P-labeling to show significant P flow from plant to plant, mediated by CMNs (e.g. , Martins and Read 1996; Tuffen et al. 2002). As with N transfer, some studies have shown P transfer among plants to be negligible (e.g., Ikram et al. 1994), but the results highlighted here show that in some systems, P transfer among plants mediated by CMNs can be substantial.

Water. Plants with deep root systems can bring deep water to the soil surface at night and distribute it among fine roots throughout shallow soil layers, through the physiological process of hydraulic lift (Caldwell et al. 1998; Simard et al. Chap. 5, this volume). This process makes water more available during the daytime to shallow roots of the plant conducting hydraulic lift and also to roots belonging to neighboring plants, potentially allowing facilitation among plant individuals of the same or different species (e.g., Dawson 1993). Studies have now shown that plants can transfer hydraulically lifted water directly to the mycelia of their mycorrhizal fungi (Querejeta et al. 2003), and that mycelia of mycorrhizal fungi can then redistribute hydraulically lifted water throughout their masses to multiple plants connected to the same CMN (Egerton-Warburton et al. 2007; Warren et al. 2008; Allen 2009). Lilleskov et al. (2009) used oxygen stable isotope analysis to show that EM sporocarps receive and transpire substantial amounts of hydraulically lifted water, either from host plant roots or via direct mycelial transport from deep water sources. Egerton-Warburton et al. (2007) used fluorescent dye and isotopic tracers, coupled with manipulations of CMNs independent of plant roots in microcosms, to show that CMNs of both AM and EM fungi could distribute hydraulically lifted water from *Quercus agrifolia* seedlings to conspecifics and to different plant species including *Salvia mellifera*, an AM plant. Warren et al. (2008) used dye tracers in a field study to show that CMNs of EM fungi can provide a conduit for hydraulic redistribution of water from adult *Pinus ponderosa* trees to nearby seedlings. Although the narrow diameter of fungal hyphae may restrict mass water flow through CMNs in some cases, the rhizomorphs of some EM fungi may be particularly well suited for this function, especially those that are hydrophobic and possess large, vessel-like internal hyphae, such as *Suillus* and *Rhizopogon* (Lilleskov et al. 2009; Agerer 2001). Thus, it is possible that mass flow of water may be more substantial through some EM fungal CMNs than through AM fungal CMNs.

These studies on N, P, and water transfer typically cannot rule out the possibility of resource leakage from roots or mycorrhizal hyphae and immediate reabsorption by the CMN, but this caveat does not alter the implications of the results, as these studies do show clearly that CMNs are significantly involved in resource transfer among plants (Wilson et al. 2006). Significant redistribution of water within a CMN, for example, regardless of whether or not it leaks out and is reabsorbed, homogenizes water availability among the shallow root systems of large and small individual plants. When water is a limiting resource, this process effectively creates facilitation of small, shallow-rooted plants by large, deep-rooted plants, mediated by CMNs. Similarly, experiments on N and P flow show clearly that CMNs are mediating significant resource transfer among plants, and additional studies demonstrate an important role for chemical signaling or chemically-mediated interactions among plants (Achatz and Rillig 2014; Song et al. 2014, 2015). These physiological mechanisms are not adequately captured by a simple consideration of the population dynamics of plants and fungi, since facilitation of one plant by another via this mechanism does not require density responses of the fungi to their host plants. Rather, these results suggest the potential for a unique influence of CMNs on plant-plant interactions via Mechanism 1.

9.1.4 Experimental Tests of CMN Effects on Plant-Plant Interactions

Testing whether CMNs affect plant ecological outcomes requires going beyond estimates of resource flows through CMNs. Measurements of plant-plant interaction strengths and/or community level consequences, such as composition and diversity, must be made and compared among treatments differing in the presence of CMN linkages, ideally under field conditions (a CMN field manipulation experiment). Moreover, claims that specific amounts of resource flow are or are not ecologically significant are difficult to justify when we have such a paucity of actual tests for CMN effects (and any associated resource flow) on the plant ecological outcomes of interest. Relatively few such experiments have been published, likely reflecting the difficulty of creating experimental treatments that differ in the presence of CMNs but not in other confounding variables. Despite these challenges, several manipulative field experiments now provide evidence that CMNs formed by mycorrhizal fungi have unique effects on aspects of green plant community ecology. The studies all have unique strengths and weaknesses, an analysis of which may help to inspire and strengthen a new generation of studies that push the field further forward. Below is such an analysis, focusing first on manipulative field experiments before discussing the implications of results from these and other types of studies.

In typical CMN field manipulation experiments, the performance of a target mycorrhizal plant (typically a seedling) is compared among several treatments. In most cases, these experiments have been designed with the possibility in mind that neighbor plant roots may have different (specifically, more negative) direct effects on seedlings than linkages with neighbor plant CMNs, and have utilized at least three treatments to estimate those two separate effects: No roots/+CMN, No roots/No CMN, and +roots/+CMN (Fig. 9.2).

Comparison of target seedling performance in a No roots/+CMN treatment versus a No roots/No CMN treatment (e.g., a trenching treatment) estimates the CMN effect of neighbor plants on a target seedling, in the absence of neighbor roots (Fig. 9.2c vs. d or b vs. d). As detailed below, a No roots/No CMN treatment may (Fig. 9.2c) or may not (Fig. 9.2b) include a control for the effect of the container used to exclude roots in the No roots/+CMN treatment. Comparison between a No roots/No CMN treatment and a +roots/+CMN treatments estimates the net effect of roots and CMNs together (Fig. 9.2a vs. b). By subtraction of these two treatment effects, the effect of roots alone can be estimated. Sometimes, a fourth treatment has been included, +roots/No CMN, which can be used to estimate the effect of CMNs in the presence of roots (when compared to the +roots/+CMN treatment) and to directly estimate the effect of roots alone (when compared to the No roots/No CMN treatment). It is important to note that in all treatments in these studies, both the target plant and the neighbor plant(s) are mycorrhizal. What is being manipulated is whether or not the mycorrhizal fungi of the target and neighbor plants are connected

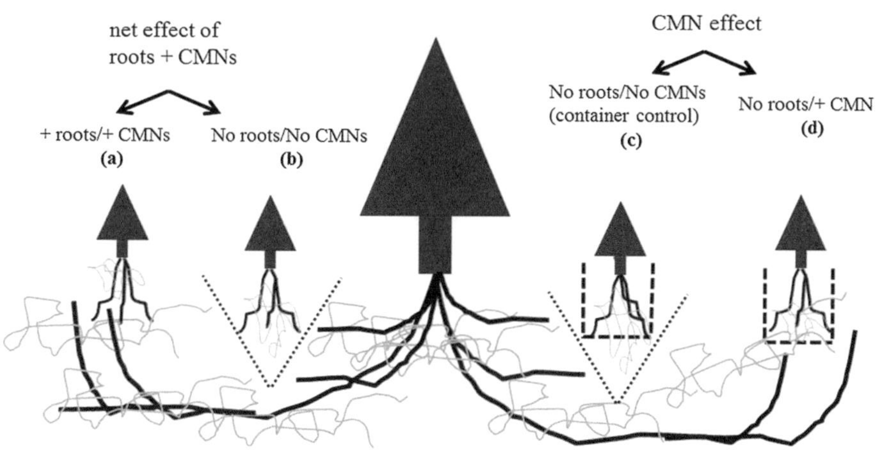

Fig. 9.2 Potential treatments in a field experiment testing separate effects on target plants of common mycorrhizal networks (CMNs) and roots associated with neighboring plants. *Solid black lines* are root, thin *gray lines* are mycorrhizal fungal mycelium, *dotted lines* are trenching treatments or solid barriers, and *dashed black lines* are containers (e.g., bags or PVC cylinders) with mesh openings large enough to allow penetration by fungal mycelium but small enough to exclude roots. **a** +roots/+CMNs. **b** No roots/No CMN. **c** No roots/No CMN (container control). **d** No roots/+CMN

in a CMN and whether or not target and neighbor plant root systems are overlapping. Below, I discuss how these treatments have been implemented, including potential strengths and weaknesses of various approaches. Table 9.1 summarizes the approaches taken in published field studies to date (of which I am aware). Note that all such field experiments so far have studied EM (not AM) fungal CMNs.

No roots/+CMN treatment. This treatment (Fig. 9.2d), when compared with a No Roots/No CMN treatment (Fig. 9.2c vs. d or b vs. d), allows estimation of the effect of neighbor plant CMNs on a target plant, in the absence of neighbor plant roots. Neighbor plant roots may have a different effect on the target plant, so it is ideal to separately estimate root effects and CMN effects. Published studies so far have created No roots/+CMN target plant plots, which lack neighbor plant roots but contain CMNs associated with neighbor plants, using mesh barriers with pore sizes between 20 and 250 μm in cylindrical or rectangular forms ranging in diameter from 10 to 20 cm and in depth from 18 to 40 cm. This pore size range, depending on the plant community, excludes most neighbor roots but allows colonization by fungi. Teste and Simard (2008) and Teste et al. (2009b) included two different treatments, one with 35 μm mesh designed to allow only individual fungal hyphae, and one with 250 μm mesh designed to also allow colonization by rhizomorphs. They observed rare instances of rhizomorphs breaking down into an unstructured form and penetrating the 35 μm mesh. Teste and Simard (2008) also argued that the thickness (not the diameter of the openings) of the mesh used in their study ($\sim$320 μm) likely prevented contact exploration types of EM fungi from colonizing these plots and forming CMNs, although it is unclear how important those contact exploration types are, relative to other types of EM fungi, in forming functional CMNs. An important consideration for this treatment is the volume of the mesh enclosure, which must be sufficient to allow unrestricted target plant root growth for the intended duration of the experiment, but which must not be too large as to be impractical to install.

No roots/No CMN treatment. Early experiments, designed to test how contact with adult tree root systems and associated CMNs may influence the EM fungal community of seedlings, used trenching or coring and subsequent insertion of impermeable barriers to separate outplanted seedlings from the roots and CMNs of neighboring adult trees (reviewed by Deacon and Fleming 1992; see also Simard et al. 1997b). In more recent experiments focused on testing effects of CMNs on plant ecology, two different approaches have been used to establish this treatment, which excludes both roots and CMNs associated with neighbor plants, but in which target plants are mycorrhizal (Fig. 9.2b, c). In one approach, bags or cylinders of the same size as those in the No roots/+CMN treatment are used, but are made of an impermeable material or covered with a mesh of sufficiently small pore size (0.45–4 μm) to exclude roots and fungi but to potentially allow some passage of water and gases (e.g., McGuire 2007; Teste et al. 2009b). An alternative approach has been to repeatedly (every 4–6 weeks) renew a conical trench around the target plant plot, severing roots and CMNs that have grown into the plot (Booth 2004; Booth and Hoeksema 2010). In both cases, target plants establish their own mycorrhizas (not linked to neighbor plants via a CMN) independently, via fungal propagules in the

Table 9.1 Characteristics of field studies used to test CMN effects on plant ecology

Author and year	No roots/+CMN	No roots/No CMN	+roots/+CMN	+roots/No CMN	Mycorrhizal community data	Resource flow data
Bingham and Simard (2012a)	35 μm mesh cylindrical bag (17 cm diam, 32 cm deep)	0.5 μm mesh cylindrical bag (17 cm diam, 32 cm deep)	No barrier	–	–	water and C (^{13}C accumulation in target plant stems)
Bingham and Simard (2013)	35 μm mesh cylindrical bag (17 cm diam, 32 cm deep)	0.5 μm mesh cylindrical bag (17 cm diam, 32 cm deep)	No barrier	–	EM fungi molecular ID (reported in Bingham, and Simard 2012b)	water and C (^{13}C accumulation in target plant stems)
Booth (2004)	44 μm mesh on pvc cylinder (15 cm diam, 18 cm deep)	1. 0.5 m diam. conical trench 2. trench around pvc cylinder	No barrier	AM seedling species planted in EM overstorey	–	–
Booth and Hoeksema (2010)	44 μm mesh on pvc cylinder (15 cm diam, 18 cm deep)	1. 0.5 m diam. conical trench 2. trench around pvc cylinder	No barrier	–	–	water and C (^{13}C accumulation in target plant leaves)
Kranabetter (2005)	250 μm mesh cylindrical bag (10 cm diam, 20 cm deep, open bottom)	4 μm mesh cylindrical bag (10 cm diam, 20 cm deep, open bottom)	No barrier	–	EM fungi morphotypes	–
McGuire (2007)	20 μm mesh pots (30 cm diam, 40 cm deep)	0.45 μm mesh pots (30 cm diam, 40 cm deep)	No barrier	–	–	–

(continued)

Table 9.1 (continued)

Author and year	No roots/+CMN	No roots/No CMN	+roots/ +CMN	+roots/No CMN	Mycorrhizal community data	Resource flow data
Teste and Simard (2008)	1. 35 μm mesh cylindrical bag (15 cm diam, 35 cm deep) 2. 250 μm mesh cylindrical bag (15 cm diam, 35 cm deep)	1. 0.5 μm mesh cylindrical bag (15 cm diam, 35 cm deep) 2. Impermeable cylindrical bag (15 cm diam, 35 cm deep)	No barrier	–	EM fungi molecular ID (reported in Teste et al. 2009b)	water (pulsing of deuterated water; reported in Schoonmaker et al. 2007)
Teste et al. (2009a)	1. 35 μm mesh cylindrical bag (15 cm diam, 35 cm deep) 2. 250 μm mesh cylindrical bag (15 cm diam, 35 cm deep)	0.5 μm mesh cylindrical bag (15 cm diam, 35 cm deep)	No barrier	–	EM fungi molecular ID	C and N (pulsing of ^{13}C and ^{15}N)

Columns 2–5 give details of how roots and/or CMNs of neighbor plants were or were not restricted from contact with target experimental plants. All experiments took place in ectomycorrhizal systems.

soil. One major advantage of the first approach (bags or cylinders the same size as in the No roots/+CMN treatment) is that once the bags or cylinders are installed, much less labor is required to maintain the treatment, whereas the trenching treatment requires a great deal of physical labor. In addition, the physical process of trenching could potentially cause disturbance of the target plant plot. On the other hand, the trenching approach has a significant advantage, which is that the volume of soil inside the trenched zone, and thus the soil volume that can be explored by the mycorrhizal fungi associated with the target plant, can be substantially larger than the mesh bag or cylinder used to create the No roots/+CMN treatment. In contrast, if an impermeable barrier or very fine mesh is used to create the No roots/No CMN treatment, then target plants and their associated mycorrhizal fungi in that treatment are restricted to explore the volume of soil inside the bag or cylinder. This situation could be problematic when comparing plant growth with the No roots/+CMN treatment, in which mycorrhizal fungi associated with target plants are theoretically free to explore a much larger soil volume. In addition, the trenching approach allows inclusion of a No roots/No CMN treatment that controls for bag/cylinder effects by growing the target seedling inside a bag or cylinder identical to the one used in the No roots/+CMN treatment (Fig. 9.2c). Regardless of how this treatment is imposed, it is possible it may select for a different subset of the mycorrhizal fungal community than other treatments allowing CMN connections, potentially confounding interpretation of comparisons of plant performance among treatments (also see below under Field CMN-manipulation Experiments).

+roots/+CMN treatment. All of the published CMN manipulation experiments reviewed here took a similar approach to establishing this treatment, which was to plant target plants into intact soil containing roots and compatible CMNs associated with neighbor plants (Fig. 9.2a). In some cases, investigators took the extra step to first impose a disturbance similar to that created in the two No Roots treatments, e.g., installing and then removing a mesh bag or cylinder, in order to control for the effects of this disturbance across the whole experiment (McGuire 2007; Teste et al. 2009b; Booth and Hoeksema 2010; Bingham and Simard 2012a).

+roots/No CMN treatment. This treatment (when compared with a +roots/+CMN treatment) potentially allows estimation of the effect of neighbor plant CMNs on a target plant, in the presence of interactions between target and neighbor plant roots. However, this treatment is the most difficult to establish in a way that does not include confounding effects. Ideally, target and neighbor plants are both mycorrhizal, and their roots and mycorrhizal fungi are intermingled, but no CMN connections form between the mycorrhizas of the two plants. The only way this treatment has been achieved in published experiments is to establish AM target plants in soil colonized by roots and incompatible CMNs of EM neighbor plants (Eason et al. 1991; Booth 2004). In theory another possible approach, if target and neighbor plants are different species, would be to inoculate target and neighbor plants with different host-specific mycorrhizal fungi. All of these approaches have the limitation of confounding the lack of CMNs with one or more other factors that may differ between this treatment and the +roots/+CMN treatment, such as differing densities of compatible mycorrhizal fungi for the target plant, differing species or

types of neighbor fungi, differing neighbor plant species or microenvironments, or differing target plant species.

Additional methodological considerations for CMN field manipulation experiments. Some studies (e.g., McGuire 2007) allowed no time for in-growth of CMNs into No roots/+CMN plots before seedlings were planted, while others allowed up to a year (e.g., Booth and Hoeksema 2010). Although both of these experiments observed positive effects of CMNs on target plant growth and/or survival, I suggest that studies allowing a longer time for establishment of the CMN before planting target plants are more likely to provide accurate estimates of the potential effect of CMNs, since these benefits may be highest when target plants are youngest and join an existing CMN. For the same reason, benefits of CMNs may be less likely to be observed when older seedlings are planted as target plants (e.g., Teste and Simard 2008) compared to when seeds or young seedlings are planted (e.g., McGuire 2007). One study compared results between both of these approaches (Teste et al. 2009b). Bingham and Simard (2012a) demonstrated how the timing of planting can dramatically affect how target seedlings respond to CMNs. Finally, one study (Kranabetter 2005) used open-bottomed bags to implement treatments, making it possible that CMNs and overstorey roots were present in both No roots/No CMN and No roots/+CMN treatments, thereby potentially masking distinct CMN effects. Open-bottomed containers for target plants may be appropriate for implementing these treatments in some systems, but investigators must take care to insure and verify that this approach does not result in confounding colonization of target plant plots by unwanted roots or mycorrhizal fungi from neighbor plants.

9.1.5 What Do Results of Previous Field, Laboratory, and Other CMN Studies Tell Us?

Field CMN-manipulation experiments. Four CMN manipulation studies conducted in EM forests have found substantially higher growth and/or survival of target seedlings in No roots/+CMN treatments compared to No roots/No CMN treatments (Booth 2004; McGuire 2007; Booth and Hoeksema 2010; Bingham and Simard 2012a), providing support for the notion that overstorey trees can have important facilitative effects on seedlings through CMNs (but see Jakobsen and Hammer, Chap. 4, this volume). Two of these studies (Booth and Hoeksema 2010; Bingham and Simard 2012a) reported data that provide a partial test for whether resource transfer (i.e., Mechanism 1) was the mechanism underlying the facilitative effect of CMNs. Booth and Hoeksema (2010) found higher survival of target *Pinus radiata* seedlings over two years in No roots/+CMN plots compared to No roots/No CMN plots, but no differences in target seedling growth, N status, or maximum photosynthetic rates, suggesting that N transfer from overstorey *Pinus radiata* trees was not responsible for facilitative effects. This study did, however, find that leaves of target seedlings connected to overstorey CMNs were significantly depleted in ^{13}C at the end of the experiment relative to non-networked seedlings. This is the opposite pattern that would be expected if the leaves of networked seedlings contained a substantial proportion of carbon transferred through the CMN from overstorey trees (Hogberg et al. 1999). Rather, this result supports the hypothesis that networked seedlings had access to and transpired more water than non-networked seedlings, as total water transpiration is correlated with depletion in ^{13}C (Farquhar et al. 1989; Fotelli et al. 2003; Querejeta et al. 2006). Moreover, most target seedling mortality occurred during the dry season. Altogether, these results suggest that CMNs associated with overstorey trees had a facilitative effect on understorey seedling dry-season survival through the redistribution of water that was hydraulically lifted by overstorey trees. In the same study, overstorey tree roots were found to have a direct negative (competitive) effect on seedling survival, but much of this competitive effect was offset by the facilitative effect of overstorey CMNs. Bingham and Simard (2012a) found that Douglas fir (*Pseudotsuga menziesii*) seedling survival under a Douglas fir canopy was dramatically higher in a No roots/+CMN treatment compared to a No roots/No CMN treatment. They measured ^{13}C natural abundance in target seedling stems, but did not find significant differences among CMN treatments.

Three field studies (Teste and Simard 2008; Teste et al. 2009b; Bingham and Simard 2013) found no difference in target Douglas fir seedling performance between No roots/+CMN treatments compared to No roots/No CMN treatments. The authors of the two earlier field studies suggested that CMN formation was limited in their No roots/+CMN treatments, and they found increases in target Douglas fir seedling performance in +roots/+CMN treatments compared to No roots/No CMN treatments, which suggests facilitation of seedlings through either roots or CMNs associated with neighbor Douglas fir individuals. One of the field studies (Teste et al. 2009b) tested for C and N transfer from Douglas fir trees to target seedlings, and one (Bingham and Simard 2013) tested indirectly for C and water transfer by measuring ^{13}C natural abundance in target seedling stems; both found little evidence of significant differences among CMN treatments in the amount of resources transferred. In the field experiment by Teste and Simard (2008), however, deuterium-labeled water was used to track water transfer, and it was estimated that target seedlings received more than 21 % of their water through hydraulic redistribution from adult neighbor trees (Schoonmaker et al. 2007). Whether or not the pathway was root-soil-root or root-CMN-root, transfer of water seems to be one of the underlying mechanisms behind this instance of seedling facilitation by trees.

Bever et al. (2010) recently argued that these CMN manipulation experiments do not distinguish between the mechanisms of resource transfer between plants versus changes in density or composition of mycorrhizal symbionts, since treatments that sever the CMN also reduce the density of mycorrhizal fungi available to the target plant and reduce resource availability to the mycorrhizas of the target plant. I disagree with this blanket assessment, and suggest that a careful examination of the details of these experiments can help distinguish between alternative mechanisms. For example, in the Booth and Hoeksema (2010) study, No roots/+CMN plots and No roots/No CMN both contained cylinders wrapped in mesh that excluded roots but allowed CMN colonization. These cylinders were allowed one year to be colonized by overstorey CMNs before young target seedlings were planted in January during the wet season and trenches were cut to sever overstorey roots and CMNs entering the No roots/No CMN plots. Planted seedlings may have encountered reduced initial densities of mycorrhizal fungi in the latter plots; however, there were no differences among treatments in seedling mortality or growth during the early phase of the experiment. Rather, treatment differences in target seedling mortality appeared later in the experiment, during two subsequent dry summers, when mycorrhizal fungal densities were likely very similar among treatments. Moreover, it seems unlikely that trenching substantially reduced resource availability to mycorrhizas associated with the target seedlings in No roots/No CMN plots (except for resources potentially transferred from overstorey CMNs), as trenches were cut well away from the target seedling cylinders, in a conical shape with a diameter at the soil surface of 0.5 m. As discussed above, stable isotope data from this experiment suggest that seedlings in the No roots/+CMN treatment survived at a higher rate during dry summers at least in part due to great access to hydraulically distributed water from CMNs. Restriction of available resources to mycorrhizas in the No roots/No CMN treatment might be more of a concern in experiments that create this treatment using impermeable or micromesh bags the same size as the mesh bags used in the No roots/+CMN treatment.

As Bever et al. (2010) rightly pointed out, however, the mechanism of resource transfer through CMNs is not mutually exclusive from that of plant-soil feedbacks through altered densities or composition of mycorrhizal symbionts and asymmetric distribution of benefits by the symbionts among different hosts (i.e., versions of Mechanisms 2 and 3) in explaining facilitation of target plants by neighbor plants (see also Simard and Durall 2004; Selosse et al. 2006), and both mechanisms may have been operating in the experiments discussed above that showed facilitation of target seedlings by CMNs of overstorey trees (Booth 2004; McGuire 2007; Booth and Hoeksema 2010; Bingham and Simard 2012a). However, none of those studies reported data on variation in mycorrhizal fungal densities or community composition among treatments. In the experiment by Booth and Hoeksema (2010) one EM fungal taxon, *Tomentella sublilacina*, was most abundant on target seedlings in all treatments, although composition of rare taxa may have differed among treatments (K.J. Hennig, unpublished data). Teste et al. (2009b) found that when Douglas fir seeds (rather than seedlings) were planted in target plant plots, ectomycorrhizal colonization on target seedlings was higher after the first growing season in the +roots/+CMN treatment, but this effect disappeared by the following growth season. In the study by Teste and Simard (2008), treatments had no effect on mycorrhizal colonization, richness, or diversity (reported in Teste et al. 2009a), whereas the field study by Bingham and Simard (2013) found that the similarity of target seedling EM fungal communities to those of neighboring adult trees was significantly altered by CMN manipulations (reported in Bingham and Simard 2012b). In future experimental studies of potential CMN effects on plant ecology, it will be essential to report companion data on how mycorrhizal fungal community composition is altered by treatments, and ideally also on variation among fungal taxa in how they affect plant growth, so that alternative mechanisms can be clearly distinguished.

Laboratory CMN-manipulation studies. Laboratory (greenhouse or growth chamber) studies across diverse systems have now shown that severing of CMN connections between plants can significantly alter outcomes of plant-plant interactions. In particular, recent experiments in AM systems support the idea that CMNs mediate antagonistic, rather than facilitative, interactions among plants (e.g., Janos et al. 2013; Merrild et al. 2013; Weremijewicz and Janos 2013). Typically, these studies have used restrictive mesh to prevent root overlap between adjacent plants and to allow mycorrhizal hyphae to form a CMN, creating a No roots/+CMN treatment, and have compared target plant performance between this treatment and a No roots/No CMN treatment in which the CMN is severed or prevented. One advantage of these laboratory experiments is that fungal community composition can be controlled, and plant and fungal densities can be either controlled or carefully monitored.

In one example from an AM system, target tomato (*Solanum lycopersicon*) seedlings were found to grow significantly larger and to attain a higher P status when CMN connections to neighbor cucumber (*Cucumis sativus*) plants were severed, suggesting that connections mediate antagonistic effects of cucumber on tomato (Merrild et al. 2013). Fungal densities and root colonization were high across treatments, and the authors argued that asymmetric distribution of P by the AM fungi forming the CMN (Mechanism 3) was responsible for the antagonistic interaction, although the possibility of interplant transfer of P or other resources through the CMN was not explored. Similar growth benefits of severing an AM CMN, despite consistent AM colonization among treatments, were found for *Eucalyptus tetrodonta* (Janos et al. 2013). In yet another AM system, intact CMNs mediated intraspecific competition between big bluestem (*Andropogon gerardii*) individuals, resulting in greater size inequality among plants interconnected by CMNs compared to those with severed CMNs (Weremijewicz and Janos 2013). In this experiment, plants with access to CMNs exhibited higher root colonization by AM fungi, supporting the idea that CMNs mediated plant-plant interactions at least partly through density effects (Mechanism 2). However, the authors argue that CMNs had access to larger volumes of soil for nutrient acquisition, compared to the AM fungi of non-networked plants, allowing higher overall productivity in systems connected by CMNs. This interpretation highlights the problem, highlighted above (under "No roots/No CMN treatment"), that emerges when CMN connections are prevented by using an impermeable container the same size as those in which CMN-connected plants are grown: In this case, the soil volume available for nutrient acquisition by plants disconnected from the CMN is much smaller than the soil volume available to CMN-networked plants, confounding effects of CMN connections per se with differences in resource availability. One laboratory study in an EM system found no difference in target Douglas fir seedling performance between No roots/+CMN treatments compared to No roots/No CMN treatments (Bingham and Simard 2011), but was notable for its effort to test for C and water transfer between Douglas fir seedlings using stable isotope pulse-labeling.

CMN-inoculation and near-planting experiments. Another type of experiment relevant to the debate about the importance of mycorrhizal networks is one in which the performance of a target plant is compared between two treatments—one in which no mycorrhizal inoculum is provided and one in which mycorrhizal inoculum is provided in the form of the mycelium associated with a companion plant (e.g., Nara 2006). In these experiments, target plant performance is consistently found to be higher when inoculum is provided by a companion plant compared to when it is not (reviewed by van der Heijden and Horton 2009). Although the inoculum in these experiments is indeed provided by the CMN associated with a neighbor plant, I suggest that these results do not provide evidence for unique CMN effects on plant ecology. Since target plant performance is compared between a +CMN treatment and a non-mycorrhizal treatment, the effect of CMN network connections per se cannot be distinguished from the simple effect of mycorrhizal inoculation. This point has been made previously by other authors (Teste and Simard 2008; Bever et al. 2010), and Teste and Simard (2008) suggest that facilitation in these experiments be termed "MN-inoculation" effects rather than direct MN effects. Certainly, these studies provide examples in which the mechanism of indirect plant-plant facilitation is altered density of the mycorrhizal fungal symbiont, as discussed by Bever et al. (2010).

More informative to the debate about unique CMN effects in which linkages per se are important is a comparison of the magnitude of benefit to a target plant when it is inoculated by a CMN versus inoculation by spores or mycelium. Such an experiment would essentially constitute a comparison between a +roots/+CMN or No roots/+CMN treatment versus a No roots/No CMN treatment as described above, in which all target plants are mycorrhizal, but the two treatments differ in the presence of CMN connections. If the mycorrhizal No roots/No CMN treatment is compared with a +roots/CMN treatment, then the relative roles of neighbor plant roots versus CMNs may not be clear. Moreover, the mechanism of interactions (Mechanisms 1–3) would not be clear without careful examination of companion data in each experiment. An example of one such experiment was provided by Kytöviita et al. (2003), who compared the growth of four herbaceous perennial plants in a greenhouse experiment across several treatments, including non-mycorrhizal target plants, target plants inoculated in isolation with spores of one of several AM fungi, or target plants inoculated via mycelial growth of AM fungi from the CMN of an established neighbor plant. They found that growth of the target plants was not improved by inoculation through the CMN compared to inoculation by spores, suggesting either no benefit of a CMN connection, a benefit that was offset by competitive effects of the neighboring plant root system, or a benefit that was offset by competition for resources through the CMN itself (as the authors argue).

In an intermediate approach between CMN manipulations and CMN inoculation studies, a few studies have effectively generated a comparison of target plant performance among +roots/+CMN, No roots/No CMN, and/or +roots/No CMN treatments by planting an EM target plant in the field near different types of neighbors and/or at different distances from neighbors (e.g., Horton et al. 1999; Dickie et al. 2002, 2005; Teste and Simard 2008). In some of these experiments, +roots/No CMN and +roots/+CMN treatments were effectively established by planting EM target plants in soil colonized by roots and either incompatible CMNs of predominantly AM neighbor plants or compatible CMNs of EM neighbor plants, respectively. Dickie et al. (2002) also created a no roots/+CMN treatment by planting their EM target plants near stumps of dead EM host plants, in soils that presumably lacked roots and CMNs associated with neighbor plants. A key difference between these experiments versus CMN inoculation experiments (reviewed by van der Heijden and Horton 2009) is that the target seedlings in all treatments had access to natural levels of compatible EM inoculum, so that an effect of neighbor CMNs was not necessarily just an inoculation effect. One potential challenge of these studies is that multiple aspects of the microenvironment had the potential to differ among treatments, so interpretation was aided by thorough measurement of microenvironmental variation. Horton et al. (1999) found substantially increased survival of EM *Pseudotsuga* target plants near EM *Arctostaphylos* neighbors compared to those near predominantly AM *Adenostoma* neighbors, and Dickie et al. (2002) found significantly increased nutrient status and growth in *Quercus* seedlings planted near EM *Quercus* neighbors compared to those near AM *Acer* neighbors. In both studies, microenvironmental differences among treatments were minimal, and data on densities and identities of EM fungi on target plants supported the hypothesis that facilitation by neighbors was mediated by altered densities and composition of EM inoculum (as hypothesized by Bever and Schultz 2005; Bever et al. 2010), although resource transfers were not estimated or ruled out, and Horton et al. (1999) also noted available soil P and moisture as potentially explaining treatment differences.

9.2 Conclusions

If we are to make efficient progress in understanding what roles mycorrhizal fungi play in mediating community dynamics and general plant ecology, we need to clearly recognize the kinds of studies that are needed to distinguish among multiple alternative hypotheses, including resource flows through CMNs and altered densities and composition of mycorrhizal fungal communities. First, we need experimental studies that clearly distinguish between effects of physical connections through CMNs per se, versus effects of CMNs on inoculum densities and composition, on strengths of interactions among plants and/or plant community composition. Although these experiments are difficult to execute in a way that allows clear inference, several examples have been published on which future efforts can build. In particular, more such field experiments are needed in AM systems. Most importantly, such studies need to be combined with companion data sufficient to distinguish among multiple types of resource flow through CMNs, as well as data to quantify altered densities, composition, and functions of mycorrhizal fungal symbionts, so that Mechanisms 1–

3 can be effectively tested. In particular, water, phosphorus, and nitrogen flow through CMNs may be more ecologically significant than attention paid to them in previous studies would indicate, and we still understand relatively little about whether or how individual fungal taxa in mixed communities may contribute differently to plant growth or the function of CMNs. Despite decades of progress, we are still in the early stages of determining the general ecological significance of common mycorrhizal networks, and significant advances will only come with careful accumulation and examination of proper evidence.

Acknowledgements Michael G. Booth was a great friend and collaborator, and many of the concepts in this chapter had their origins in conversations with him. The author was supported by National Science Foundation grant DEB-1119865.

References

Achatz M, Rillig MC (2014) Arbuscular mycorrhizal fungal hyphae enhance transport of the allelochemical juglone in the field. Soil Biol and Biochem 78:76–82

Agerer R (2001) Exploration types of ectomycorrhizae-a proposal to classify ectomycorrhizal mycelial systems according to their patterns of differentiation and putative ecological importance. Mycorrhiza 11(2):107–114

Allen MF (2009) Bidirectional water flows through the soil-fungal-plant mycorrhizal continuum. New Phytol 182:290–293

Bethlenfalvay GJ, Reyes-Solis MG, Camel SB, Ferrera-Cerrato R (1991) Nutrient transfer between the root zones of soybean and maize plants connected by a common mycorrhizal mycelium. Physiol Plant 82:423–432

Bever JD (2002) Negative feedback within a mutualism: host-specific growth of mycorrhizal fungi reduces plant benefit. Proc R Soc B 269:2595–2601

Bever JD (2003) Soil community feedback and the coexistence of competitors: conceptual frameworks and empirical tests. New Phytol 157:465–473

Bever JD, Schultz PA (2005) Mechanisms of arbuscular mycorrhizal mediation of plant-plant interactions. In: Dighton J, White J, Oudemans P (eds) The fungal community: its organization and role in the ecosystem. CRC Press, New York

Bever JD, Dickie IA, Facelli E, Facelli JM, Klironomos J, Moora M, Rillig MC, Stock WD, Tibbett M, Zobel M (2010) Rooting theories of plant community ecology in microbial interactions. TREE 25:468–478

Bidartondo MI (2005) The evolutionary ecology of mycoheterotrophy. New Phytol 167:335–352

Bingham MA, Simard SW (2011) Do mycorrhizal network benefits to survival and growth of interior Douglas fir seedlings increase with soil moisture stress? Ecol Evol 1:306–316

Bingham MA, Simard S (2012a) Ectomycorrhizal networks var. glauca trees facilitate establishment of conspecific seedlings under drought. Ecosystems 15:188–199

Bingham MA, Simard SW (2012b) Mycorrhizal networks affect ectomycorrhizal fungal community similarity between conspecific trees and seedlings. Mycorrhiza 22:317–326

Bingham MA, Simard SW (2013) Seedling genetics and life history outweigh mycorrhizal network potential to improve conifer regeneration under drought. For Ecol Manag 287:132–139

Booth MG (2004) Mycorrhizal networks mediate overstorey-understorey competition in a temperate forest. Ecol Lett 7:538–546

Booth MG, Hoeksema JD (2010) Mycorrhizal networks counteract competitive effects of canopy trees on seedling survival. Ecology 91:2294–2302

Caldwell MM, Dawson TE, Richards JH (1998) Hydraulic lift: consequences of water efflux from the roots of plants. Oecologia 113:151–161

Dawson TE (1993) Hydraulic lift and water use by plants: implications for water balance, performance and plant-plant interactions. Oecologia 95:565–574

Deacon JW, Fleming LV (1992) Interactions of Ectomycorrhizal Fungi. In: Allen MF (ed) Mycorrhizal functioning: an integrative plant-fungal process. Chapman and Hall, New York, pp 249–300

Dickie IA, Koide RT, Steiner KC (2002) Influences of established trees on mycorrhizas, nutrition, and growth of *Quercus rubra* seedlings. Ecol Monogr 72:505–521

Dickie IA, Schnitzer SA, Reich PB, Hobbie SE (2005) Spatially disjunct effects of co-occurring competition and facilitation. Ecol Lett 8:1191–1200

Eason WR, Newman EI, Chuba PN (1991) Specificity of interplant cycling of phosphorus: the role of mycorrhizas. Plant Soil 137:267–274

Egerton-Warburton LM, Querejeta JI, Allen MF (2007) Common mycorrhizal networks provide a potential pathway for the transfer of hydraulically lifted water between plants. J Exp Bot 58:1473–1483

Farquhar GD, Ehleringer JR, Hubick KT (1989) Carbon isotope discrimination and photosynthesis. Ann Rev Plant Physiol Plant Mol Biol 40:503–537

Fellbaum CR, Mensah JA, Cloos AJ, Strahan GE, Pfeffer PE, Kiers ET, Bücking H (2014) Fungal nutrient allocation in common mycorrhizal networks is regulated by the carbon source strength of individual host plants. New Phytol 203:646–656

Fitter AH, Hodge A, Daniell TJ, Robinson D (1999) Resource sharing in plant-fungus communities: did the carbon move for you? TREE 14:70

Fotelli MN, Rennenberg H, Holst T, Mayer H, Gessler A (2003) Carbon isotope composition of various tissues of beech (*Fagus sylvatica*) regeneration is indicative of recent environmental conditions within the forest understorey. New Phytol 159:229–244

Francis R, Read DJ (1994) The contributions of mycorrhizal fungi to the determination of plant community structure. Plant Soil 159:1–25

He X, Critchley C, Ng H, Bledsoe C (2004) Reciprocal N ($^{15}NH_4$+ or $^{15}NO_{3-}$) transfer between non-N2-fixing *Eucalyptus maculata* and N2-fixing *Casuarina cunninghamiana* linked by the ectomycorrhizal fungus *Pisolithus* sp. New Phytol 163:629–640

He XH, Critchley C, Ng H, Bledsoe C (2005) Nodulated N-2-fixing *Casuarina cunninghamiana* is the sink for net N transfer from non-N-2-fixing *Eucalyptus maculata* via an ectomycorrhizal fungus *Pisolithus* sp using $^{15}NH_4$+ or $^{15}NO_3-$ supplied as ammonium nitrate. New Phytol 167:897–912

He X, Xu M, Qiu GY, Zhou J (2009) Use of ^{15}N stable isotope to quantify nitrogen transfer between mycorrhizal plants. J Plant Ecol 2:107–118

Hogberg P, Plamboeck AH, Taylor AFS, Fransson PMA (1999) Natural C-13 abundance reveals trophic status of fungi and host-origin of carbon in mycorrhizal fungi in mixed forests. PNAS 96:8534–8539

Horton TR, Bruns TD, Parker VT (1999) Ectomycorrhizal fungi associated with *Arctostaphylos* contribute to *Pseudotsuga menziesii* establishment. Can J Bot 77:93–102

Ikram A, Jensen ES, Jakobsen I (1994) No significant transfer of N and P from *Pueraria phaseoloides* to *Hevea brasiliensis* via hyphal links of arbuscular mycorrhiza. Soil Biol Biochem 26:1541–1547

Janos DP, Scott J, Aristizábal C, Bowman DMJS (2013) Arbuscular-mycorrhizal networks inhibit *Eucalyptus tetrodonta* seedlings in rain forest soil microcosms. PLoS ONE 8(2):e57716

Julou T, Burghardt B, Gebauer G, Berveiller D, Damesin C, Selosse MA (2005) Mixotrophy in orchids: insights from a comparative study of green individuals and nonphotosynthetic individuals of Cephalanthera damasonium. New Phytol 166:639–653

Kranabetter J (2005) Understory conifer seedling response to a gradient of root and ectomycorrhizal fungal contact. Can J Bot 83:638–646

Kytöviita MM, Vestberg M, Tuomi J (2003) A test of mutual aid in common mycorrhizal networks: established vegetation negates benefit in seedlings. Ecology 84:898–906

Leake JR, Johnson D, Donnelly DP, Muckle GE, Boddy L, Read DJ (2004) Networks of power and influence: the role of mycorrhizal mycelium in controlling plant communities and agroecosystem functioning. Can J Bot 82:1016–1045

Lilleskov EA, Bruns TD, Dawson TE, Camacho FJ (2009) Water sources and controls on water-loss rates of epigeous ectomycorrhizal fungal sporocarps during summer drought. New Phytol 182:483–494

Martins MA, Read DJ (1996) The role of the external mycelial network of VA mycorrhizal fungi. II A study of phosphorus transfer between plants interconnected by a common mycelium. Rev de Microb 27:30–35

McGuire KL (2007) Common ectomycorrhizal networks may maintain monodominance in a tropical rain forest. Ecology 88:567–574

Merrild MP, Ambus P, Sr Rosendahl, Jakobsen I (2013) Common arbuscular mycorrhizal networks amplify competition for phosphorus between seedlings and established plants. New Phytol 200:229–240

Molina R, Massicotte HB, Trappe JM (1992) Specificity phenomena in mycorrhizal symbioses: community-ecological consequences and practical implications. In: Allen MF (ed) Mycorrhizal functioning: an integrative plant-fungal process. Chapman and Hall, New York, pp 357–423

Moyer-Henry KA, Burton JW, Israel DW, Rufty TW (2006) Nitrogen transfer between plants: a 15N natural abundance study with crop and weed species. Plant Soil 282:7–20

Nara K (2006) Ectomycorrhizal networks and seedling establishment during early primary succession. New Phytol 169:169–178

Newman EI (1988) Mycorrhizal links between plants–their functioning and ecological significance. Adv Ecol Res 18:243–270

Pfeffer PE, Douds DD, Bücking H, Schwartz DP, Shachar-Hill Y (2004) The fungus does not transfer carbon to or between roots in an arbuscular mycorrhizal symbiosis. New Phytol 163:617–627

Querejeta JI, Egerton-Warburton LM, Allen MF (2003) Direct nocturnal water transfer from oaks to their mycorrhizal symbionts during severe soil drying. Oecologia 134:55–64

Querejeta JI, Allen MF, Caravaca F, Roldan A (2006) Differential modulation of host plant delta C-13 and delta O-18 by native and nonnative arbuscular mycorrhizal fungi in a semiarid environment. New Phytol 169:379–387

Robinson D, Fitter A (1999) The magnitude and control of carbon transfer between plants linked by a common mycorrhizal network. J Exp Bot 50:9–13

Schoonmaker AL, Teste FP, Simard SW, Guy RD (2007) Tree proximity, soil pathways and common mycorrhizal networks: their influence on the utilization of redistributed water by understory seedlings. Oecologia 154:455–466

Selosse MA, Richard F, He XH, Simard SW (2006) Mycorrhizal networks: des liaisons dangereuses? TREE 21:621–628

Shen Q, Chu G (2004) Bi-directional nitrogen transfer in an intercropping system of peanut with rice cultivated in aerobic soil. Biol Fert Soils 40:81–87

Simard SW, Durall DM (2004) Mycorrhizal networks: a review of their extent, function, and importance. Can J Bot 82:1140–1165

Simard SW, Perry DA, Jones MD, Myrold DD, Durall DM, Molina R (1997a) Net transfer of carbon between ectomycorrhizal tree species in the field. Nature 388:579–582

Simard SW, Perry DA, Smith JE, Molina R (1997b) Effects of soil trenching on occurrence of ectomycorrhizas on *Pseudotsuga menziesii* seedlings grown in mature forests of *Betula papyrifera* and *Pseudotsuga menziesii*. New Phytol 136:327–340

Simard SW, Beiler KJ, Bingham MA, Deslippe JR, Philip LJ, Teste FP (2012) Mycorrhizal networks: mechanisms, ecology and modelling. Fungal Biol Rev 26:39–60

Song YY, Ye M, Li C, He X, Zhu-Salzman K, Wang RL, Su YJ, Luo SM, Zeng RS (2014) Hijacking common mycorrhizal networks for herbivore-induced defence signal transfer between tomato plants. Scientific reports, 4

Song YY, Simard SW, Carroll A, Mohn WW, Zeng RS (2015) Defoliation of interior douglas-fir elicits carbon transfer and stress signalling to ponderosa pine neighbors through ectomycorrhizal networks. Scientific reports 5

Teste FP, Simard SW (2008) Mycorrhizal networks and distance from mature trees alter patterns of competition and facilitation in dry Douglas-fir forests. Oecologia 158:193–203

Teste FP, Simard SW, Durall DM (2009a) Role of mycorrhizal networks and tree proximity in ectomycorrhizal colonization of planted seedlings. Fungal Ecol 2:21–30

Teste FP, Simard SW, Durall DM, Guy RD, Jones MD, Schoonmaker AL (2009b) Access to mycorrhizal networks and roots of trees: importance for seedling survival and resource transfer. Ecology 90:2808–2822

Tuffen F, Eason WR, Scullion J (2002) The effect of earthworms and arbuscular mycorrhizal fungi on growth of and 32P transfer between *Allium porrum* plants. Soil Biol Biochem 34:1027–1036

van der Heijden MGA, Horton TR (2009) Socialism in soil? The importance of mycorrhizal fungal networks for facilitation in natural ecosystems. J Ecology 97:1139–1150

Warren JM, Brooks JR, Meinzer FC, Eberhart JL (2008) Hydraulic redistribution of water from *Pinus ponderosa* trees to seedlings: evidence for an ectomycorrhizal pathway. New Phytol 178:382–394

Weremijewicz J, Janos DP (2013) Common mycorrhizal networks amplify size inequality in *Andropogon gerardii* monocultures. New Phytol 198:203–213

Whitfield J (2007) Underground networking. Nature 449:136–138

Wilson GWT, Hartnett DC, Rice CW (2006) Mycorrhizal-mediated phosphorus transfer between tallgrass prairie plants *Sorghastrum nutans* and *Artemisia ludoviciana*. Funct Ecol 20:427–435

Zimmer K, Meyer C, Gebauer G (2008) The ectomycorrhizal specialist orchid *Corallorhiza trifida* is a partial myco-heterotroph. New Phytol 178:395–400

Subject Index

T.R. Horton (ed.), *Mycorrhizal Networks*,
Ecological Studies 224, DOI 10.1007/978-94-017-7395-9

Taxonomic Index

T.R. Horton (ed.), *Mycorrhizal Networks*,
Ecological Studies 224, DOI 10.1007/978-94-017-7395-9

MIX
Papier aus verantwortungsvollen Quellen
Paper from responsible sources
FSC® C105338

If you have any concerns about our products,
you can contact us on
ProductSafety@springernature.com

In case Publisher is established outside the EU,
the EU authorized representative is:
Springer Nature Customer Service Center GmbH
Europaplatz 3, 69115 Heidelberg, Germany

Printed by Libri Plureos GmbH
in Hamburg, Germany